Embedding New Technologies into Society

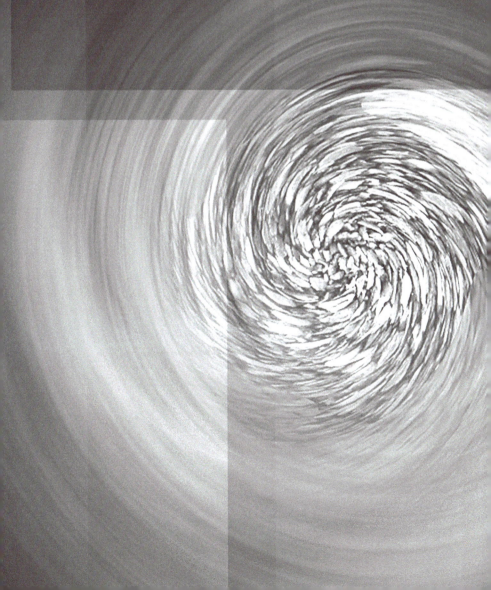

Embedding New Technologies into Society

A Regulatory, Ethical and Societal Perspective

edited by
Diana M. Bowman
Elen Stokes
Arie Rip

Published by

Pan Stanford Publishing Pte. Ltd.
Penthouse Level, Suntec Tower 3
8 Temasek Boulevard
Singapore 038988

Email: editorial@panstanford.com
Web: www.panstanford.com

British Library Cataloguing-in-Publication Data
A catalogue record for this book is available from the British Library.

Embedding New Technologies into Society: A Regulatory, Ethical and Societal Perspective

Copyright © 2017 Pan Stanford Publishing Pte. Ltd.

All rights reserved. This book, or parts thereof, may not be reproduced in any form or by any means, electronic or mechanical, including photocopying, recording or any information storage and retrieval system now known or to be invented, without written permission from the publisher.

For photocopying of material in this volume, please pay a copying fee through the Copyright Clearance Center, Inc., 222 Rosewood Drive, Danvers, MA 01923, USA. In this case permission to photocopy is not required from the publisher.

ISBN 978-981-4745-74-1 (Hardcover)
ISBN 978-1-315-37959-3 (eBook)

Printed in the USA

Contents

1. **Introduction** — 1
   Diana M. Bowman, Arie Rip, and Elen Stokes
   1.1 Novelty and Indeterminacy — 3
   1.2 Anticipation and Tentative Governance — 5
   1.3 Change, Re-Interpreted and Re-Negotiated — 7
   1.4 This Collection — 8
   1.5 In Conclusion — 12

   ### Part 1: Variety in the Governance of Newly Emerging Technologies

2. **Reflexive Co-Evolution and Governance Patterns** — 17
   Harro van Lente and Arie Rip
   2.1 Introduction — 17
   2.2 Actual Reflexive Co-Evolution of Technology in Society — 19
   2.3 Early Warning as a Governance Pattern — 22
   2.4 Bridging the Gap between Technology Development and Society — 24
   2.5 Governance by Orientation: The Discourse of Grand Challenges — 26
   2.6 In Conclusion — 28

3. **Regulatory Governance Approaches for Emerging Technologies** — 35
   Bärbel Dorbeck-Jung and Diana M. Bowman
   3.1 Introduction — 35
   3.2 Promising Regulatory Governance Approaches — 39

	3.2.1	Models of Technology Regulation	39
	3.2.2	Regulatory Models for Nanotechnologies	40
3.3	Reviewing Nano Regulatory Governance: The Story so Far		42
	3.3.1	Introduce Evaluation Frame of Cases: Some Effectiveness Questions	42
	3.3.2	Governance Arrangements to Limit Exposure to Nanoparticles in the Workplace	43
	3.3.3	Nanocosmetics within the European Union	47
3.4	Lessons Learned, and Moving Forward		52

4. Society as a Laboratory to Experiment with New Technologies — 61

Ibo van de Poel

4.1	Introduction		61
4.2	Technology as Social Experiment		64
4.3	A Brief History of the Idea		65
	4.3.1	Society as Laboratory	65
	4.3.2	Engineering as Social Experiment	67
	4.3.3	Social Experiments in Strategic Niche Management	68
	4.3.4	Social Experiments in Social Science	68
4.4	Characteristics of Social Experiments		69
	4.4.1	An Experiment in Society	69
	4.4.2	An Experiment on Society	69
	4.4.3	An Experiment by Society	70
	4.4.4	Responsible Experimentation	71
4.5	Uncertainty, Learning and Experimentation		72
	4.5.1	Uncertainty	73
	4.5.2	The Need for Experimentation	75
	4.5.3	Learning-by-Experimentation	77

4.6	An Example: Sunscreens with Titanium Dioxide Nanoparticles	80
4.7	Conclusions: Towards Responsible Experimentation	83

5. Care and Techno-Science: Re-Embedding the Futures of Innovation — 89

Christopher Groves

5.1 Introduction	89
5.2 Care and Innovation	91
5.2.1 Needs, Attachments and Care	91
5.2.2 Expectations and Performing Futures	95
5.3 Disembedding: Generic Futures in Nano- and Biotechnology	98
5.4 Care-Full Re-Embedding: Performing Concrete Futures	102
5.5 Conclusion	106

6. Division of Moral Labour as an Element in the Governance of Emerging Technologies — 115

Arie Rip

7. Ethical Reflexivity as Capacity Building: Tools and Approaches — 131

Clare Shelley-Egan and Federica Lucivero

7.1 Introduction	131
7.2 Institutional Requirements and Opportunities to Engage in Ethical Reflexivity	133
7.3 Ethical Reflexivity and Building Capacity	137
7.4 Building Capacities: Supportive Tools and Approaches	140
7.4.1 Articulating Reflexivity	140
7.4.2 Imagining and Discussing Societal Relevance	142
7.4.3 Performing Reflexivity	147
7.5 Conclusions and Discussion	148

Part 2: Promises, Politics and Particularities of Nanotechnologies

8. The Demand Side of Innovation Governance: Demand Articulation Processes in the Case of Nano-Based Sensor Technologies   159

Haico te Kulve and Kornelia Konrad

8.1 Introduction	159
8.2 Distributed Processes of Demand Articulation	161
8.2.1 Demand Articulation as Part of Innovation Governance	161
8.2.2 Dynamics in Distributed Demand Articulation Processes	164
8.2.3 Interventions in Demand Articulation and Tools to Support Them	167
8.3 Demand Articulation of New Sensor Applications for the Drinking Water and Food and Beverages Sectors	169
8.3.1 Sensors and Characteristics of the Drinking Water and Food and Beverages Sectors	169
8.3.2 Fit and Stretch Strategies in Demand Articulation Processes	170
8.3.3 Sensors for Monitoring Water Quality in the Distribution Network: Fit or Stretch the Monitoring Regime?	171
8.3.4 Sensors for Monitoring Food Quality: Stretching Commercial Relationships along the Value Chain	173
8.3.5 Certification and Standardisation in Societal Embedding Articulation: Fit or Stretch	174
8.4 Supporting Demand Articulation Processes via CTA Workshops	175
8.5 Conclusions	178

9. Evolving Patterns of Governance of, and by, Expectations: The Graphene Hype Wave   **187**

Kornelia Konrad and Carla Alvial Palavicino

9.1 Introduction	187
9.2 Governance of, and by, Expectations	190
9.2.1 Modes of Governance	190
9.2.2 Intentional and De Facto Governance	193
9.3 The Graphene Hype Wave	195
9.3.1 A Graphene Hype Emerging in the Science Space	195
9.3.2 Graphene Moves into the Policy and Media Space	198
9.3.3 Graphene Moving into the Market Space and the Hype Becoming Reflexive	202
9.3.4 The Emergence of Concerns Besides the Promises and the Call for Standardisation and New Definitions of Graphene	205
9.4 Conclusion	207

10. Transactional Arrangements in the Governance of Emerging Technologies: The Case of Nanotechnology   **219**

Evisa Kica and Ramses A. Wessel

10.1 Introduction	219
10.2 The Transnationalisation of Nanotechnology Governance	226
10.3 Transnational Governance Arrangements Generally and Their Attributes	228
10.4 The Governance of Nanotechnology: A Typology of Transnational Governance Arrangements	233
10.4.1 ISO Technical Committee on Nanotechnology	233
10.4.2 OECD Working Party on Manufactured Nanomaterials	236
10.4.3 International Risk Governance Council	239

            10.4.4 International Council on Nanotechnology   241
            10.4.5 Intergovernmental Forum on Chemical
                    Safety   242
    10.5   Conclusion   244

11. **Co-Regulation of Nanomaterials: On Collaborative Business Association Activities Directed at Contributing to Occupational Health and Safety** 259

    Aline Reichow

    11.1   Introduction   259
    11.2   Approach to Business Association Activities   264
            11.2.1 Business Associations in Germany   266
            11.2.2 The Chemical Industry Association   268
    11.3   Discussion: Effective Nanomaterials OHS
            Regulation   276
    11.4   Conclusions: An Outlook to the Future of
            Nanomaterials OHS Regulation   279

    Part 3: LOOKING TO THE FUTURE OF DISRUPTIVE TECHNOLOGIES

12. **The 'Metamorphosis' of the Drone: The Governance Challenges of Drone Technology and Border Surveillance** 299

    Luisa Marin

    12.1   Introduction: Let's Face It! They Are Here to Stay 299
    12.2   Drones and Border Surveillance   304
    12.3   The US Experience of Border Surveillance
            with Drones   307
            12.3.1 The American Drone   307
            12.3.2 Legal Framework for Border Surveillance
                    and Privacy in the US   309
    12.4   The Deployment of Drone Technology into Border
            Surveillance in the EU   313
            12.4.1 Border Surveillance in the EU: Frontex   313
            12.4.2 Frontex and Drones   315

12.5		Issues, Ethical and Regulatory Challenges Underlying the Political Choice of Deploying Drones for Border Surveillance	317
	12.5.1	Policy Issues Underlying the Deployment of Drones in Border Surveillance	318
	12.5.2	Ethical and Regulatory Questions	321
	12.5.3	Legal Framework for Border Surveillance and Privacy in the EU	325
12.6	Conclusion		329

13. On the Disruptive Potential of 3D Printing   335

Pierre Delvenne and Lara Vigneron

13.1	Introduction	335
13.2	A Brief Introduction to 3D Printing	338
13.3	3D Printing in the Industrial Sector	340
13.4	3D Printing in the Biomedical Industry	341
	13.4.1 3D Printing of Biomedical Instruments and Implants for Patients	341
	13.4.2 Additive Bio-Manufacturing	343
13.5	3D Printing in the Non-Industrial Domains	344
13.6	Discussion	347

14. Advanced Materials and Modified Mosquitoes: The Regulation of Nanotechnologies and Synthetic Biology   357

Diana M. Bowman, Elen Stokes, and Ben Trump

14.1	Introduction	357
14.2	Regulations: Past and Present	360
	14.2.1 Common Narratives of Early-Stage Emerging Technology Risk Governance	361
	14.2.2 Nanotechnology Regulation in the EU and Australia	362
14.3	Synthetic Biology: The Next Evolutionary Technology	366
	14.3.1 Synthetic Biology: Early Steps to Regulation and Governance	367

		14.3.2	Synthetic Biology: The Challenges for Regulators	368
	14.4	Synthetic Biology in the Environment		370
		14.4.1	Genetically Modified Mosquitoes as Population Control: Initial Trials	371
		14.4.2	Critical Response to Modified Mosquito Field Trials	372
		14.4.3	Synthetic Biology and Mosquitoes: Health Concerns and Regulatory Challenges	373
		14.4.4	Synthetic Biology and Mosquitoes: Looking Forward	375
	14.5	Conclusions		377

Index 385

Chapter 1

Introduction

Diana M. Bowman,[a] Arie Rip,[b] and Elen Stokes[c]

[a]*Sandra Day O'Connor College of Law and
School for the Future of Innovation in Society, Arizona State University,
111 E Taylor St, Phoenix, AZ 85004, USA*
[b]*Department of Science, Technology and Policy Studies,
School of Management and Governance,
University of Twente, 7500 AE Enschede, The Netherlands*
[c]*Birmingham Law School, University of Birmingham,
Edgbaston, Birmingham, B15 2TT, UK*

Diana.Bowman@asu.edu, a.rip@utwente.nl, e.stokes@bham.ac.uk

> *In science fiction, space and time warps are a commonplace.
> They are used for rapid journeys around the galaxy,
> or for travel through time.
> But today's science fiction, is often tomorrow's science fact.*
> —Stephen Hawking [1]

There have been at least two watershed moments in the modern governance of emerging technologies, linked to issues of the embedding of those technologies in society. First, the tribulations of genetic modification (GM) technology, especially in agriculture in the 1990s, dealt a blow to the progressivist perspective of technologists and technology promotors. There were of course earlier debates about technology and society, such as in the

---

Embedding New Technologies into Society: A Regulatory, Ethical and Societal Perspective
Edited by Diana M. Bowman, Elen Stokes, and Arie Rip
Copyright © 2017 Pan Stanford Publishing Pte. Ltd.
ISBN 978-981-4745-74-1 (Hardcover), 978-1-315-37959-3 (eBook)
www.panstanford.com

nuclear sector as regards issues of safety and radioactive waste. Yet, even in the aftermath of such controversies, newly emerging technologies might have continued to have the benefit of the doubt and there was still room for promises of progress. The progressivist perspective may still be prevalent, but the promotors are more prudent now. Their 'social licence to operate' could no longer be seen as automatic, if it ever was. Second, the advent of nanoscience and nanotechnology in the early 2000s prompted new calls for specially adapted modes of governance. 'Responsible development' became a touchstone of law and policy in the field, which later evolved into the principle of Responsible Research and Innovation (RRI).

Of course, one reason why technological development may be contentious is that it is difficult to know where the 'optimal' balance lies between innovation and caution. Such questions are made all the more complex by issues of scientific and political uncertainty, and cross-generational, cross-national and cross-cultural differences in the values attached to technological progress. The idea that there can be a single 'correct' approach to new technologies has long been criticized as being unrealistic and undesirable. That is not to say, however, that there are not better and worse ways of dealing with technological advance, and embedding it in society.

This edited collection attempts to take stock of the governance of emerging technologies. It does so, not by offering a comprehensive overview of the governance arrangements for all major new technologies, but by identifying some of the main themes running through a small number of examples. For the most part, the collection draws on, and analyses examples from nanotechnology.[1] The edited volume highlights interesting and important issues as encountered by the contributors in their studies of, and experiences with, various applications of nanotechnology. But the book also looks forward, by commenting on the evolving patterns of technology governance and regulation, and by extending the analysis to other emerging technologies. Although the chapters vary in their approaches to technology, they share a central concern with the following issues.

---

[1]The edited volume grew out of the work being done in the Technology Assessment sub-program of the Dutch national nanotechnology R&D consortiums NanoNed and NanonextNL. See [2].

1.1 Novelty and Indeterminacy

Many aspects of emerging technologies are, by definition, new and uncertain, raising questions about whether and to what extent 'existing' governance arrangements continue to be fit for purpose.[2] The prospect of regulatory gaps often prompts exercises in forecasting, to create an 'anticipatory' evidence-base as it were, although the robustness of such policy-relevant knowledge (and its underpinning practices and assumptions) has frequently been called into question. Collingridge [4] introduced the idea of the 'dilemma of control', to describe the trade-off between regulating a technology when it is in its infancy (and then face the problem of a lack of evidence) and regulating a technology after it has become embedded in society (by which time it has become difficult to exert any influence over the pace and trajectory of the technology's further development). The dilemma cannot be 'solved' in any simple way, but it might involve strategies of 'hedging' and 'flexing' [5], and greater reflexivity.[3] The challenge, as perceived by policymakers and regulators, is to identify frameworks of governance that are stable, yet flexible, rooted in experience but also anticipatory, controlling and facilitative. Achieving this in conditions of severe uncertainty—and indeterminacy—is no easy task.

Indeterminacy goes deeper than uncertainty. Whereas uncertainty implies a lack of information, indeterminacy describes situations in which causal chains, networks and processes are open and defy prediction. Indeterminacy also creates a potentially greater space for projections about technologies, such as big promises about their future performance and societal effects—these promises give rise to particular visions of technoscientific progress. The early years of nanotechnology were full of references

---

[2]"Technological revolutions', philosopher James Moor reminds us, 'do not arrive fully mature' [3]. They take time to emerge, evolve and become embedded in society. Initially, the possibilities seem almost limitless; a new technology could develop in many different ways, along different trajectories, and directions.

[3]This is taken up in Constructive Technology Assessment (CTA) by arguing that the harshness of Collingridge's dilemma is softened when considering that there are informal and formal assessments throughout the development of an emerging technology. In other words, instead of one big dilemma, there are many small dilemmas that can be addressed in concrete contexts. One should still be concerned about the path dependencies that might arise.

to the expected 'new industrial revolution', and to the 'alleviation of so many earthly ills' as the United States (US) Undersecretary of Commerce phrased it [6]. Such imagery may lead to developmental 'hype-cycles'—sometimes the promises are so far-fetched that they can only end in disappointment [7]. At the same time, there is also the possibility that 'big promises' (e.g. concerning the advantages of the nanoparticles and nanostructures) may trigger 'big concerns' (e.g. about health and environmental hazards). Again, it is important to maintain a sense of complexity, and to not assume that the answer lies solely in more or better risk assessments.

Historically, there has been a tendency to think about technology in terms of its associated 'effects', 'hazards', and 'risks' [8]. Over the years, various contributions from science-and-technology studies have highlighted the problems with such narrow framing of technology, and have shown how exercises in 'black-boxing' can conceal previously unacknowledged, but nonetheless important, subtleties and complexities. In this spirit, there is growing recognition in the academic and policy literatures that the 'facts' of technology may be contested and contingent, and that ambiguity can arise because technology cannot be separated from its constitutive social and political relations. The upshot is that technology amounts to more than a list of benefits and adverse consequences; there is more at play, including deeper social and political considerations, visions and values, risk-benefit calculations, trade-offs, and diverse knowledge and experiences, all of which can contribute to understandings of technological development. These are not static, nor are they consistent within one population, let alone across populations.

The interface between technology and policy-making is itself a source of uncertainty, not only in the sense that there are not always clear scientific answers directing policy action, but also because policy rules and practices themselves bring about novelty and change. This is particularly so where high-level policy commitments are made to becoming the 'world's leading knowledge-based economy' and to creating the 'Innovation Union' [9:5] by ensuring that innovative ideas are turned into new products and services.

Technological innovation rarely occurs independently of normative policy goals. Rather, law and policy can provide the stimulant for rapid technological change, opening up new areas of uncertainty as a result. Equally, law and policy may act as a disincentive to technology investment and innovation, closing off potential pathways before they can begin to be explored.

1.2 Anticipation and Tentative Governance

There is a strong tradition in the policy literature of calls for 'evidence-based' decision-making, which is usually taken to mean that policies are to be underpinned by scientific facts—although, understandably, it bears a more fraught relationship with technologies so new that evidence is in short supply or deeply contradictory. More recently, it has become associated with the 'Better Regulation' agenda in the European Union (EU) [10] and individual Member States. For example, a United Kingdom Government report, titled *Modernising Government*, stated that

> This Government expects more of policy makers. More new ideas, more willingness to question inherited ways of doing things, better use of evidence and research in policy making and better focus on policies that will deliver long term goals (cited in [11:90]).

Invariably, new technologies, especially those bringing more disruptive forms of change, pose a challenge to the goals of Better Regulation. At the very least, new and disruptive technologies make it difficult to determine what sort of regulation would be 'better' or 'worse' in the circumstances. The question is how policy may be 'evidence-based' in conditions of partial knowledge and even ignorance.

The aim, therefore, is not just to act early but to act in ways that allow for reflection, revision and a degree of 'self-confrontation'. This has been captured by the notion of tentative governance [12]. This means—among other things—developing tools and approaches within the public and private sectors that are attuned to their own limits, contradictions, prior commitments, and potential roles in shaping new technological development.

Whereas the social sciences may in the past have been slow to provide reflexive analysis, the emergence of new technology is now seen as providing an opportunity for re-evaluation and for reflexively incorporating key insights into the technology's regulation and development. If nothing else, new technology should give us pause for thought about the status quo of existing (not just prospective) arrangements for adapting to, and coping with, a changing technological world. But, one must ask, how often do policymakers, regulators and other relevant stakeholders stop to assess and evaluate such arrangements? Whose responsibility is it to engage in such activities? And what are the practical implications of doing so?

The extent to which law and regulation provide opportunities for reflexivity in this context is much debated. For one thing, it is a well-rehearsed argument that law struggles to keep up with technological innovation. If law marches with technological progress, it is 'in the rear and limping a little'.[4] The problem has been described as the potential 'disconnect' [13] and/or 'pacing problem' [14] between technological development and regulatory frameworks, where there are genuine questions as to whether a particular new technology falls within the letter and spirit of existing regulatory provisions. Where there are plausible gaps, law—in the form of legislative intervention—is often said to be too slow to respond. It is a well-documented feature of law-making processes that they can be unwieldy and bureaucratically cumbersome mechanisms for dealing with rapidly evolving and contested technological futures, and with the wider social implications lying outside conventional issues of 'risk'. At a fundamental level, law is concerned with resolving uncertainty and it deploys various tools to achieve that goal including, for example, the burden of proof and evidentiary presumptions to bridge the gap between knowns and unknowns. These traditional legal norms can become less straightforward to apply when confronted with novel and open-ended circumstances. As Richard Posner observes:

> Law is the most historically oriented, or if you like the most backward-looking, the most 'past-dependent,' of the professions... It is suspicious of innovation, discontinuities, 'paradigm shifts', and the energy and brashness of youth [15:573].

Given the potential limits of conventional legal responses to new technology, there have been calls for new approaches to governance. Although the idea of 'new governance' is not a settled one, it tends to signal a shift away from hierarchical, command and control regimes to more varied techniques of regulation.

---
[4]Justice Windeyer, *Mt Isa Mines v. Pusey* (1970) 125 CLR 383.

These techniques are usually described as less rigid and less prescriptive than those deployed under traditional regulatory frameworks—and, because of this, they may lend themselves to the more experimental and more reflexive governance of emerging technologies.

1.3 Change, Re-Interpreted and Re-Negotiated

A focus on governance opens up space both within, and beyond, traditional structures and institutional frameworks for further analysis. A recurring theme in this collection is change, be it technological, scientific, social, or political change, and the governance frameworks, principles and behaviours that emerge or are adapted as a consequence. There is a sense that governance may be better suited than traditional legal approaches to dealing with change and uncertainty because it is less committed to specific, uniform outcomes (e.g. such as those contained in precise legal standards), and entails more agile and process-oriented techniques (e.g. using more inclusive and discursive approaches to decision-making). It is in these governance spaces that policies and practices on new technology may be more openly re-interpreted and re-negotiated as experiences unfold.

In this sense, technological development can bring about more than a one-off change; it can induce multiple shifts throughout its social embedding. One example might be the initial polarization of responses to a particular hi-tech product during its early stage of development or commercialization, but a subsequent softening of views as the product becomes more 'mainstream'. The opposite is, of course, also possible. The rhetoric of technoscientific innovation is often framed in sharply dualistic terms—technology is either a saint or sinner, revolutionary or evolutionary, continuous or discontinuous, liberating or enslaving. Other dualisms that infuse the debate include: stagnation or change, heedlessness or precaution, individual or collective, regulated or unregulated, public or private, global or local, poverty or wealth, potential or actual, control or chaos, power or helplessness, upstream or downstream, expert or lay, transparent or opaque, reversible or irreversible.

While not unique to technology, such dichotomies have the potential to obfuscate the issues and short-circuit central debates

[16]. Over time, as a technology becomes embedded, such dualisms are less easily sustained, since issues, patterns and relations emerge in more complex and kaleidoscopic ways. For example, in the case of nanotechnology, the traditional division of moral labour—between on the one hand technology developers and promoters, and on the other hand regulatory agencies and concerned groups—has become blurred, paving the way for better integrated approaches to decision-making. By developing insights into the complexities of processes of and approaches to embedding new technologies into society, and by feeding those insights back into governance arrangements, the overall aim is to build-in prolonged moments of reflexivity.

1.4 This Collection

There is a plethora of newly emerging and potentially disruptive technologies. Nanotechnologies. Synthetic biology. Additive manufacturing (AM) or three-dimensional (3D) printing. Unmanned aerial vehicles. And gene editing using CRISPR-Cas9 techniques. Each of these technological domains was, at one time or another, found only in the imagination of science fiction writers, on the pages of their manuscripts, and in the creativity of movie director's minds as they sought to bring these visions to the big screen. And while travel through time may still fall within the realm of science fiction, certain other technological possibilities, as chapters of this volume illustrate, are part of today's science fact.

What they also build on is what we, as a collective, can learn from consequences of the 'magic mineral' asbestos, the widespread use of ozone-depleting chlorofluorocarbons, environmental exposure to polychlorobiphenyls, and the human health and environmental hazards of lead in paint and petrol [17]. This does not, of course, mean that mistakes will no longer be made in determining suitable regulatory and governance responses to techno-scientific innovation. For one thing, Sir John Meurig Thomas reminds us of the fallibility of predictions of scientific and technological advances [18]. This does not make the attempt to predict the future a purely academic exercise, but it does highlight that many important discoveries are often influenced by a range of (often unpredictable) external factors, including political, commercial, societal and ethical pressures. Consequently, new technologies and their products will experience periods of under- and over-regulation [19], although we might not be in a position to make such judgments without the benefit of hindsight. This raises important questions about how to approach and evaluate the regulatory, ethical and social dimensions of a given technology as, and when, it emerges.

The chapters in this collection consider these questions; they are divided into three parts. *Part 1—Variety in the Governance of Newly Emerging Technologies* (Chapters 2–7) examines governance issues arising from newly emerging technologies generally. In Chapter 2, van Lente and Rip examine the co-evolution of science, technology and science in order to identify recurrent governance patterns for emerging technologies. The authors highlight the significance of 'early warnings' and 'early signalling' as aids to policymakers and technology developers, as well as the orientation towards Grand Challenges in the policy discourse.

In Chapter 3, Dorbeck-Jung and Bowman explore different modes of governance—precautionary, anticipatory and responsive—that have been employed in addressing real work regulatory challenges posed by nanotechnologies, and the uncertainties thereof. Their chapter dwells on the notion of effectiveness, and the inability to measure effectiveness in an environment characterized by an absence of (known) harms.

Notions of uncertainty, innovation and potential hazards are threads that similarly run through van de Poel's chapter (Chapter 4). He observes that society is 'the laboratory in which new technologies are tried out'. Van de Poel suggests that as it is not possible to mitigate all potential risks associated before a new technology and/or its products enter the market, we should embrace the concept of experimentation once a technology reaches the market and is embedded within it. Careful and deliberate design of experiments allows for real world experimentation, the results of which can then be employed to address the very challenges presented by the technology.

The notion of the *disembedded* future, one that 'tames' an uncertain future by imposing certain logics of decision at the expense of a broader ethics of care, is explored by Groves in Chapter 5. Developing more 'care-full' practices, accordingly to Groves, depends on, among other things, making space for the virtues and goals of RRI. With RRI now a cross-cutting theme of EU policy in areas of science and technology, an understanding of how ethical and moral issues play into innovation, and its embedding in society, is fundamental to ensuring the goals of RRI are achieved.

In Chapter 6, Rip introduces a novel element in de-facto governance, the division of moral labour, as it specifies roles and responsibilities that have been settled over time. As is clear from the discussion in Chapter 5 (Groves) and Chapter 7 (Shelley-Egan and Lucivero), it can also a productive focus for discussing the governance of new technologies. Rip discusses how the roles and responsibilities are being articulated for nanotechnology.

Shelley-Egan and Lucivero (Chapter 7) focus on the notion of ethical reflexivity and responsibility as applied to emerging technologies. In their chapter the authors ask whether, and how, RRI and institutional reflexivity might be built into scientific practice, notwithstanding the absence of formal regulation (particularly, hard law) to that effect. Again, as with other chapters in the book, the focus is very much on developing more ethically sensitive ways of effectively governing a technology without resorting to traditional legal mechanisms.

Part 2—Promises, Politics and Particularities of Nanotechnologies looks explicitly at the applications of nanotechnology, in order to tease out the different lessons from, and perspectives of, the technology's entry into the market. As this section of the book discusses, governance arrangements (broadly defined) for nanotechnology-based products have now emerged and are not confined to the nation or supranational level; they are now also part of the transnational landscape.

The focus of Chapter 8 is on nano-based sensor technologies. In this chapter, te Kulve and Konrad explore the governance tools that have developed by producers and users of the product as a result of their ongoing interactions; so-called 'demand-side' governance. Their chapter illustrates a deepening and widening of new governance spaces, including those that are developed from the ground up through less formal means, and that go beyond the obvious domains of state-centred rule-making.

In their chapter, Konrad and Palavicino (Chapter 9) make the argument that expectations or imaginaries of the future, and not just technologies per se, may themselves be governed. This might happen, for example, through the discursive politics of innovation, as a particular technological future may be framed in terms that define its benefits broadly but its hazards narrowly.

The volume then shifts to an examination by Kica and Wessel (Chapter 10) of public-private and private governance schemes involving international standard-setting bodies and non-governmental organizations. Their chapter raises important questions relating not just to where governance activities occur and by whom, but also about the normative basis of such actions. How *should* governance work be distributed?

A common theme among the chapters in this part of the book is the need for collaboration between relevant stakeholders. In her chapter, Reichow (Chapter 11) continues to build on this premise by looking at the ways that business associations, firms and the government have shaped, and are shaping, the occupational health and safety regulatory framework addressing nanomaterials. By looking specifically at two jurisdictions—Germany and the US—Reichow is able to show the level of influence that can be wielded by non-state actors. Moreover, her chapter provides a clear illustration of the variety of governance tools that might be deployed in response to nanotechnology, including information exchange between state and private actors.

Together, these four chapters provide insight into less conventional approaches to the development of governance regimes, and how important and influential non-state actors can be. Such practices shall no doubt continue to evolve along with the emergence of new technologies; how successfully different actors are will depend, so it would seem, on the level of power they have at their discretion, who they seek to influence, and the public values and societal expectations associated with the technology and its products.

Part 3—Looking to the Future of Disruptive Technologies speculates as to just how society may respond to the next wave of emerging technologies. In Chapter 12, Marin focuses on the metamorphosis of the drone; a technology that was developed by, and for, the military that has now found its way into the civilian world. Within the context of this chapter, Marin explores the ethical, legal and social issues created by the use of drone technologies for border surveillance activities in the US and the EU. The analysis presented by Marin paints a picture of rapidly decreasing privacy, in which the deployment of drones 'changes border surveillance and makes it more pervasive and subtle'. She argues that drones should be considered a 'game changer' within the context of such applications, and that careful consideration must be given to the ways that data may be collected and used.

AM, or so-called 3D printing, has been described as a disruptive technology. In Chapter 13, Delvenne and Vigneron explore the history and use of AM/3D printing today, and paint a detailed picture of the way in which the technology could be employed across all societal domains. The chapter speaks to the seemingly limitless possibilities offered by the technology, and the ways in which AM/3D printing could transform systems of trade, alter global political dynamics, and shift power in ways that we can barely comprehend at this time. Their chapter challenges the reader to contemplate who the potential winners and losers are likely to be, and the ramifications thereof to individuals and society more generally.

In the final chapter (Chapter 14), Bowman, Stokes and Trump explore the tendency and implications of resorting to existing regulatory regimes when faced with a new technology, in this case a particular application of synthetic biology. Here, the authors examine the potential risks and benefits of relying on the regulatory *status quo,* and ask if there are better approaches to bringing a technology to market—especially one that involves such a high degree of scientific uncertainty.

1.5   In Conclusion

In this edited volume, we are taking stock of what has been learned about governance of emerging and possibly disruptive technologies. There has been a lot of learning in practice, including some trial and error. Our contributors build on that, but have also tried to identify important issues and approaches that deserve to be further developed and applied. We invite our readers to think with us, and join in this further development and application.

References

1. Hawking, S. (undated) *Space and Time Warps*. Available at: http://www.hawking.org.uk/space-and-time-warps.html.
2. Rip, A., and H. van Lente (2013) Bridging the Gap between Innovation and ELSA: The TA Program in the Dutch Nano-R&D Program NanoNed, *NanoEthics*, **7**(1), 7–16.
3. Moor, J. H. (2005) Why We Need Better Ethics for Emerging Technologies, *Ethics and Information Technology*, **7**, 111–119.
4. Collingridge, D. (1980) *The Social Control of Technology* (Frances Pinter, London).
5. Collingridge, D. (1983) Hedging and Flexing: Two Ways of Choosing under Ignorance, *Technological Forecasting and Social Change*, **23**(2), 161–172.
6. Bond, P. J. (2005) Responsible Nanotechnology Development. In: Swiss Re (ed.), *Nanotechnology: Small Size-Large Impact?* (Swiss Re, Zurich).
7. Rip, A. (2006) Folk Theories of Nanotechnologists, *Science as Culture*, **15**(4), 349–365.
8. Macnaghten, P., M. B. Kearnes, and B. Wynne (2005) Nanotechnology, Governance, and Public Deliberation: What Role for Social Sciences?, *Science Communication*, **27**(2), 268–291.
9. European Commission (2010) *Communication from the Commission on Europe 2020: A Strategy for Smart, Sustainable and Inclusive Growth*, COM (2010) 2020 Final (European Commission, Brussels).
10. European Commission (2010) *Communication from the Commission to the European Parliament, the Council, the European Economic and Social Committee and the Committee of the Regions on Smart Regulation in the European Union*, COM (2010) 543 Final (European Commission, Brussels).
11. Solesbury, W. (2002) The Ascendancy of Evidence, *Planning Theory and Practice*, **3**(1), 90–96.
12. Kuhlmann, S., P. Stegmaier, K. Konrad, and B. Dorbeck-Jung (forthcoming 2017) Tentative Governance in Emerging Science and Technology—Conceptual Introduction and Overview, *Research Policy*.
13. Brownsword, R., and M. Goodwin (2012) *Law and the Technologies of the Twenty-first Century: Text and Materials* (Cambridge University Press, Cambridge).

14. Marchant, G. E., B. R. Allenby, and J. R. Herkert (eds.) (2011) *The Growing Gap Between Emerging Technologies and Legal-Ethical Oversight* (Springer, New York).
15. Posner, R. A. (2000) Past-Dependency, Pragmatism, and Critique of History in Adjudication and Legal Scholarship, *The University of Chicago Law Review*, **67**(3), 573–606.
16. Hodge, G. A., A. D Maynard, and D. M. Bowman (2013) 'Nanotechnology: Rhetoric, Risk and Regulation' (2013), *Science & Public Policy*, sct029.
17. Harremoës, P. (ed.) (2001) *Late Lessons from Early Warnings: the Precautionary Principle 1896–2000* (European Environmental Agency, Copenhagen).
18. Thomas, J. M. (2001) Predictions. *Notes and Records of the Royal Society of London*, 105–117.
19. Ludlow, K., D. M. Bowman, and D. D. Kirk (2009) Hitting the Mark or Falling Short with Nanotechnology Regulation?, *Trends in biotechnology*, **27**(11), 615–620.

Part 1

Variety in the Governance of Newly Emerging Technologies

Part 2

Chapter 2

Reflexive Co-Evolution and Governance Patterns

Harro van Lente[a] and Arie Rip[b]

[a]*Department of Technology and Social Studies,*
Maastricht University, Minderbroedersberg 4-6,
6211 LK Maastricht, The Netherlands
[b]*Department of Science, Technology and Policy Studies,*
School of Management and Governance,
University of Twente, PO Box 217, 7500 AE Enschede, The Netherlands
h.vanlente@maastrichtuniversity.nl, a.rip@utwente.nl

2.1 Introduction

This chapter is an attempt to broaden the notion of governance so as to be able to include a larger variety of patterns of governance and governance arrangements than is usual in the literature, at least as emerging science and technology are concerned. To do so, we introduce the notion of ongoing co-evolution of science, technology and society, a notion that has become quite common in science and technology studies and economics of innovation, explicitly [1–3], or in related terminology like mutual shaping of science, technology and society [4] and co-production [5].

Embedding New Technologies into Society: A Regulatory, Ethical and Societal Perspective
Edited by Diana M. Bowman, Elen Stokes, and Arie Rip
Copyright © 2017 Pan Stanford Publishing Pte. Ltd.
ISBN 978-981-4745-74-1 (Hardcover), 978-1-315-37959-3 (eBook)
www.panstanford.com

Typically, the development of new technologies is a process in which science, technology and society co-evolve. It does not just involve the creation of a new artefact with promising performance that is sent out into society. There are interdependencies from the beginning, in opportunities and constraints, and in anticipation on them. Researchers, firms and governmental actors can be aware of such interdependencies and may act upon their understanding of the process. In that sense, the co-evolution is reflexive. There are lots of intended and unintended inputs to this reflexive co-evolution of science, technology and society. The net effect is that certain patterns become visible in the interactions, for example the emergence of a dominant design [6, 7], or the accommodation to societal expectations and rules from government agencies [8]. These patterns of interaction have tentative rules and structures, which may institutionalize and then guide subsequent actions and interactions.

The key step we make is that such more or less institutionalized patterns constitute governance—not governance arranged by one or another actor, but governance in, and through, the pattern.[1] General examples of such governance patterns include an industry structure or a sociotechnical regime [10]. It is de facto governance [11], and it can be recognized for what it is and does, and how it is referred to, and can be reflexively used to guide choices. There can also be attempts to modify or modulate it [3].

What we do in this chapter is flesh out the notion of governance patterns that emerge in the reflexive co-evolution of science, technology and society. We discuss three examples of governance patterns and how they emerge, including the promises and warnings that create structures-to-be-filled-in-by-agency [12]. We can also consider the nature and desirability of the governance pattern, even if only speculatively when its institutionalization is still at an early stage.

We focus on the examples of

(1) early warning,

---

[1]This can be compared to how Mintzberg [9] viewed intentional (and often top-down) strategy in firms and other organisations as part of overall de facto, or in his words 'pattern' or 'emergent' strategies that are out there already.

(2) bridging of gaps between technology and society, and
(3) orientation towards Grand Challenges,

because in these cases the nature of the emerging technologies is a key element of how society attempts to re-orders itself in relation to those technologies.

The advantage of calling it a governance pattern, over and above it being an instance of the sociological phenomenon of institutionalization, is that questions of legitimacy and productivity can be raised and used to evaluate the quality of the pattern: Is this pattern a good governance pattern? Such a question appears to require standards against which the governance pattern and its eventual justifications can be evaluated. But these are not pre-given. They become articulated in, and through, the reflexive co-evolutionary processes. This last observation is particularly important for newly emerging technologies where eventual performance and societal embedding are indeterminate, but where there are already issues of governance; one example would be the debate about new possibilities for human enhancement promised by nanotechnology and pushed for by some promoters. It might well happen that our values about human enhancement will change because of the new technological options—another thread in the co-evolution of science, technology and society [13].

2.2 Actual Reflexive Co-Evolution of Technology in Society

An important insight from historical and socio-economic studies of technological change is that 'technology' and 'society' are deeply intertwined [4]. Instead of deciding between technological determinism (the assumption that technology as such shapes societal change) or societal determinism (assuming that the eventual shape of technology can be explained from social forces), it is their entanglement and co-evolution that should be studied. Indeed the term 'co-evolution' has become popular. There are, for instance, studies of the co-evolution of the design of artefacts and their use, which show how the eventual shape of an artefact can be understood as a result of interactions with actual and intended users [14]. Others have studied the co-evolution of novel products and markets, and demonstrated how the developers of new products seek to create new markets but, at the same time, are constrained by such emerging new markets [6].

Another important lesson relates to the co-evolution of organizations and their competitive environment: In responding to environments, organizations also change those environments. In general, the co-evolution of technology and institutions [2] and evolution of socio-technical regimes [10] is an important perspective in the social studies of science and technology. The idea of co-evolution has also been applied to innovation policies and practices and the theories of innovation policies [15], and to science and technology development and attempts to modulate it [3].

The additional aspect of co-evolution is its reflexivity. Arguably, reflexivity is a feature of late-modern societies that became apparent in the second half of the 20th century [16]. Giddens [17] argues that modernity is shot through with reflexivity: Modern institutions will not just act according to their role and position, but will also reflect on them to ensure their survival and to improve their efficacy. Reflexivity plays out at the level of individuals as well, and they might take up concepts and claims of social science about social actors and interactions. This is where our analysis of governance patterns can come in. It describes actors and the outcome of their interactions, but can then be used by the actors themselves; this is what Giddens has called 'double hermeneutics' [17].

Reflexive co-evolution thus happens all the time, with more or less explicit reflexivity. There are macro-level patterns, such as the modern settlement in general [17–19]. Other patterns relate to technological change, for example the equation of technology and progress since World War II [20], the contestation of modernity [21], the fertile ground for technological promises and hype-disappointment cycles [22].

Patterns also occur at the meso-level, which is where our cases are mainly located. The phenomenon of innovation journeys [23] is an important example for the analysis of newly emerging technologies:

The reason the innovation journey for industrial product and process innovations has a recognizable and recurrent pattern is the way industrial innovation has become institutionalized in our societies. The pattern is reproduced because it is seen as the natural way to do things and because it remains productive [24:213].

When such patterns are reflected upon as forms of de facto governance, this overlaps with reflections on governance. Established forms of governance, characterized as markets, hierarchies and networks, do not just passively exist, but are designed, contested and negotiated in particular ways. Thus, there is a governance of governance, or 'meta-governance' as discussed by Jessop [25].

For the governance of governance, also in the case of newly emerging technologies, one entrance point is to trace co-evolutions over time. The patterns manifested in the emergence of new technologies become governance structures in themselves, and these build on each other. The patterns are not just repetitive in their structuring of developments, but are also used by actors as ways (reflexively) to understand the situation and act accordingly. This is visible in the way responses to the challenge of newly emerging technologies in general show a certain progression [26].

In the 1970s and 1980s, new technologies such as recombinant DNA and biotechnology in general, or nuclear power, were subjects of societal contestation. The exchange between proponents and critics led to particular dynamics of polarization, with the repetition of moves and establishment of repertoires. In the 1990s, the pattern of polarization shifted to a combination of the general interest in innovation and the parallel emergence of Ethical, Legal and Social Issues (ELSI) studies—with the Human Genome Project as an exemplar.

With the identification of nanotechnology as a new frontier, a next stage has emerged where co-evolution of technology and society is becoming more reflexive. The different relevant actors, ranging from technology developers, policy makers, insurance companies to NGOs and other civil society groups, now operate from the acknowledgement of the interdependences between research activities, scientific fields, funding opportunities and societal visions. There are increasing interactions, manifesting in public dialogues and calls for 'responsible development', which are now common in nanotechnology (see Section 2.4). Over the last 30 years or so, the pattern of contestation of new technology has been accompanied by the co-construction and co-production of embedding in society. Actually, the novel handling of nanotechnology may become a model for how society can recognize and manage an emerging technology. The distributed governance patterns that emerge for nanotechnology will be available for other emerging technologies too, as one sees happening already for synthetic biology [27].

2.3 Early Warning as a Governance Pattern

The notion of 'early warning' emerged in the early 1960s in the debate around Rachel Carson's book *Silent Spring* [28]. Carson's book is now seen as the exemplar of early warning, and used to indicate the legitimacy of early warnings in other contexts.[2] In retrospect, earlier examples of early warning have been recognized, such as the case of asbestos which has become being well known because of the award of liability payments to victims and the decision that insurance companies had to cover them, once the cause of health damage was established and accepted.[3]

There is a recurrent pattern associated with early warning: resistance to early warning by those with vested interests, linked to attempts to create doubts about the plausibility of the warning, and then gradual shifts in position when the early warning is getting a hearing. The debate about the reduction of the ozone layer after the early warning by Rowland and Molina in 1974 is a clear example [32]. A similar pattern of early warnings, resistance by the industry and eventual adaptations was visible in the contestations about safety of the American automobile, as detailed by Penna and Geels [33].

---

[2]Exemplar as in a Kuhnian paradigm, or a regime of early warning. This is clear in how Harremoës [29] can write about 'late lessons from early warning'.
[3]An additional point is that the example of asbestos played an important role in the discussion about health risks of carbon nanotubes with a shape similar to that of asbestos fibres. It led re-insurance company Swiss Re to issue a warning [30, 31] that created legitimacy for the concerns about health risks of nanoparticles.

We emphasized the 'warning' aspect here, but the element of 'early' in the development of emerging technologies is important for the dynamics as well, so one might want to speak of 'early signalling' which includes promises as well as warnings. At an early stage, the situation is indeterminate: The nature of the technology and the dimensions of performance, let alone its embedding in society, are unknown. All claims can be contested, whether these are promises, warnings or claims about the impossibility or irrelevance of the new technology. In other words, a new technology opens up the existing order, at least in the realm of anticipation, and in that space old and new voices are heard, claims are articulated [34], and positions are negotiated—including negotiation of the legitimacy of contestation at all.

While the basic pattern of contestation remains, and is unavoidable in a sense because of the indeterminacy of the situation, shifts have occurred. For one thing, there is now an expectation that there will be opponents to a new technology. More important as a de facto governance pattern is that it is no longer contested that there might be undesirable side effects and impacts, so early-warners cannot be accused of being against progress (although that still occurs in concrete cases, for example in the debate about risks of genetically modified organisms (GMOs)). The need for some early warning was one of the arguments to have parliamentary technology assessment offices, explicitly in the United States. The new legitimacy of early warning supported the acceptance of the precautionary principle and the concurrent shift in the burden of proof about possible risks.

There is, however, a downside as well. If every early warning is accepted just because it is an early warning, the door is opened to eccentric and ridiculous claims.[4] Giving all warning claims a hearing would undermine the productivity of this governance pattern, so there must be thresholds even if these can never be unequivocal. But this is clearly a different situation from the one where concerned early-warners had to go to great effort to get a hearing at all.

---

[4]An example is the campaign against vaccination for influenza, when one argument was that with the vaccine nano-robots would be injected which would allow governments to monitor and control people at a distance.

There is more to say, for example about the role of visionaries and prophets (including prophets of doom), but also about whistle blowers and their protection. What we have discussed here is sufficient, however, to show that a new governance pattern has emerged and is stabilized.

2.4 Bridging the Gap between Technology Development and Society

For nanotechnology, but also earlier for biotechnology (but then more tentatively), there are attempts to bridge the gap between technology development and society—which to some extent are two different worlds. Figure 2.1 offers a visualization of the various activities that occur and have become more or less accepted.

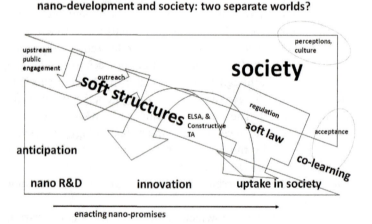

Figure 2.1 Bridging the gap between nanotechnology and society.

This visualization includes some strong assumptions, including a rather harmonious view of society full of good citizens interested in engagement and learning about the new technology.[5] While this occurs, there is also contestation, there

---

[5]There is also an assumption that the new technology is full of promises. This is visible with researchers and with policy makers (who have to fill their portfolios). Industrialists are not always so keen because of the open-ended character of the promises, and so-called waiting games occur [35].

are conflicting interests and perspectives, and struggles about the directions in which to go. The point of the visualization, however, is to indicate interactions and how these get entangled in what we call here 'soft structures' like consultations and codes of conduct for science and technology actors.

What could not be visualized in Fig. 2.1 is the increasing interest and activities of technology actors in anticipating societal embedding. Given earlier experiences with newly emerging technologies, particularly biotechnology, which included setbacks with respect to the promises, concerns about impacts and criticisms of the directions of development, the need to do better is translated into activities to 'do it right from the very beginning' [36, 37]. There is, we add, an illusion of control here—what you do in the beginning will make everything alright in the end—and in practice, too rapid a recourse to better communication rather than actually doing things right. But it does show a shift in thinking, and this has created further openings for approaches like Constructive Technology Assessment [38] and Real Time Technology Assessment [39]. The recent policy discourse of Responsible Research and Innovation (RRI) [40], where science and technology actors are asked to take on some responsibility for the embedding of their technologies in society, also builds on this shift, and constitutes a further step: Anticipation becomes a requirement.[6]

Anticipation on eventual societal embedding is precarious, not just because of the many unknowns (and unknown unknowns), but also because technology developers have at best partial control of the societal embedding of their technologies [41]. Thus, there must be alliances with other actors; some of this is visible in exercises of co-construction, even co-production of new technology.

While many of these activities are still tentative, one can see some patterns emerge and these might coalesce into a new regime of 'managing technology in society'. One can speculate about the shape this governance pattern will take [11]. There is definitely emphasis on horizontal interactions, with some authority. Thus, responsibilities for better technology in a better society will be delegated to interactions between technology actors and

---

[6]Rip [8] argues that spaces for anticipation in general are becoming a characteristic of technology development.

accredited society actors (stakeholders, NGOs). This is a move away from traditional representative democracy, where the responsibilities for a better society lie with Parliament and Government. In other words, there is an element of neo-corporatism. With its challenge to existing divisions of moral labour [42], RRI, when implemented, will reinforce this trend. This need not be frowned up if it is reflexive neo-corporatism. Maybe this is the best we can do in managing new technologies in society [43].

2.5 Governance by Orientation: The Discourse of Grand Challenges

Innovation, as well as innovation policies, are changing, partly because of challenges of technoscientific developments and partly because of the need that is felt to orient R&D also towards broader societal issues than economic competitiveness. In this respect, the recent policy discourse of Grand Challenges is interesting, as such, and because it may create governance patterns rather than rely on steering to achieve a mission.

A starting point is the European Union's *Lund Declaration*, which opens with a clarion call: 'European research must focus on the Grand Challenges of our time moving beyond current rigid thematic approaches' [44:1]. This thinking continues in Horizon 2020, the successor of the EU's Seventh Framework Program. Typically, the challenges, often characterized by a short label like 'the ageing society' to indicate the overall thrust, refer to improving the quality of life in terms of health, mobility or security; often sustainability is added as umbrella term. The Lund Declaration already notes that this involves new forms of coordination, between European and national policies as well as between public and private actors. The discourse of Grand Challenges is not an exclusively EU affair: The US 'Strategy for American Innovation' also refers to societal challenges [45].

The discourse now also permeates national and agency policy frameworks in EU Member States. The language used sometimes resembles that of a war to be fought, as earlier in the US 'War on Cancer' [46]. According to Daimer *et al.* the focus on Grand Challenges indicates a normative turn in innovation policies and the emergence of a new rationale for policy making: 'orientation failure' [47:222]. This phrasing refers to the notion of 'market failure', the dominant legitimation of pro-active technology and innovation policy since World War II. The idea of 'market failure' is that government intervention is only justified if the level and nature of R&D investments and activities on the basis of market mechanisms is sub-optimal. Increasingly, and since the 1990s supported by studies of innovation systems, governments have tried to improve the framework conditions for innovation [48]. There can be system failure, however, for example, when the innovation system shows structural deficits, such as poor interaction between different actors in the system. The next step now appears to be a concern about the direction of innovation, an orientation failure as it were.

The Grand Challenges can also pick up on the promises of emerging technologies. Conversely, new technologies link up with societal needs, as is clearly visible in the recent R&D programs for nanotechnology (e.g. the nanotechnology program of UK Research Councils (UKRC) led by the Engineering and Physical Sciences Research Council (EPSRC), and NanoNextNL in The Netherlands), which refer to problems of health, energy, water etc. to which nanotechnology R&D should contribute. Such programs also involve industry and other private actors to achieve societally important goals. This ambition neatly aligns with the wish for 'valorization': the commercial uptake and use of research outcomes. Researchers now anticipate such claims and are willing to indicate how much their research contributes to Grand Challenges [49].

Thus, in the recent discourse of Grand Challenges and the first attempts at implementation, general societal goals are introduced to pull innovation in specific directions. In other words, we see demand-pull techno-governance without market demand. Interestingly, while the Grand Challenge discourse started in discussions of government science and technology policy, there are now scenarios where the government steps back in favour of consortiums in which it still plays some role, and where big charitable foundations come in as well, with their administrative flexibility and their avowed goal of serving the public interest [50]. While Nordmann's 'Converging Technologies for the European Knowledge Society' [51] proposed broad articulation of directions in which to go, because converging technologies are crosscutting and can address a variety of goals, by now the reference to the Grand Challenges is more important than broad discussion.

Apart from the question of implementation, what is important to note about the Grand Challenges are the shift in language, and the new modes of handling technology and society. At the moment, Grand Challenges are new policy-speak, but the issues are important. They may well require a cultural shift in the research and innovation system, in agendas and ways of working, rather than checking whether research proposals address a specified Grand Challenge. Thus, overall orientation is more important than dedicated steering. Taken together, we see a possible pattern of governance, with emerging but not yet stabilizing institutions and soft structures.

2.6 In Conclusion

We offered a general argument and three cases of institutionalized patterns in reflexive co-evolution of technology and society. Such patterns have emerged and stabilized as a result of ongoing interactions and learning—without conscious or forceful planning—and thus should be seen as de facto governance patterns. We discussed three such de facto governance patterns as they occur around new technologies. One, about early warning, is fully stabilized (but evolving further), the second, about bridging the worlds of technology and society, is in the process of institutionalizing, and the third, about orientation towards Grand Challenges, is still emerging.

The patterns overlap: Scientists, technologists, industrialists, policy makers, and civil society actors are in situations where early warnings have to be responded to while there is also an expectation that work will address Grand Challenges. There may even be competing claims about what to do.

Our evaluation indicated the difficulties of giving all early warnings a hearing: It proffers legitimacy, but decreases productivity. Or when we pointed out that interactions of technology developers with stakeholders and civil society groups about technological developments, when outcomes would have some authority, would add neo-corporatist elements to our democracies. In general, such evaluations should focus on the actual development of the governance pattern, including the path dependencies that emerge.[7]

There is more to say. All governance patterns in the co-evolution of technology and society have to live with two key tropes: the trope (or figure) of technology coming in from the outside (as in the common metaphor of impact of technology on society), and the more recent trope of inclusive governance. The latter term was introduced by the European Commission in the early 2000s; the idea is strong independently of the particular label that is used. Both add a narrative element to the governance patterns that can be quite strong in how it orients actors.

What we also see is how the narrative of progress through technology, now enlightened by anticipation, remains strong. Actors may realize that these are narratives, and thus become more reflexive about them, and about the governance patterns more generally. One can compare this with macro-level trends, in particular reflexive modernization as diagnosed by Beck and Giddens. Beck in particular emphasizes blurring of boundaries between the grand dichotomies of modernity [16]. In the three cases we discussed we see some blurring, and new institutionalization, for instance through bridging structures between technology development and society. Blurring and bridging, and the related recognition of hybridity, are important to create openings for reflexive change. But then, inevitably, there will be some institutionalization and the path dependencies that go with it. Thus, in addition to welcoming the changes, it is important to evaluate the governance patterns that arise. That requires anticipation on further evolution of the pattern, an important element of meta-governance.

References

1. Nelson, R. R., and R. Winter (1982) *An Evolutionary Theory of Economic Change* (Harvard University Press, Cambridge).

---

[7]Rip [42] does this for the present interest, in the European Union and more broadly, in RRI, and calls for the inclusion of dedicated reflexive moments in the developments.

2. Nelson, R. R. (1994) The co-evolution of technology, industrial structure, and supporting institutions, *Industrial and Corporate Change*, **3**(1), 47–63.
3. Rip, A. (2006) A co-evolutionary approach to reflexive governance– and its ironies. In: Voß, J.-P., D. Bauknecht, and R. Kemp (eds.) *Reflexive Governance for Sustainable Development* (Edward Elgar, Cheltenham), pp. 82–100.
4. Bijker, W. E., and J. Law (eds.) (1992) *Shaping Technology/Building Society. Studies in Sociotechnical Change* (MIT Press, Cambridge).
5. Jasanoff, S. (2004) *States of Knowledge: The Co-Production of Science and the Social Order* (Routledge, London).
6. Utterback, J. M., and W. J. Abernathy (1975) A dynamic model of product and process innovation, *Omega*, **3**(6), 639–656.
7. Anderson, P., and M. Tushman (1990) Technological discontinuities and dominant designs: A cyclical model of technological change, *Administrative Science Quarterly*, **35**(4), 604–635.
8. Rip, A. (2012) The context of innovation journeys, *Creativity and Innovation Management*, **21**(2), 158–170.
9. Mintzberg, H. (1994) *The Rise and Fall of Strategic Planning* (Basic Books, London).
10. Geels, F. W. (2005) *Technological Transitions and System Innovations. A Co-Evolutionary and Socio-Technical Analysis* (Edward Elgar, Cheltenham).
11. Rip, A. (2010) De facto governance of nanotechnologies. In: Goodwin, M., B.-J. Koops, and R. Leenes (eds.) *Dimensions of Technology Regulation* (Wolf Legal Publishers, Nijmegen), pp. 285–308.
12. Van Lente, H., and A. Rip (1998) Expectations in technological developments: An example of prospective structures to be filled in by agency. In: Disco, C., and B. J. R. van der Meulen (eds.) *Getting New Technologies Together* (Walter de Gruyter, Berlin), pp. 195–220.
13. Swierstra, T., and A. Rip (2007) Nano-ethics as NEST-ethics: Patterns of moral argumentation about new and emerging science and technology, *NanoEthics*, **1**, 3–20.
14. Fleck, J. (2000) Artefact activity: The coevolution of artefacts, knowledge and organization in technological innovation. In: Ziman, J. (ed.) *Technological Innovation as an Evolutionary Process* (Cambridge University Press, Cambridge), pp. 248–266.

15. Smits, R. E., S. Kuhlmann, and P. Shapira (eds.) (2010), *The Theory And Practice Of Innovation Policy, An International Research Handbook* (Edward Elgar, Cheltenham).
16. Beck, U., W. Bonβ, and C. Lau (2003) The theory of reflexive modernization. problematic, hypotheses and research programme, *Theory, Culture & Society*, **20**, 1–33.
17. Giddens, A. (1991) *Modernity and Self-Identity. Self and Society in the Late Modern Age* (Polity, Cambridge).
18. Beck, U., A. Giddens, and S. Lash (1994) *Reflexive Modernization. Politics, Tradition and Aesthetics in the Modern Social Order* (Polity, Cambridge).
19. Latour, B. (1991) *Nous n'avons jamais été modernes. Essai d'anthropologie symétrique* (La Découverte, Paris).
20. Sarewitz, D. (1996) *Frontiers of Illusion: Science, Technology and Politics of Progress* (Temple University Press, Philadelphia).
21. Schot, J., and A. Rip (2010) Inventing the power of modernization, In: Schot, J., H. Lintsen, and A. Rip (eds.) *Technology and the Making of the Netherlands. The Age of Contested Modernization, 1890–1970* (MIT Press, Cambridge).
22. Van Lente, H., C. Spitters, and A. Peine (2013) Comparing technological hype cycles: Towards a theory, *Technological Forecasting and Social Change*, **80**(8), 1615–1628.
23. Van de Ven, A. H., D. E. Polley, R. Garud, and S. Venkataraman (1999) *The Innovation Journey* (Oxford University Press, Oxford).
24. Rip, A. (2010) Processes of technological innovation in context—and their modulation. In: Steyart, C., and B. van Looy (eds.) *Relational Practices, Participative Organizing* (Emerald Advanced Series in Management, Bingley), pp. 199–217.
25. Jessop, B. (2009) From governance to governance failure and from multi-level governance to multi-scalar meta-governance. In: *The Disoriented State: Shifts in Governmentality, Territoriality and Governance* (Springer, Amsterdam), pp. 79–98.
26. Rip, A., and H. van Lente (2013) Bridging the gap between innovation and ELSA: The TA program in the Dutch nano-R&D program NanoNed, *NanoEthics*, **7**(1), 7–16.
27. Douglas, C. M. W., and D. Stemerding (2013) Governing synthetic biology for global health through responsible research and innovation, *Systems and Synthetic Biology*, **7**(3), 139–150.

28. Carson, R. (1962) *Silent Spring* (Houghton Mifflin Harcourt).
29. Harremoës, P. (ed.) (2001) *Late Lessons from Early Warnings: The Precautionary Principle 1896–2000* (European Environment Agency, Copenhagen).
30. Swiss Re (2004) *Nanotechnology: Small Matter, Many Unknowns* (Swiss Reinsurance Company, Zürich).
31. Swiss Re Centre for Global Dialogue (2007) *The Risk Governance of Nanotechnology: Recommendations for Managing a Global Issue*, Conference Report (Swiss Reinsurance Company, Zürich).
32. Dotto, L., and H. Schiff (1978) *The Ozone War* (Doubleday, Garden City).
33. Penna, C. C. R., and F. W. Geels (2012) Multi-dimensional struggles in the greening of industry: A dialectic issue life-cycle model and case study, *Technological Forecasting and Social Change*, **79**, 999–1020.
34. van Lente, H., and J. I. van Til (2008) Articulation of sustainability in the emerging field of nanocoatings, *Journal of Cleaner Production*, **16**(8), 967–976.
35. Parandian, A., A. Rip, and H. te Kulve (2012) Dual dynamics of promises and waiting games around emerging nanotechnologies, *Technology Analysis & Strategic Management*, **24**(6), 565–582.
36. Roco, M. C. (2001) International strategy for nanotechnology research, *Journal of Nanoparticle Research*, **3**(5–6), 353–360.
37. Krupp, F., and C. Holliday (2005) Let's get nanotech right, *Wall Street Journal*, **14**, B2.
38. Schot, J., and A. Rip (1997) The past and future of constructive technology assessment, *Technological Forecasting and Social Change*, **54**, 251–268.
39. Guston, D. H., and D. Sarewitz (2002) Real-time technology assessment, *Technology and Society*, **24**, 93–109.
40. Owen, R., J. Bessant, and M. Heintz (eds.) (2013) *Responsible Innovation* (John Wiley, London).
41. Jasper Deuten, J., A. Rip, and J. Jelsma (1997) Societal embedding and product creation management, *Technology Analysis & Strategic Management*, **9**(2), 131–148.
42. Rip, A. (2014) Past and future of responsible research and innovation, *Life Sciences, Society and Policy*, **10**, 17.
43. Fisher, E., and A. Rip (2013) Responsible innovation: Multi-level dynamics and soft intervention practices. In: Owen, R, J. Bessant, and

M. Heintz (eds.) *Responsible Innovation: Managing the Responsible Emergence of Science and Innovation in Society* (John Wiley & Sons, Chichester), pp. 165–183.

44. Council of the European Union (2009) *Lund Declaration* (Swedish Presidency of the Council of the European Union, Brussels).

45. White House (2011) *A Strategy for American Innovation: Securing Our Economic Growth and Prosperity* (National Economic Council, Council of Economic Advisers, and Office of Science and Technology Policy, Washington, D. C.).

46. Rettig, R. A. (1977) *Cancer Crusade: The Story of the National Cancer Act of 1971* (Princeton University Press, Princeton).

47. Daimer, S., M. Hufnagl, and P. Warnke (2012) Challenge-oriented policy-making and innovation systems theory: Reconsidering systemic instruments. In: Fraunhofer ISI (ed.) *Innovation System Revisited: Experiences from 40 Years of Fraunhofer ISI Research* (Fraunhofer Verlag, Stuttgart), pp. 217–234.

48. Larédo, P., and P. Mustar (eds.) (2001) *Research and Innovation Policies in the New Global Economy: An International Comparative Analysis* (Edward Elgar, Cheltenham).

49. Bos, C, A. Peine, and H. van Lente (2013) Articulation of sustainability in nanotechnology: Funnels of articulation. In: K. Konrad, C. Coenen, A. Dijkstra, C. Milburn, and H. van Lente (eds.) *Shaping Emerging Technologies: Governance, Innovation, Discourse* (AKA Verlag, Heidelberg), pp. 231–242.

50. Kuhlmann, S., and A. Rip (2014) *The Challenge of Addressing Grand Challenges. Report to the European Research and Innovation Area Board* (University of Twente, Enschede).

51. Nordmann, A. (2004) *Converging Technologies: Shaping the Future of European Societies* (Office for Official Publications of the European Communities, Luxembourg).

Chapter 3

Regulatory Governance Approaches for Emerging Technologies

Bärbel Dorbeck-Jung[a] and Diana M. Bowman[b]

[a]Department of Public Administration,
Faculty of Behavioural Management and Social Science,
University of Twente, 7500 AE Enschede, The Netherlands
[b]Sandra Day O'Connor College of Law and
School for the Future of Innovation in Society,
Arizona State University, 111 E Taylor St Phoenix, Arizona 85004, USA

b.r.dorbeck-jung@utwente.nl, diana.bowman@asu.edu

3.1 Introduction

> *If a man will begin with certainties, he shall end in doubts; but if he will be content to begin with doubts he shall end in certainties.'*
> — Sir Francis Bacon

Technology and technological innovations are—as history reminds us—tightly coupled with societal change. The emergence of new technologies and the products associated with their commercialization may, for example, be driven in response by a societal need, or through entrepreneurial foresight. In both cases, the emergence of the technology will be paired with

---

Embedding New Technologies into Society: A Regulatory, Ethical and Societal Perspective
Edited by Diana M. Bowman, Elen Stokes, and Arie Rip
Copyright © 2017 Pan Stanford Publishing Pte. Ltd.
ISBN 978-981-4745-74-1 (Hardcover), 978-1-315-37959-3 (eBook)
www.panstanford.com

questions regarding its utility, its potential impacts and social—and therefore market—acceptance. Questions regarding how to facilitate beneficial innovation, and how to safeguard constitutional concerns (amongst which are the protection of intellectual property rights and safety), will similarly accompany the emergence of any new technology, and its applications, into today's market. Technological innovation operates, so to speak, with a 'social license'. This means, as Brownsword and Somsen have argued, that

> it falls to politicians and regulators, and ultimately to the law, to set the limits of technological innovation, to co-ordinate the assessment and management of risk, to design procedures for public participation, and to set the terms of compensatory responsibility. At the same time, though, it also falls to the regulators and to the law to establish a governance environment that is supportive of desirable technological innovation and that ensures that benefits are fairly shared [1:2].

According to the focus on co-regulation in the current governance movement, the 'social license' is interpreted as a shared responsibility of public and private regulators.[1] This view inspires the regulatory governance approaches that will be discussed within the context of this chapter.

For the inventor, manufacturer, commercialiser and public safety regulator of any new technology, support of beneficial development and minimisation of harms will contain multiple dimensions of uncertainty related to the effects, development and governability of the technology; the degree of which will differ depending on the question itself, the stage of the life-span, and the breadth and nature of the technology itself. Such uncertainties are not in themselves new; they are, instead, inherent with the emergence of any new technology.

What is arguably new, however, is the potential scope and impact that these 'uncertainties' may have, in part, due to the level of sophistication and complexity now associated with any new, emerging technology. But this is only part of the story. Another significant component is the nature of our society today; we live in a highly connected and integrated global environment, underpinned by the rapid movement of people and goods. As such, today's social laboratory, in which emerging technologies

---

[1]Governance can be defined as 'a system of co-production' of norms and goods where the co-producers are different kinds of actors [2].

are placed, is, for the most part, much larger, more integrated, and complex than what it was even a century ago. So too are the potential consequences—positive and negative—of any such uncertainties should they transpire.

Also new is the current effort to take the social license of technologies seriously at early stages of the development. Under the umbrella of so-called 'Responsible Research and Innovation' (RRI), a myriad of proposals have been brought forward to develop frameworks for guiding technological development and innovation effectively, and legitimately, under circumstances in which multiple uncertainties exist [3–5]. These initiatives aspire to address any negative perceptions of a technology before it comes to fruition, as well as to support beneficial development. They propose anticipatory and responsive governance approaches that can contribute to connect regulation to fast moving technological development. For nanotechnologies, such an approach has been proposed by Levi-Faur and Comaneshter [6]. In their words,

> ...there are not many examples that fit so well with, or speak so clearly to, the notion of the 'risk society' as the challenges of nanotechnology. Moreover, unlike other cases where the discussion of the associated risks has followed the development of new technologies, the discussion of the proper regulatory framework for this new technology advances hand in hand with the development of the technology itself [6:150].

In acknowledging that risk and regulation are omnipresent with the emergence of any new technology, the authors go on to suggest that such forms of innovation enable society to rethink what we mean, and how we frame, regulatory governance, and the objectives that such structures should address [6].

Nanotechnologies are particularly interesting for the discussion of regulatory governance approaches to emerging technologies. A significant volume of literature examining the 'fit' of current governance structures for nanotechnologies, and nano-based products more specifically, has been published in recent years. These analyses and critiques have, in general, become increasingly sophisticated and specific in the questions posed by the authors, and the process by which they are framed [7–11]. The growing volume of production, and entry of nano-based products into the market (see, for example, [12]), has also shifted many of these analyses from the theoretical to a more empirical evaluation [13].

This chapter explores effectiveness aspects of two governance arrangements that have been recently launched in the European Union (EU) to align nanotechnologies to requirements of the responsible research and innovation agenda. Emerging technologies do not materialise in a governance vacuum. Pre-market research, development and testing, for example, are subject to—at least in most jurisdictions—a myriad of governance requirements. This may include, for example, occupational health and safety (OH&S) requirements, compliance with certain industry/government standards, and general law obligations, such as the law of torts. Entry of the technology and/or its products onto to market will similarly attract, or trigger, a multitude of governance arrangements, which will include components of hard and soft regulation. It is the applicability and the effectiveness of these regimes for managing the uncertainties associated with the latest wave of emerging technologies that have been increasingly questioned in relation to emerging technologies [14–16].

The first governance arrangement that is discussed in this chapter serves to handle OH&S in the research, development, production and use of nanomaterials. The second example focuses on the arrangements that have been put into place to address uncertainties associated with cosmetics containing certain types of nanomaterials. These two examples have been selected because they are prominent in the current discourse on how to cope with regulatory problems of emerging technologies, in which the theoretical or perceived uncertainties continue to dominate debate. Due to the differences between the governance arrangements and approaches, as well as to different conditions in regulatory practice, the effectiveness issues may be different in the two examples, but they may also be similar. Therefore it can be assumed that the evaluation of the case studies will provide interesting insights that trigger our imagination on how to regulate emerging technologies effectively.

The chapter is structured as follows. Section 3.2 deals with promising regulatory governance approaches to emerging technologies; in doing so, it lays the foundation for the two examples of nano-specific regulation. Section 3.3 explores effectiveness issues of governance arrangements on exposure to nanoparticles in the workplace (benchmarks for exposure limits) and cosmetic products containing nanoparticles ('nanocosmetics'). Finally, tentative lessons are drawn with regard to effective regulation of emerging technologies with complex uncertainties in Section 3.4.

3.2 Promising Regulatory Governance Approaches

The objective of Section 3.2 is to examine the different regulatory governance models that could be employed to address the uncertainties associated with emerging technologies. Here we focus on the theory of technology regulation, as well as on regulatory strategies and some theoretical models that have been developed specifically to address the challenges of nanotechnologies.

3.2.1 Models of Technology Regulation

In the history of technology regulation we see that responses to earlier emerging technologies spread across a continuum, flowing from absolute prohibition by the state (for instance, human reproductive cloning), to almost no specific regulation (for instance, the Internet, see [17]). Traditionally, development of regulatory models has been evidence-based. Regulators usually wait before introducing new legislation until there is robust knowledge on the characteristics and effects of social phenomena [1, 18]. Traditional approaches build on 'inherited regulation' [19]. When no specific legislation is established by the legislature, the emerging technology and its products are still regulated across the whole of their life-cycle through the same instruments and regimes as their conventional equivalents.

For more than 15 years, the evidence- and risk-based, reactive model to technology regulation has been subject to scholarly critique and contested by those in favour of more precautionary approaches. In the EU, the precautionary principle is enshrined in the Treaty (Article 191(2) *Treaty on the Functioning of the European Union*[2]) and requires EU policy to aim at a high level

---
[2]Consolidated version of the *Treaty on the Functioning of the European Union* [2010] OJ C83/47.

of environmental protection. This has been interpreted as an obligation on EU institutions to take protective steps, even if the risks cannot be fully demonstrated because of the insufficiency of the scientific data (see, for example, [20–22]). Given the potential for a mismatch between the technology and the existing regulatory regime, and the resulting periods of under- and over-regulation and/or regulatory lag, regulatory scholars questioned the short and long-term wisdom of the do-nothing approach [19, 23, 25].

Proposing alternatives such as, for example, responsive regulation [6, 16, 24–27], co-regulation [28], reflexive regulation [29], experimental regulation [30], and meta-regulation [13, 31, 32], have been articulated over the past two decades. Under these frameworks and theoretical constructions, regulatory scholars have continued to build on, and refine, their insights on governance theory. These alternatives seem to be promising to take the social license of technological innovation seriously under conditions of multiple uncertainties (e.g. co-regulation). They appear to respond to the current movement of responsible research and innovation, which shifts the focus from risk regulation to the support of beneficial (technological) innovation [4].

To manage the delicate balance between promising development and potential risks Mandel [27] recommended a particular governance system for emerging technologies. This model includes improving data gathering and sharing, filling newly created or exposed regulatory gaps, incentivizing strong corporate stewardship, enhancing agency expertise and co-ordination, providing for regulatory adaptability and flexibility, and achieving substantial, diverse stakeholder involvement.

3.2.2 Regulatory Models for Nanotechnologies

To cope with the regulatory challenges posed by nanotechnologies regulatory strategies have now been developed and applied in the United States (US), the EU, Australia, New Zealand and Canada. A number of theoretical models have been proposed by leading commentators. This section illustrates an approach that is prevailing in regulatory policies. Thereafter some leading theoretical proposals are examined which draw on the critique of this approach.

Most governments have taken similar incremental regulatory approaches to nanotechnologies. This approach involves activities such as testing the adequacy of the existing regulatory regimes and revising them only if regulatory gaps are detected [33–35]. In some countries this approach is complemented through co- or meta-regulatory initiatives. Meta-regulation like the *Code of Conduct for Responsible Nanosciences and Nanotechnologies Research* launched by the European Commission in 2008 [36] aims to steer the process of self-regulation of nanotechnological business and research organisations. Public regulators seem to expect that they can benefit from the collaboration with private regulators that have established specific nano-related guidance [11, 37]. Most of the nano-specific regulatory initiatives focus on soft regulation. The launching of voluntary data reporting schemes,[3] the incorporation of definitions of 'nanomaterials' into policy documents,[4] and through the stimulation of corporate stewardship and public dialogue, governments appear to follow the approach articulated by Mandel [27].

The prevailing incremental or 'muddling through' strategy,[5] which is not embedded in a coherent regulatory governance system, has received critical comments of leading scholars. As Marchant and Wallach have observed,

> to date, governance of emerging technologies has generally proceeded in a piecemeal approach, in which government agencies and developers of soft law programs propose new oversight initiatives one piece at a time, with little regard to other initiatives affecting the same technology [40:9].

In order to address these deficiencies, Marchant and Wallach [40] suggest that greater coordination is required between the different national bodies and agencies. Such an approach would reduce duplication, assist in identifying gaps and uncertainties, while simultaneously promoting innovation. A new regulatory body would not be needed for this, Marchant and Wallach [40] argue. Rather, coordination should take place through, in their view, an independent Issues Manager, who would

---

[3]See, for example, the Voluntary Reporting Scheme of the United Kingdom's Department for Environment, food and Rural Affairs [38].
[4]See Commission Recommendation of 18 October 2011 on the definition of nanomaterial (2011/696/EU) (EC, 2011).
[5]See Lindblom's characterization of the incremental approach [39].

act 'more like an orchestra conductor in trying to harmonize and integrate the various governance approaches that have been implemented or proposed by others' [40:9].

This role would be complemented through the formation of a more formal, yet still independent, body—the Governance Coordinating Committees (GCC) [40]. The role of a GCC would not be to regulate; rather, to 'monitor, manage and modulate an emerging technology' [40:20]. This would, however, including overseeing and coordinating regulatory approaches to the technology and its products, as well as a range of broader activities around innovation activities, public engagement, acting as clearing house for information, etc. A similar body, the Scanning Probe Agency, an institution of permanent vigilance related to nanotechnologies was previously proposed by Gammel et al. [41].

Against this backdrop, a number of commentators have suggested the need for a more integrated regulatory approach that reflects both the global nature of trade and the porous state of jurisdictional boarders in relation to human and environmental health risks. Abbott et al. [42, 43] have, for example, proposed the creation of a *Framework Convention on Nanotechnologies*. Such instruments are inherently flexibility in nature, and it is this characteristic, combined with its ability to gradually harden obligations on member states, that makes it ideal for addressing a technology that is underpinned by so many uncertainties [42].

3.3 Reviewing Nano Regulatory Governance: The Story so Far

3.3.1 Introduce Evaluation Frame of Cases: Some Effectiveness Questions

The two EU governance arrangements that are explored in this section are unique because they include nanospecific provisions that have not been established in other countries or regions. The nanospecific regulations are either part of soft regulation (exposure limits) or legislation (nanocosmetics), the latter of which came fully into effect on 11 July 2013.[6] At this stage, it

---

[6]Regulation (EC) No 1223/2009 of the European Parliament and of the Council of 30 November 2009 on cosmetic products.

seems to be particular useful to get insights into potential effectiveness problems.

Generally, regulatory effectiveness is focused on the degree to which regulation achieves certain policy goals [44]. Regulatory scholars assume that the achievement of policy goals requires that the regulated parties follow defined rules and/or models of behaviour [45, 46]. Rule compliance depends on whether the regulated parties *can* follow the rules ('capacity') and whether they *are willing* to do so ('willingness') [47, 48]. Regulatory governance of emerging technologies, however, requires a broader perspective on effectiveness than rule compliance. Under conditions of multiple uncertainties, and at this early stage of implementation, the strength of regulation seems to be primarily to stimulate the reflection of the involved social actors [29] and to convince them to take their regulatory responsibilities seriously ('responsibilisation', see [49]). Science, technology and society (STS) scholars remind us of rules being useful even when they function as a conversational tool [50, 51]. Hence, the exploration of the two examples focuses on the process in which rule compliance is embedded including responsibilisation activities.

3.3.2 Governance Arrangements to Limit Exposure to Nanoparticles in the Workplace

According to EU Directives, employers are obliged to care for all aspects of safety and health in the workplace according to the state of science.[7] With regard to hazardous substances, the legal duty to care includes the obligation to limit the employees' exposure following generally acknowledged values ('Occupational Exposure Limits' (OELs) and 'Derived No-Effect

---

[7]See, Directive 89/391/EEG (Council Directive of 12 June 1989 on the Introduction of Measures to Encourage Improvements in the Safety and Health of Workers at Work, OJ. L 183 (29-06-1989)), pp. 1–8, Directive 98/24/EG Council Directive 98/24/EC of 7 April 1998 on the protection of the health and safety of workers from the risks related to chemical agents at work (fourteenth individual Directive within the meaning of Article 16(1) of Directive 89/391/EEC) OJ L 131 (5. 5. 98) pp. 11–13, and Directive 2004/37/ EG (Directive 2004/37/EC of the European Parliament and of the Council of 29 April 2004 on the protection of workers from the risks related to exposure to carcinogens or mutagens at work (Sixth individual Directive within the meaning of Article 16(1) of Council Directive 89/391/EEC) OJ (29.6.2004) L 229, pp. 23–34.

Levels' (DNELs); see, for example, [52]). The intentional or unintentional exposure to free nanoparticles of any type in the workplace creates certain difficulties in relation to the duty to comply with this obligation due to two crucial uncertainties. First, since there is no evidence at this time as to whether exposure to free any or specific types and/or families of nanoparticles are harmful to human health, we do not know whether the legal obligation applies. Clarification is still needed whether nanomaterials, or which types of free nanoparticles under what specific conditions, should be characterised as hazardous substances.

If we treat all nanomaterials as hazardous substances—which is problematic and is not supported by the scientific literature—we are confronted with a further difficulty. This is that risk management procedures and processes that are specific to nanomaterials have not yet been established; there is not, for example, such a thing as a 'gold standard' for nanospecific risk management. Although a few nanospecific OELs and DNELs, as well as risk assessment methods and exposure measurement devices, have been proposed, legal acknowledgment is still lacking.

To better implement a precautionary approach in the EU, some member states (including the United Kingdom (UK), Germany and the Netherlands) have developed provisional benchmarks that can guide employers to limits the employees' exposure to nanoparticles in companies were nanomaterials are produced or used [52].

This section discusses the Dutch governance approach. In this context, we focus on certain exposure benchmarks, known as the 'Nano Reference Values' (NRVs), that were recommended by the Dutch Minister of Social Affairs and Employment to the Dutch Parliament in 2010 and 2012.[8] This example is particularly interesting because it has gained the attention of the media

---

[8]See letters of the Minister to Parliament on 10 August 2010 (G&VW/GW/2010/14925) and of 11 December 2012 (G&VW/GW/2012/7864). Nano Reference Values (NRVs) are benchmarks that provide provisional limits of the employees' exposure to nanoparticles until OELs and DNELs can be established. The NRV defines a maximum, generic level (benchmark) for the concentration of nanoparticles at the workplace which takes background particles into account. Usually, NRVs focus on the exposure of engineered nanoparticles. Engineered nanoparticles are distinguished from background and process-generated nanoparticles [52]. NRVs are the outcome of standard setting 'by analogy'. For carbon nanotubes, for example, a provisional fibre concentration is proposed, based on an exposure risk ratio for asbestos.

and relevant stakeholders beyond national borders.[9] It is a rich example because it builds on, refines and extends the earlier UK and German proposals. Moreover, data on the potential use of NRVs have been published [53], which provide us with insights on potential effectiveness problems.[10]

The Minister's 2010 and 2012 recommendations on NRVs are one of the outcomes of an anticipatory, collaborative and responsive governance approach that has been developed in the Netherlands to safeguard OH&S aspects of employees working in an environment in which nanomaterials are produced and/or used. The approach to developing the NRVs was based on political and scientific consensus. To achieve such a consensus in 2008 the Minister requested employers, their organisations and trade unions ('social partners') to jointly develop a strategy to address potential nanoparticle exposure within the workplace. In 2010, the Minister responded to the social partners' recommendation on developing NRVs by commissioning a scientific organisation with a technical review of this instrument, as well as by funding another project on the further development of NRVs and the evaluation of governance aspects. On the basis of the recommendation to follow the conclusions of the scientific reports, the Minister proposed in 2012 that employers use the developed NRVs. He also encouraged the Labour Inspectorate to integrate the NRVs into enforcement activities.

The Minister's 2010 and 2012 recommendations are an example of *soft regulation* by the state. The recommendations involve certain rules of conduct (e.g. to use NRVs as a provisional tool and as pragmatic warning levels which are part of the

---

[9]The Dutch benchmarks build on the BSI, NIOSH, Safe Australia and German proposals. Since they have been launched in 2012 they have 'travelled' to Finland and to the European Union.

[10]The evaluation was conducted in the context of a pilot study that was funded by the Dutch Ministry of Social Affairs and employment [53]. To get insights into the capacity and willingness to use the NRVs in practice 12 producers and users (companies), 25 Dutch key stakeholders (government, business associations, trade unions), 10 German key stakeholders (government, companies, business associations, assurance organisation and 2 European key stakeholders (business association and scientific expert) were interviewed (in total 49 interviews). In September 2013 and November 2013, the current use of NRVs in practice was explored with a document analysis and 2 interviews with the Dutch and the German key stakeholder. The discussion of potential effectiveness problems in this section builds on the empirical research.

state of science, as well as to include them in the enforcement activity sheet). The use of NRVs is voluntary, yet not 'without commitment'. While there is a legislative obligation on employers to ensure workplace safety, there is no requirement that employers follow NRVs. Employers may choose to use scientifically acknowledged alternatives, such as 'control banding', which does not primarily focus on quantitative measurement.

To date, NRVs are have yet to be adopted in the workplace, or in enforcement activities. According to the broad understanding of effectiveness that is proposed in this chapter, the fact that employers do not use NRVs yet to limit the exposure to nanoparticles does not mean that the soft instrument has no meaning in practice. Since the Dutch approach has been discussed in EC, and has been adopted in Finland, it appears to have had some effect. A review of empirical research [53], and recent literature, indicates that the capacity of employers to use NRVs is problematic, mainly because of the contestation of equipment and methods to measure exposure to nanoparticles. In this context one difficulty is to distinguish engineered nanoparticles from process-generated and background nanoparticles [52]. An additional problem is that the scientific community had not yet reached agreement on the appropriate method for standard measurement [54]. Complex measurement requires sophisticated knowledge and skills that do not seem to be available in small and medium sized companies [55]. Hence, an important condition for rule following appears not to have been fulfilled in practice.

With regard to the second compliance condition—the willingness to apply the benchmarks—companies appear at this time to be ambivalent. Most of them contend that NRVs are 'better than nothing'. Although they would prefer certain legally established occupational exposure limits, some of the surveyed companies appreciate 'the temporary certainty' of the NRVs. Of those companies surveyed, there was an overwhelming preference for legislation on nanomaterials because of the perceived certainty that is provided, especially in relation to the need for further investment.[11] Main interests to apply NRVs are to enhance the company's reputation and to build trust at the work floor. To date, rewards and sanctions to apply the NRVs in practice

---

[11]This conclusion is confirmed by other studies on nanotechnological regulation [56, 57].

are lacking. Some of the surveyed companies wondered whether the Labour Inspectorate is knowledgeable in this field.

In sum, this example indicates that, at present, there are substantial shortcomings with the implementation of the NRVs in relation to compliance. Although informed employers, as part of a survey, appeared willing to employ and use the NRVs in their workplace, in practice it appeared that they may be hindered to do so by the limited number of acknowledged measurement devices and methods. The uncertainties and provisionality of the NRVs appears to be contrary to the employers' demand for uncontested instruments and legal certainty.

3.3.3 Nanocosmetics within the European Union

Moves by the European Parliament and Council in 2009 to differentiate cosmetic products containing nanomaterials from those containing more conventional macro scale particles for regulatory purposes provides an excellent example of anticipatory governance at work. The push to create additional regulatory provisions for those products containing nanomaterials occurred within the context of the broader recast of the regulatory regimes for cosmetic products within the EU [58, 59], against the backdrop of increasing debate—within the scientific and broader communities—on the potential risks that some nanomaterials may pose to consumers when incorporated into topically applied products [60–62]. The fact that, even today, there is limited consensus as to that nature of any such risks (if, indeed, they exist at all), and the nature of the exposure pathway [61], is illustrative of the success that advocates for nano-specific regulation had within the debates.

In this section of the chapter, we explore the notion of 'effectiveness' within the context of the Cosmetic Regulation (Regulation (EC) No 1223/2009[12]). In exploring effectiveness within this context, it is pertinent to note that no conclusive evidence of harm to an individual as a consequence of wearing nano-based cosmetics has been identified at this time. However, despite this absence of harm to date, uncertainty exists over the potential risks posed by some families of nanomaterials when topically applied over time. Moreover, nor is it clear what any

---
[12]OJ L 342 of 22.12.2009.

such harm would look like should it occur. Such fundamental uncertainties go to the very heart of the question of effectiveness when looking at the issue from a policy perspective—for example, from preventing harm to the user. Parallel policy goals also include promoting EU market competitiveness through nano-based cosmetics, promoting innovation more generally, and providing greater choice to the consumer. However, the question is also underpinned by the broader notion of effectiveness articulated above in relation to specifically the degree to which industry is taking these obligations and responsibilities seriously despite, or in the face of, such uncertainties.

As such, we must look at effectiveness at this time from the following standpoint: what is it that the Parliament and Council are seeking to achieve by virtue of the inclusion of nano-specific provisions in the Cosmetic Regulation? Are such objectives possible or likely given the nature of the provisions? And is industry seemingly willing to comply with the legal obligations created by the instrument despite the uncertainties surrounding topic exposure to nanomaterials at this time? Any such conclusions can only be tentative at this time, given that Regulation (EC) No 1223/2009 was only fully implemented in July 2013.

The intent of the European Parliament (EP) and Council (EC) is articulated in somewhat aspiration forms in the Preamble of the Cosmetic Regulation. Clause 29, for example, stresses public health and economic need for agreement, across countries, in relation to a universally accepted definition of 'nanomaterials'. Clause 30, in contrast, focuses on scientific risk:

> At present, there is inadequate information on the risks associated with nanomaterials. In order to better assess their safety the SCCS should provide guidance in cooperation with relevant bodies on test methodologies which take into account specific characteristics of nanomaterials.

Concern over safety is further echoed in Clause 35; the call for the Scientific Committee on Consumer Safety (SCCS) to provide 'opinions where appropriate on the safety of use of nanomaterials in cosmetic products', is further underpinned by the EP's and EC's desire for greater transparency and information sharing between the regulator, the scientific committees, and industry.

The substantive requirements relating to nanomaterials are set out in the Articles of the Regulation (see Articles 2, 13, 16 and 19). These provisions *only* apply to those products that contain nanomaterials as an ingredient. As such, the first—and arguably most important—trigger when considering the effectiveness of the legal regime is the definition itself, as it is this definition that establishes the framework for which industry must then follow. For the purposes of Cosmetic Regulation (Regulation (EC) No 1223/2009) as

> an insoluble or biopersistant and intentionally manufactured material with one or more external dimensions, or an internal structure, on the scale from 1 to 100 nm;...(Article 2(1)(k)).

While this definition is consistent with the classes of nanomaterials that were identified by the SCCP [61] as those that are more likely to give rise to concern when used in such products, the adoption of such a narrow definition would appear to purposely exclude the many families of soluble and/or biodegradable nanoparticles that can, and have, been used in cosmetic products. Nanomaterials that also fall outside of the 1–100 nm size range, despite still exhibiting novelty, similarly fall outside the scope of the definition. Manufacturers of products containing such nanomaterials (if considered as such) will not be required to comply with the nano-specific provisions set out in, for example, Articles 16 and 19. Regulation (EC) No 1223/2009 recognises that the definition may need to be updated in line with the evolving state of the scientific art (as set out in Recital 29 of the Regulation).[13]

As such, the narrow definition adopted by the Parliament and Council would appear to purposely exclude many of the nanomaterials that are being used in commercial cosmetic products today including, for example, biodegradable nanosomes.[14]

---

[13]Recital 29 of Regulation (EC) No 1223/2009 specifically states that, '...agreement on a definition in appropriate international fora. Should such an agreement be reached, the definition of nanomaterials in this Regulation should be adapted accordingly'.

[14]According to Nohynek et al. [63] nanosomes are a globular vesicle formulation, with at least one dimension at the nanoscale, manufactured through traditional techniques for the purposes of skin penetration. They usually contain active ingredients, and improve stability of said ingredients during the penetration process.

This, in our view, appears to undermine the push by the Parliament and Council towards greater transparency around the use of nanotechnologies in consumer products generally, and cosmetic products more specifically. This is especially true in relation to the labelling requirements (as discussed below), and that ability for consumers to exercise so-called informed choice, as a result of the labelling provisions.

Article 13 addresses notification requirements that must be submitted by the responsible parties to the Commission. Pursuant to Article 13(1)(f), the electronic dossier must contain information regarding the 'presence of substances in the form of nanomaterials...'. Additional notification requirements are further set out in Article 16(3). These notification requirements only apply to those that fall within the definition set out in Article 2(1)(k). This provision reflects the Parliament's desire for enhanced information sharing between industry and the regulator, as well as the provision of an information base on which decisions over the need for scientific opinions may be made (see also Article 16(4)–(6)). The effectiveness of this provision shall, at least in part, depend on several elements. These include, for example, the technical capacity of the responsible parties to accurately measure particle size (i.e. does the responsible party have the necessary infrastructure to accurately measure the particles?). We would argue that this, at least in part, may be dependent on how well they are supported they supported in taking their duty to provide information seriously. Effectiveness may also depend upon the actions of the EC in terms of how they catalogue, and utilises, the provided information.

The narrow definition adopted in Regulation (EC) No 1223/2009 has the ability to undermine, albeit to varying degrees, this notification requirement, and therefore the effectiveness of the regime. As hypothesised by Bowman et al. [64], there is a possibility that products that would currently trigger the nanomaterial notification requirement (and subsequent requirements) could be purposely reformulated so as to no longer trigger this regulatory requirement. For example, whereas a concealer (a flesh-coloured cosmetic product designed to cover blemishes and dark under-eye circles) containing titanium dioxide at 90 nm would be subject to the notification requirement, the same concealer containing titanium dioxide particles at

140 nm would not—despite potentially exhibiting the same functionality. Issues around functionality, performance, cost and competitive advantage are likely to be the key drivers in any such decision, suggesting that a decision to avoid this requirement would be inherently complex and not simply about the notification provision.

Article 19 sets out the labelling requirements for those products generally, and includes a specific requirement for those products containing nanomaterials. Pursuant to 19(1)(g),

> ...All ingredients present in the form of nanomaterials shall beclearly indicated in the list of ingredients. The names of such ingredients shall be followed by the word `nano' in brackets...

This requirement would appear to be fundamental to promoting transparency between industry and the consumer, and the market more generally. How effective it is in assisting consumers to make an informed choice about the product remains unclear at this time. Informed choice is dependent *inter alia* on the capacity and willingness to make comparisons between different products, as well as time, knowledge and a willingness to make comparisons between different products [64–66]. Without these, the word 'nano' by itself is unlikely to significantly advance and/or promote informed decision making by itself.

The effectiveness of the public catalogue of nanomaterials as a communications and/or a transparency tool for consumers—as required under Article 16(10)(a)—is similarly dependent on more than its simple existence. While the catalogue, and even the annual status report (pursuant to Article 16(10)(b)), are useful apparatuses for overcoming the transparency deficit generally (but primarily between policymakers, the regulator and the market), one has to ask just how likely it is that consumers will be aware of, drawn to, and utilise such information? Such action requires the consumer to not only be aware, but to be proactive in their search for information. In short, the onus is on the consumer to actively seek out the information.

In sum, in the absence of categorical scientific evidence of harm to humans as direct consequence of using nano-based cosmetics, the EP and EC opted to take a precautionary approach to the regulation of such products. This approach may be viewed, in our opinion, as a cautionary tale—given the divergent scientific options and acknowledged scientific uncertainty—as well as an exercise of political muscle. The provisions have been designed to promote safety (or, from a different perspective, enable the Commission to mitigate harm in the shortest possible time should such evidence arise), as well as promote transparency. Such motives are not unique to nanotechnologies. However, the narrow definition adopted in the Regulation, and broader questions relating to the appropriateness of conventional risk assessment paradigms for nanomaterials, raise questions about the overall effectiveness of the requirements.

3.4 Lessons Learned, and Moving Forward

The precautionary, anticipatory and responsive approaches presented in the context of this chapter are two 'real world' examples of regulatory action taken by responsible actors within the realms of EU governance. The models adopted in each case appear to correspond to the calls and proposals outlined by influential scholars and commentators in terms of mechanisms for dealing with the uncertainties of emerging technologies. However, as highlighted by Section 3.3, each approach brings with it a number of effectiveness issues; our analysis shows that neither approach can be consider to be a perfect solution for the problems presented. The question must therefore be asked as to whether such approaches are really promising when it comes to regulatory practice? Or, should we—at least at this time—consider such action to be more symbolic in nature?

Both questions give rise to many more questions, but few answers. Does this mean, for example, that the best way forward is to go back to the traditional evidence- and risk-based model of regulation and wait until there is more certainty on the effects of technologies and governance frameworks? We would argue no. As illustrated by the EU's employment of the precautionary principle, a lack of certainty is not a reason, at least by itself, to be paralysed in relation to policy or regulatory developments [21]. This is especially true given that many of the questions and uncertainties associated with engineered nanomaterials are unlikely to be resolved in the short term. If safety regulators must wait for that solid, and undisputed, evidence-base before they are able to act, the regulatory lag is likely to be significant and on-going. As we have already seen, across multiple domains and jurisdictions, evidence-based regulation on fast moving, and evolving, technologies are often outdated at the time of their implementation, or shortly thereon after. Moreover, the emergent stage of technologies, with a high degree of flexibility, and a low attachment to the status quo, can present a unique opportunity to bring together diverse stakeholders and to incentivise responsible research and innovation [27].

Some more general remarks on the two examples presented within the context of this chapter are as follows. In relation to the NRVs, we would argue that it is likely that incentives like governmental rewards, but also pressures of a knowledgeable Labour Inspectorate, could enhance the uptake and use of the benchmarks going forward. This case also reminds us that regulators have to cope with industry demands for regulatory certainty; for many within the private sector, a lack of clear guidance and/or regulation provides its own uncertainty, which can hinder development and commercialization of products.

Accordingly, we would argue that this need for certainty, and guidance, suggests that the way forward is not to depend heavily on forms of soft regulation within this field. Rather, an appropriate balance or combination of legislation (which provides legal certainty) and soft regulation (which provides flexibility) that is adaptive to changing insights on the emerging technologies, should be developed. Given the huge gap between the industry's demand for regulatory certainty, and the persistent uncertainties about the governability of the technology, the OH&S example illustrates the need for regulators to avoid, wherever possible, contested implementation tools. To gain trust across all stakeholder groups, regulators must not only be proactive in their dissemination of guidance documents and information, but also better rely on assessment methods that are underpinned by commonly accepted processes.

In a similar vein, the EU's approach to regulating nanocosmetics—which is to be adopted by New Zealand in the coming years—suggests that regulators cannot afford to ignore, or leave questions relating to narrow scientific definition and the appropriateness of conventional risk assessment paradigms, unanswered. To do so is to likely attract the disapproval and discontent of not only those who are bound by the regulatory regime, but also the broader scientific and stakeholder community. We would also argue that while the labelling mechanism introduced for nanocosmetics is an important step for promoting transparency within the industry, the mechanism is not a means to an end. Rather, it is a first step. And without further opportunities for public engagement and deliberation, the value of the label is limited; one could argue, for example, that without educating the public as to what the label means, and allowing them to then exercise an informed choice over their purchasing, the label by itself appears to be more symbolic than meaningful.

In sum, this chapter has sought to draw from current developments and approaches to regulating an emerging technology, and consider their effectiveness in light of the prevailing incremental regulatory strategy. The case studies selected here suggest that such a strategy is only promising if it involves activities and mechanisms that create, and continue to support commitment to regulation. While it is clear that, at least in the short term, much of the regulatory and policy style will be driven by a 'muddling through' approach, there is still an opportunity to incorporate sophisticated tools and strategies into such a style. The 'social license' framework articulated by Brownsword and Somsen [1] calls for politicians, policy makers and regulators to use a sophisticated and responsive co-regulatory approach for technology innovation. Such an approach, which is adaptive and flexible, seems the way forward to cope with the difficult task of guiding technological innovation both in the short and longer term.

References

1. Brownsword, R., and H. Somsen (2009) Law, innovation and technology: Before we fast forward—a forum for debate, *Law, Innovation and Technology*, **1**, 1–73.
2. Bartolini, S. (2011) New modes of European governance: An introduction. In: Héritier, A., and M. Rhodes (eds.) *Modes of Governance in Europe* (Palgrave, London), pp. 1–18.

3. Roco, M. C., B. Harthorn, D. Guston, and P. Shapira (2011) Innovative and responsible governance of nanotechnology for societal development. In: *Nanotechnology Research Directions for Societal Needs in 2020* (Springer, Amsterdam), pp. 561–617.
4. Owen, R., P. Macnagten, and J. Stilgoe (2012) Responsible research and innovation: From science in society to science for society, with society, *Science and Public Policy*, **39**, 751–760.
5. Von Schomberg, R. (2013). A vision of responsible innovation. In: Owen, R., M. Heintz, and J. Bessant (eds.) *Responsible Innovation* (John Wiley, London), pp. 51–74.
6. Levi-Faur, D., and H. Comaneshter (2007) The risks of regulation and the regulation of risks: The governance of nanotechnology. In: Hodge, G., D. Bowman, and K. Ludlow (eds.) *New Global Regulatory Frontiers in Regulation: The Age of Nanotechnology* (Edward Elgar, Cheltenham), pp. 149–165.
7. European Commission (2008) *Regulatory Aspects of Nanomaterials* (European Commission, Brussels).
8. European Commission (2012) *Communication on the Second Regulatory Review on Nanomaterials* (European Commission, Brussels).
9. Ludlow, K., D. M. Bowman, and G. A. Hodge (2007) *Final Report: Review of Possible Impacts of Nanotechnology on Australia's Regulatory Frameworks* (Monash Centre for Regulatory Studies, Monash University Melbourne).
10. Royal Commission on Environmental Pollution (2008) *Twenty-Seventh Report Novel Materials in the Environment: The Case of Nanotechnology* (Royal Commission on Environmental Pollution, The Stationery Office, Norwich).
11. Hodge, G. A., D. M. Bowman, and A. D. Maynard (eds.) (2010) *International Handbook on Regulating Nanotechnologies* (Edward Elgar, Cheltenham).
12. Project on Emerging Nanotechnology (2013) *Consumer Products An Inventory of Nanotechnology-Based Consumer Products Currently on the Market*, Project on Emerging Nanotechnologies. Available at: http://www.nanotechproject.org/inventories/consumer/.
13. Dorbeck-Jung, B. R., and C. Shelley-Egan (2013) Meta-regulation and nanotechnologies: The challenge of responsibilisation within the European Commission's code of conduct for responsible nanosciences and nanotechnologies research, *NanoEthics*, **7**, 55–68.

14. Marchant, G. E., B. R. Allenby, and J. R. Herkert (eds.) (2011) *The Growing Gap between Emerging Technologies and Legal-Ethical Oversight: The Pacing Problem* (Springer, Amsterdam).
15. Marchant, G. E., K. Abbott, and B. R. Allenby (eds.) (2013) *Innovative Governance Models for Emerging Technologies* (Edward Elgar, Cheltenham).
16. Dorbeck-Jung, B. R. (2011) Soft regulation and responsible nanotechnological development in the European Union: Regulating occupational health and safety in the Netherlands, *European Journal of Law and Technology*, **2**(3), 1–14.
17. Hodge, G. A., D. M. Bowman, and K. Ludlow (eds.) (2007) *New Global Frontiers in Regulation: The Age of Nanotechnology* (Edward Elgar, Cheltenham).
18. Ladeur, K. H. (2003) The introduction of the precautionary principle into EU law: A Pyrrhic victory for environmental and public health law? Decision-making under conditions of complexity in multi-level political systems, *Common Market Law Review*, **40**(6), 1455–1479.
19. Stokes, E. (2012). Nanotechnology and the products of inherited regulation, *Journal of Law and Society*, **39**(1), 93–112.
20. European Commission (2000). *Communication on the Precautionary Principle* (COM(2000) 1, 2.2.2000) (European Commission, Brussels).
21. Fisher, E. (2007) *Risk Regulation and Administrative Constitutionalism* (Hart Publishing, Oxford and Portland Oregon).
22. Fischer, E. (2002) Precaution, precaution everywhere: Developing a 'common understanding' of the precautionary principle in the European Community, *Maastricht Journal of European Comparative Law*, **8768**, 7–28.
23. Ludlow, K., D. M. Bowman, and D. Kirk (2009) Hitting the mark or falling short with nanotechnology regulation?, *Trends in Biotechnology*, **27**(11), 615–620.
24. Ayres, I., and J. Braithwaite (1992) *Responsive Regulation. Transcending the Deregulation Debate* (Oxford University Press, Oxford).
25. Braithwaite, J. (2006) Responsive regulation and developing economies, *World Development*, **34**, 885, 887–888.
26. Baldwin, R., and J. Black (2008) Really responsive regulation, *The Modern Law Review*, **71**(1), 59–94.
27. Mandel, G. N. (2009) Regulating emerging technologies, *Law, Innovation and Technology*, **1**, 75–92.

28. Gunningham, N., P. Grabosky, and D. Sinclair (1998) *Smart Regulation. Designing Environmental Policy* (Oxford University Press, Oxford).
29. Teubner, G. (1983) Substantive and reflexive elements in modern law, *Law & Society Review*, **17**(2), 239–285.
30. Sabel, C. F., and J. Zeitlin (eds.) (2010) *Experimentalist Governance in the European Union: Towards a New Architecture* (Oxford UP, Oxford).
31. Parker, C. (2007) Meta-regulation: Legal accountability for corporate social responsibility? In: McBarnet, D., A. Voiculescu, and T. Campbell (eds.) *The New Corporate Accountability: Corporate Social Responsibility and the Law* (Oxford University Press, Cambridge), pp. 207–241.
32. Coglianese, C., and E. Mendelson (2010) Meta-regulation and self-regulation. In: Cave, M., R. Baldwin, and M. Lodge (eds.) *The Oxford Handbook of Regulation* (UP, Oxford, New York), pp. 146–168.
33. Bowman, D. M., and K. Ludlow (2013) Assessing the impact of a 'for government' review on the nanotechnology regulatory landscape, *Monash Law Review*, **38**(3), 168–212.
34. Haum, R., U. Petschow, and M. Steinfeldt (2004) *Nanotechnology and Regulation within the Framework of the Precautionary Principle* (IOW, Berlin).
35. European Commission (2008). *Regulatory Aspects of Nanomaterials' and the Accompanying Commission Staff Document 'Summary of Legislation in Relation to Health, Safety and Environment Aspects of Nanomaterials, Regulatory Research Needs and Related Measures* (European Commission, Brussels).
36. European Commission (2008) *Recommendation on a Code of Conduct for Responsible Nanoscience and Nanotechnologies Research* (COM 7.2.2008) (European Commission, Brussels).
37. Bowman, D. M., and G. A. Hodge (2008) A big regulatory tool-box for a small technology, *NanoEthics*, **2**(2), 193–207.
38. Defra (2006) *Department for Environment, Food and Rural Affairs Voluntary Reporting Scheme.* Available at: http://archive.defra.gov.uk/environment/quality/nanotech/documents/vrs-nanoscale.pdf.
39. Lindblom, C. E. (1959) The science of muddling through, *Public Administration Review*, **19**, 79–88.
40. Marchant, G., and W. Wallach (2013) Governing the governance of emerging technologies. In: Marchant, G., K. Abbott, and B. Allenby

(eds.) *Innovative Governance Models for Emerging Technologies* (Edward Elgar, Cheltenham), pp. 136–152.

41. Gammel, S., A. Lösch, and A. Nordmann (2010) A 'scanning probe agency' as an institution of permanent vigilance. In: Goodwin, M., B. J. Koops, and R. Leenes (eds.) *Dimensions of Technology Regulation* (Wolf Legal Publishers, Nijmegen), pp. 125–146.

42. Abbott, K., G. E. Marchant, and D. J. Sylvester (2006) A framework convention for nanotechnology?, *Environmental Law Reporter*, **36**, 10931–10942.

43. Abbott, K., D. J. Sylvester, and G. E. Marchant (2010) Transnational regulation of nanotechnology: Reality or romanticism?. In: Hodge, G. A., D. M. Bowman, and A. D. Maynard (eds.) *International Handbook on Regulating Nanotechnologies* (Edward Elgar, Cheltenham) pp. 525–544.

44. Opschoor, H., and K. Turner (1994) *Economic Incentives and Environmental Policies: Principles and Practice* (Kluwer Academic Publishers, Dordrecht).

45. Griffiths, J. (1999). Legal knowledge and the social working of law: The case of euthanasia. In: van Schooten, H. (ed.) *Semiotics and Legislation* (D. Charles Publications, Amsterdam), pp. 81–108.

46. Griffiths, J. (2003) The social working of legal rules, *Journal of Legal Pluralism and Unofficial Law*, **48**, 1–84.

47. Havinga, T. (2006). Private regulation of food safety by supermarkets, *Law & Policy*, **28**(4), pp. 515–533.

48. Karlsson-Vinkhuyzen, S. I., and A. Vihma (2009) Comparing the legitimacy and effectiveness of global hard and soft law: An analytical framework, *Regulation & Governance*, 3(4), 400–420.

49. Shamir, R. (2008) The age of responsibilization: On market embedded morality, *Economy and Society*, **37**, 1–19.

50. Black, J. (2001) Decentering regulation: Understanding the role of regulation and self-regulation in a 'post-regulatory' world, *Currant Legal Problems*, **54**, 103–147.

51. Von Schomberg, R. (2007) From the ethics of technology to the ethics of knowledge assessment. In: *The Information Society: Innovation, Legitimacy, Ethics and Democracy in Honor of Professor Jacques Berleur sj* (Springer, New York), pp. 39–55.

52. Van Broekhuizen, P. (2012) *Nano Matters, Building Blocks for a Precautionary Approach* (IVAM/UvA, Amsterdam).

53. Van Broekhuizen, P., and B. R. Dorbeck-Jung (2012) Exposure limit values for nanomaterials: Capacity and willingness of users to apply a precautionary approach, *Journal of Occupational and Environmental Hygiene*, **10**(1), 46–53.
54. Abbott, L. C., and A. D. Maynard (2010) Exposure assessment approaches for engineered nanomaterials, *Risk Analysis*, **30**, 1634–1644.
55. Reichow, A., and B. R. Dorbeck-Jung (2013) Discovering specific conditions for compliance with soft regulation related to word with nanomaterials, *NanoEthics*, **7**, 83–92.
56. Groves, C., et al. (2011) Is there room at the bottom for CSR? Corporate social responsibility and nanotechnology in the UK, *Journal of Business Ethics,* **101**(4), 525–552.
57. Stokes, E. (2013) Demand for command: Responding to technological risks and scientific uncertainties', *Medical Law Review*, **21**(1), 11–38.
58. Gergely, A., and L. Coroyannakis (2009) Nanotechnology in the EU cosmetics regulation, *Household and Personal Care Today*, **3**, 28–30.
59. Bowman, D. M., G. van Calster, and S. Friedrichs (2010) Nanomaterials and the regulation of cosmetics, *Nature Nanotechnology*, **5**(2), 92.
60. Scientific Committee on Consumer Products (2007) *Opinion on Safety of Nanomaterials in Cosmetic Products* (Health & Consumer Protection Directorate-General, Brussels).
61. Scientific Committee on Consumer Safety (2012) *Guidance on the Safety Assessment of Nanomaterials in Cosmetics* (Health & Consumer Protection Directorate-General, Brussels).
62. Scientific Committee on Emerging and Newly Identified Health Risks (2009) *Risk assessment of products of nanotechnologies* (Directorate-General for Health & Consumers, Brussels).
63. Nohynek, G. J., J. Lademann, C. Ribaud, and M. S. Roberts (2007) Grey goo on the skin? Nanotechnology, cosmetic and sunscreen safety, *CRC Critical Reviews in Toxicology*, **37**(3), 251–277.
64. Bowman, D. M., G. van Calster, and S. Friedrichs (2010) Nanomaterials and regulation of cosmetics, *Nature nanotechnology*, **5**(2), 92–92.
65. Throne-Holst, H., and P. Strandbakken (2009) Nobody told me I was a nano-consumer: How nanotechnologies might challenge the notion of consumer rights, *Journal of Consumer Policy*, **32**(4), 393–402.
66. D'Silva, J., and D. M. Bowman (2010) To label or not to label: It's more than a nano-sized question, *European Journal of Risk Regulation,* **1**, 420.

Chapter 4

Society as a Laboratory to Experiment with New Technologies

Ibo van de Poel

Department of Values, Technology and Innovation,
School of Technology, Policy and Management
Technical University Delft, Jaffalaan 5,
2628 BX Delft, The Netherlands

i.r.vandepoel@tudelft.nl

4.1 Introduction

The introduction of any new technology into society raises several challenges. One challenge is to appropriately deal with the expected and potential benefits and disadvantages. From a societal point of view, the benefits of new technologies should outweigh their risks, and this raises the challenge how we can predict benefits and hazards. A second challenge is to successfully embed technology in society. New technologies may require new organizational forms and new forms of regulation or other institutions. Technology and society need to be mutually adapted to successfully introduce new technology in society. Third, new technology might create new moral questions and dilemmas.

These challenges may be especially hard to deal with in the face of uncertainty. Uncertainty may arise with respect to all three challenges. Hazards and benefits are often not just uncertain in the rather straightforward sense that we can only express them as probabilities; often we even cannot meaningfully assign probabilities to potential hazards and benefits; or we even do not know all possible hazards and benefits, as in the case of 'unknown unknowns'.

Embedding technology in society may also give raise to large uncertainties. Often, it is not clear beforehand what organizational or institutional forms are most appropriate for a new technology or how a technology should be adjusted to fit existing organizations and institutions. The involved actors may be uncertain about their roles and responsibilities. As a consequence, a technology may never get off the ground, for example, in a waiting game where all actors wait for each other to take the first step in adopting the technology and, by doing so, reduce uncertainty [1]. Alternatively, when a technology nevertheless does develop it may be that no one in particular is responsible or takes responsibility for undesirable consequences, in a so-called problem of many hands, so that, as a consequence, these consequences are less likely to be prevented [2].

Also dealing with moral questions and dilemmas may be troubled by uncertainty [3, 4]. This may be uncertainty about what moral questions will arise, which is often hard to predict beforehand, but also normative uncertainty about what is desirable.

Of course not every new technology raises all three challenges and not all technologies raise these challenges to the same degree, and the mentioned uncertainties may play a larger role in one technological domain than in another. Still, there are a number of technologies that face all these challenges and in which uncertainty of consequences, institutions, and moral issues plays a large role. One example is nanotechnology. Nanotechnology is not one technology, but rather an enabling technology that enables a large range of nanotechnological applications and innovations in other domains. Still, even if we focus on specific applications, the mentioned challenges are easy to identify. Products that contain nanoparticles have raised the suspicion of being dangerous to human health and the environment. Although it is known that not all nanoparticles are dangerous and that their hazards depend on, for example, their exact shape and how humans or the environment are exposed to them, in many cases the exact hazards are hard to assess before these products enter the market [5, 6]. One example are titanium dioxide particles in sunscreens, an example that I will discuss in more detail later in this chapter.

There is also a lot uncertainty with respect to social embedding, especially in relation to regulation and social acceptance. Normative uncertainty is especially large in the case of human enhancement that might become possible through nanotechnology (and related technological developments). There is strong moral disagreement about the desirability of human enhancement, especially about the desirability to enhance the cognitive, and particularly the moral capabilities of human beings [7–9].

Most current approaches try to deal with the aforementioned challenges by anticipating the future embedding of technology in society and its consequences. This is most obvious in more traditional forms of risk assessment and technology assessment (TA). But also constructive technology assessment (CTA) fits the more general aim of TA approaches that has been described as 'to reduce the human costs of trial and error learning in society's handling of new technologies, and to do so by *anticipating* potential impacts...' [10:251; emphasis added]. Also an approach like the development of techno-moral scenarios to deal with future ethical issues is based on anticipation [11], as is the precautionary principle as way to deal with uncertainty [12].

The aim of this chapter is to describe an alternative approach that is based on experimentation rather than anticipation as a way to deal with the sketched challenges for embedding new technology in society. In this approach, society becomes a kind of laboratory to experiment with new technologies. I start this chapter with sketching the approach by briefly summarising the main theoretical strands on which it builds and by highlighting four main aspects of it (Section 4.2). In the next section, I argue that the uncertainties inherent in embedding new technologies in society may be of such a kind that they cannot be reduced before a technology is actually introduced in society. Reducing uncertainty may therefore require experimentation (rather than just anticipation). I also distinguish learning-by-experimentation as a mode of learning in addition to learning-by-anticipation and learning-by-doing that tries to avoid the pitfalls of both other modes of learning. After an illustration of the approach with an example from the field of nanotechnology, some brief conclusions are presented at the end of the chapter.

4.2 Technology as Social Experiment

The basic idea of the approach I sketch in this chapter is that society has become the laboratory in which new technologies are tried out. Consequently, the introduction of a new technology into society may be called a social experiment. This experiment is *social* in three different meanings of the term. First, it is an experiment that is done *in society*. Unlike traditional experiments that are done in a laboratory or in another confined and contained setting, it takes place in society, or at least in some part of society. Second, it is an experiment *on society*. Not only in the sense that society may be affected by (negative) effects of the experiment but also in the sense that the experiment is not restricted to the technical operation of the technology but also relates to its social consequences, risks, societal embedding and normative dimensions. Third, it is an experiment done *by society*. With that I mean to say that these experiments do not usually have a single experimenter as in traditional experiments. Experimentation is often not controlled by a single experimenter but rather an institutional pattern that arises from the actions of many actors. This is not to say that experimentation cannot be deliberate or that we cannot change the conditions under which such social experiments are done, but doing so is more complicated than just convincing a single experimenter.

It should be noted that what I call social experiments are different from standard scientific experiments in at least two important respects. First, these are often implicit experiments, i.e. experiments not called by name. The claim is that if a technology is introduced in society, it is *de facto* an experiment whether it is recognised or not as such. Obviously, if experiments are done unknowingly, the ones being experimented on may ask questions about the acceptability of such experiments, an issue to which I return below.

Second, social experiments are usually uncontrolled experiments. In science, experiments are controlled in the sense that the experimenter typically controls (and varies) the independent variables to study the effect on the dependent variables [13]. Such control is usually absent in social experiments. Still, social experiments may be properly called experiments, I will argue, because they allow to try out certain technologies in society and to learn from it.

Below I will further elaborate these points. I do so by first sketching some of the theoretical strands on which the idea of society as laboratory to experiment with technology builds. After that I recapitulate the three characteristics of social experiments mentioned above, and add a fourth point about responsible experimentation.

4.3 A Brief History of the Idea

The idea of society as a laboratory to experiment with new technologies is not new. Below, I briefly sketch four relevant strands of literature on which it builds or at least could build. These strands have developed more or less independently and they emphasise different aspects of the idea and also employ different conceptualisations of what a social experiment with technology exactly is.

4.3.1 Society as Laboratory

The publication that is usually referred to as introducing the idea of society as a laboratory for experimentation with technology is the article by Krohn and Weyer [14] published in *Science and Public Policy* in 1994. In this publication, the authors mainly focus on the unpredictability of certain risks, like a nuclear meltdown, a plane crash or toxic effects of a new chemical, before a technology is introduced in society. They present four reasons why we can often not reliably predict risks before a technology is introduced in society. First, laboratory and field test circumstances often are only partly representative for the real circumstances in which products are used. Second, hazards may be due to long-term, cumulative or interactive effects which are hard to study in the laboratory. Third, systems may be characterised by nonlinear system dynamics and emergent behaviour and, fourth, we may overlook certain potential hazards due to ignorance.

Krohn and Weyer [14], and their colleagues do not speak about social experiments, but rather about real-world experiments. They thus stress the aspect that these experiments take place in society (rather than in the traditional laboratory) but less the aspect that it is done on and by society. They stress that 'real-life experiments...presuppose the existence of an organised research; otherwise, it would be more appropriate to speak of hazardous evolutionary change' [14:176]. Sometimes, however, organised research may only be established after, for example, an accident has happened.

The notion of technology as a form of social experimentation has been applied to several domains and cases including waste facilities [15], urban studies [16], genetically modified crops [17], engineering research laboratories [18], ecological design [19] and sunscreens with nanoparticles [5]. In a 2007 report the European Expert Group on Science and Governance again stressed the importance of the notion:

> [W]e are in an unavoidably experimental state. Yet this is usually deleted from public view and public negotiation. If citizens are routinely being enrolled without negotiation as experimental subjects, in experiments which are not called by name, then some serious ethical and social issues would have to be addressed [20:68].

Three things stand out in this quote. First, that social experiments with technology are to some extent unavoidable, an idea that goes back to Krohn and Weyer; second, that experiments are often not called by name; and, third, that such experiments raise serious questions about their social and moral acceptability.

As noted by the European Expert Group, members of the public may well feel that experimenting on society, i.e. on them, in order to develop a technology is morally unacceptable. Indeed, the introduction of new technology is sometimes described as an experiment to denote its undesirability. Prince Charles, for example, has called the development of GM foods by large corporations a 'gigantic experiment with nature and the whole of humanity which has gone seriously wrong' [21].

However, once one recognises, as the European Expert Group does, that experimenting in and on society is sometimes unavoidable, it is not so obvious that *all* social experiments with new technology are unacceptable. Rather, the question becomes under what conditions such experiments may be considered morally acceptable. The European Expert Group does not answer this question but other authors have formulated tentative answers that I will discuss below.

4.3.2 Engineering as Social Experiment

In engineering ethics, the notion of engineering as experiment has been proposed by Martin and Schinzinger [22]. There seems to be no connection to the discussion mentioned in the previous section. Still, also here the focus is on the risks of new technologies and the unavoidability of some form of experimentation due to uncertainty. Martin and Schinzinger are particularly interested in the moral consequences of engineering as social experimentation. They propose the principle of informed consent to deal with these. Engineers should inform the public about their experiments and ask them for their consent to carry out such experiments.

Informed Consent is indeed one of the main ethical principles to judge the acceptability of experiments with humans. It says that human subjects should be fully informed about risks and expected benefits. They should also freely and knowingly consent to participation in the experiment. It is, however, questionable whether it feasible and sensible to apply the notion of informed consent to social experiments.

It has been suggested that the principle might be too strict, especially if risks and benefits are collective rather than individual [23]. In bioethics, the principle of informed consent has also been criticised for neglecting other ethical concerns like justice [24]. Elsewhere I have therefore proposed to replace the informed consent condition by other conditions for responsible experimentation, which I discuss below.

4.3.3 Social Experiments in Strategic Niche Management

The notion of 'social experiment' has also been used in the literature on strategic niche management [25]. Strategic niche management is based on a quasi-evolutionary theory of technological development. Technological variations in many cases cannot survive the selection environment; they first need to mature in niches. It is within these niches that social experimentation takes place. Social experiments allow for learning, both technical and institutionally, which may help to improve the technology and so they may be a stepping-stone to survival outside the niche.

4.3.4 Social Experiments in Social Science

In social science, different notions of social experiments have been developed. I will briefly mention two schools of thought and their notion of social experiment: the Chicago School of Sociology and the tradition of experimental sociology and its application to social reform that was especially influential in the United States between the 1960 and 1980s and which was established by Campbell. They two hold quite different, and partly opposing, views on social experiments.

The Chicago School has stressed the idea of society as laboratory in which continuously social experiments are going on [cf. 26]. The following quote from Small is typical:

> All the laboratories in the world could not carry on enough experiments to measure a thimbleful compared with the world of experimentation open to the observation of social science. The radical difference is that the laboratory scientists can arrange their own experiments while we social scientists for the most part have our experiments arranged for us [27:188].

Experiments, in this view, can be uncontrolled and they are taken place all the time, whether we are aware of it or not.

The tradition of experimental sociology that was successful in establishing a large number of experiments on social reform to evaluate public policy interventions between the 1960s and 1980s holds a quite different notion of experiments [28, 29]. In this tradition, the importance of randomisation and control groups is stressed in order to be able to draw reliable conclusions from experiments. In other words, to count as a proper social experiment, certain methodological preconditions had to be met, in particular, randomisation and the use of control groups.

4.4 Characteristics of Social Experiments

4.4.1 An Experiment in Society

All theoretical strands discussed share the idea of experimenting in society, or society as laboratory to experiment with technologies; only in Strategic Niche Management the emphasis is more on experiments in niches, but these are also part of society, one could argue.

Experimenting in society is important because if offers possibilities for learning and knowledge gathering that traditional forms of experimenting do not offer. In particular, it offers the possibility of learning under real-life, or near real-life circumstances. The important point here is not just that we *can* experiment in society with technology but that we sometimes *must* do so (possibly, in addition to experiments in more contained settings). The reason for this is that the uncertainty inherent in technological development can often not be reduced before a technology is actually introduced in society. In Section 4.5, I will say a bit more on why social experiments may be unavoidable to reduce uncertainty, but for now the important point is that a first idea of the approach is that social experiments with technology not only can take place in society but sometimes *must* be done in society in order to reduce uncertainty.

4.4.2 An Experiment on Society

A second characteristic of social experiments with technology is that they are experiments on certain parts of society. Such experiments are done not only to learn about technology in a narrow sense but also to learn about the social embedding of technology. In line with the three challenges that I have distinguished in the introduction, one could argue that in a social experiment, one learns (or should aim to learn) about impacts, institutional embedding and normative issues.

If the social experiment has a clear experimenter, for example, the government or a company introducing a technology, that experimenter may deliberately try to learn from the experiment. However, as we have seen, social experiments may be done implicitly or not have a single experimenter. Also in such cases, often learning takes place, although often more diffuse and less deliberate. Users may, for example, learn to better use a technology; or the government or the public may learn that they are certain unexpected drawbacks of a technology.

Learning is important because it helps to better deal with a technology—in particular, to better deal with impacts and, for example, to reduce negative impacts. Learning also helps to better embed a technology in society, or to better deal with the normative issues that a technology raises, such as privacy concerns of new surveillance technology.

4.4.3 An Experiment by Society

The third characteristic of social experiments, that they are done by society, is a bit more controversial, especially when it comes to the question whether there can be experiments that are not recognised as such by the involved actors. The European Expert Group clearly suggests the possibility of such experiments by speaking of experiments that are not called by name; also in the Chicago School of Sociology, experiments are not necessarily recognised as such; Small's quote suggests that any interesting social phenomenon might count as an experiment whether the actors involved see it as an experiment or not. At the other end of the specter, the Campbell tradition suggests that experiments need to meet certain methodological requirements to count as experiments. Krohn and Weyer are less strict, but also for them experiments require an organised research process.

The position I want to defend here sits in between these two extremes. I think we should be careful not to call all social phenomena experiments because then the notion loses much of its meaning and strength; on the other hand, we should allow for a somewhat looser or less strict meaning than common in science or philosophy of science. In particular, we should allow,

I think, for the possibility of uncontrolled experiments as experiments in society can usually not be controlled by experimenters.

The idea of uncontrolled experiments is not so uncommon as it may seem. Several authors have introduced notions of experiments that are uncontrolled [cf. 30]. Examples are the notion of 'design experiment' [31, 32] and of 'democratic experiment' [33]. In both cases, it concerns experiments in which a design or policy is tried out with the partial aim to learn from it and to improve the design or policy. This is similar to how I use the notion of technology as social experiment: the experiment is uncontrolled in the sense that no variables are controlled to find cause-effect relations. Still it is an experiment in the sense that something is tried out in society—in this case: the introduction of a new technology—with the partial aim to learn from this introduction and so to improve the further introduction of that technology in society.

4.4.4 Responsible Experimentation

The fourth point is not really a characteristic of social experiments with technology but rather a concern that is raised by doing social experiments with technology, and this concern is the more pressing if indeed, as I have claimed, sometimes reducing uncertainty requires an experiment in society rather than in a more refined or contained setting. The concern is that experimenting with technology in society raises some serious ethical and social questions as has been articulated by the European Expert Group. Addressing these concerns requires responsible forms of experimentation.

Responsible experimenting has to do with both epistemological and ethical concerns. The epistemological concerns have to do with setting up experiments in such a way that we can learn most from them or that the resulting knowledge is as reliable and as relevant as possible. The ethical concerns have to do with the fact that social experiments might negatively impact society (and nature).

Table 4.1 Possible conditions for responsible experimentation

1. Absence of other reasonable means for gaining knowledge about hazards
2. Monitoring
3. Possibility to stop the experiment
4. Consciously scaling up
5. Flexible set-up
6. Avoid experiments that undermine resilience of receiving 'system'
7. Containment of hazards as far as reasonably possible
8. Reasonable to expect social benefits from the experiment
9. Experimental subjects are informed
10. Approved by democratically legitimised bodies
11. Experimental subjects can influence the setup, carrying out and stopping of the experiment
12. Experimental subjects can withdraw from the experiment
13. Vulnerable experimental subjects are either not subject to the experiment or are additionally protected, and
14. A fair distribution of potential hazards and benefits

In earlier publications, I have tried to develop a tentative set of conditions under which a social experiment with new technology might be called responsible [5, 6]. Table 4.1 lists what such a set of conditions might look like. It should be noted that I have replaced the traditional criterion of 'informed consent' by a set of conditions (conditions 9–12) that do not guarantee individual informed consent but that still address the moral value underlying informed consent, i.e. moral autonomy or respect for persons. Also the other conditions are related to moral values that have been articulated in bioethics for the acceptability of experiments with human subjects [e.g. 34, 35], i.e. non-maleficence (my conditions 1–7), beneficence (my condition 8), and justice (my conditions 13 and 14).

4.5 Uncertainty, Learning and Experimentation

Successfully embedding new technology in society is complicated by uncertainty and requires learning to deal with, and eventually overcome uncertainties. The nature of the uncertainties involved, however, is often of such a kind that they cannot be fully reduced before a technology is introduced in society. For this reason, experimentation may be unavoidable to reduce certain uncertainties.

Below, I first say a bit more on the type of uncertainties that are involved in embedding new technology in society. I then make an argument that some of these uncertainties cannot be reduced before a technology is introduced in society. In the final subsection, I distinguish what I call learning-by-experimentation from two other modes of learning, i.e. learning-by-doing and learning-by-anticipation.

4.5.1 Uncertainty

To see to what extent uncertainties can be reduced before a technology is actually introduced in society, it is useful to distinguish between different kinds of uncertainty. A first distinction that is useful to make relates to what we are uncertain about. I propose to distinguish here between the following three types of uncertainty:

- Impact uncertainty, i.e. uncertainty about the social impacts (hazards and benefits) of a new technology.
- Institutional uncertainty, i.e. uncertainty about what institutions will evolve in conjunction with a new technology or what institutions are most appropriate for successfully embedding a technology in society. This includes uncertainty about responsibilities and roles.
- Normative uncertainty, i.e. uncertainty about what is desirable.

These three types of uncertainty are, at least roughly, the correlates of the three challenges that I have identified for the introduction of new technology in society: the challenge of predicting and appropriately dealing with the expected impacts of a technology, the challenge of successfully embedding a technology in society and the challenge of recognising and dealing with the relevant ethical issues.

In addition to the distinction with respect to what we are uncertain about, we can distinguish different types of uncertainty in particular between epistemological uncertainty and indeterminacy. Epistemological uncertainty is uncertainty that is due to a lack of knowledge; conversely, indeterminacy refers to the situation in which the causal chains towards the future are still open so that it is indeterminate what exactly will happen. Typically, doing something might reduce indeterminacy. The indeterminacy of human behaviour is, for example, resolved once a person acts. At the moment a person has acted, his (previous) action is no longer indeterminate, even if it remains true that he might have acted otherwise.

What makes the distinction between epistemological uncertainty and indeterminacy important for our current purpose is that they differ in how they can be reduced. Epistemological uncertainty can usually be reduced by research and other modes of investigation. Such uncertainty can at least be partly, although usually not fully, be reduced before a technology is introduced in society. Indeterminacy, on the other hand, is only resolved once something is done. Indeterminacy can therefore not be reduced before a technology is introduced in society. This is in fact a main reason, although not the only one, that the introduction of technology in society always has an experimental character.

Although epistemological uncertainty can often in principle be reduced before a technology is introduced in society, in practice this often turns out to be difficult or practically impossible. Here it is relevant to distinguish between different levels of epistemological uncertainty:

- Statistical uncertainty: We speak of statistical uncertainty if we know what might happen (the scenarios) and we know the probability of each scenario. We do not, however, know with certainty what scenario will happen as we only know the likelihood of each scenario.
- Scenario uncertainty: We speak about scenario uncertainty if we know the scenarios that might occur but cannot meaningfully attach probabilities to each of the scenarios. In case of scenario uncertainty, we thus know the possible outcomes but not their probabilities.
- Recognised ignorance: In the case of recognised ignorance there are certain things we do not know (e.g. we do not know all possible scenarios) and we are aware (i.e. we

know) that we do not know these things. This is also sometimes described as 'known unknowns'.
- Unrecognised ignorance: In the case of unrecognised ignorance we do not know certain things but we are unaware of our ignorance. We might even believe that we know something while we actually do not. This category is also known as 'unknown unknowns'.

4.5.2 The Need for Experimentation

Successfully introducing a new technology into society requires the reduction of uncertainty with respect to potential impacts, institutional embedding and normative issues. It is worthwhile to see to what extent each of these uncertainties can be reduced before a technology is actually introduced in society.

Impact uncertainty is often conceived as epistemological uncertainty that leads regularly to a call for more research into, for example, risks before a technology is introduced in society. There are, however, two fundamental limits to such an approach. First, the type of uncertainty involved in predicting impacts is not just epistemological uncertainty but also indeterminacy. What impacts a technology will have will often, at least partly, depend on how people use it, regulate it or otherwise react to it. History is full of examples of technologies like the telephone and Internet that are used in quite different ways than beforehand expected or foreseen. Second, the level of epistemological uncertainty may sometimes be so high that it becomes practically impossible to reliably predict or anticipate impacts before a technology is introduced in society. While tools have been developed to deal with statistical uncertainty (like probabilistic risk assessment) and with scenario uncertainty (like scenarios,), dealing with (recognised and especially unrecognised) ignorance is much harder and often impossible before a technology is actually introduced in society.

Institutional uncertainty can be partly epistemological, as we might not know what the best institutions are to embed a technology in society. Laws and regulations, for example, tend to lag behind technological development. One reason is that new technologies may have unexpected consequences or raise new ethical issues or concerns that were not foreseen at the time of their introduction. One example is the 'right to be forgotten' that the European Union recently established to deal with some of the privacy issues raised by the Internet.

Tort or liability law may be able to deal with unexpected consequences to some extent, as it offers victims protection against risks arising from negligent or reckless behaviour of, for example, companies introducing new technology. However, they do usually not offer protection against risks that could not reasonably have been foreseen. For example the EU directive for product liability holds producers liable for damage if a product 'does not provide the safety which a person is entitled to expect, taking all circumstances into account' [36: article 6.1]. However, the producer is not liable if he could not have foreseen that the product was unsafe given 'the state of scientific and technical knowledge at the time' [36: article 7 (e)]. The directive thus does not offer protection in cases of what I have above called ignorance.[1]

Institutional uncertainty also has a large indeterminacy component. Often the problem with embedding especially very new technologies is that actors are uncertain about how others will act and react. As a consequence, they may wait for others to take the initiative and to reduce uncertainty (waiting games) or certain responsibilities may be taken up by nobody (problem of many hands). Part of the solution here may be coordination between the actions, roles and expectations of the various actors involved. More generally, successful embedding of a technology in society requires an alignment of technology and institutions. Such an alignment, however, is not a matter of finding the right institutions for a given technology; rather, it is a process of mutual adaption of technology and its institutional environment technology and society co-evolve and this co-evolution can often not be reliably predicted beforehand as it is indeterminate.

Normative uncertainty with respect to new technology might at least take three forms. First, we may be uncertain about the normative issues, such as issues of privacy, security, safety, sustainability, justice, freedom, identity, or democracy, which

---

[1]However, article 15.1 allows countries to deviate from 7(e) and to establish tort laws that assume liability even if the lack of safety could not have been foreseen given the state of scientific and technological knowledge at the time, so that the actual protection may be better than the quote suggests.

a new technology raises. Second, we may be uncertain about how to apply existing moral rules, principles, values and norms to these problems; third, we may be uncertain about the moral rules, values and principles themselves.

Normative learning might require experimentation, i.e. the trying out of norms and values in new situations to see whether they are able to deal with those situations. This idea has especially been elaborated in John Dewey's pragmatic philosophy. Dewey sees existing normative principles, norms and values merely as tools to deal with normative problems [37: chapter 7]. Although such tools have proven their usefulness and adequacy in the past, they may not be the best to deal with new problems. Dewey suggests that sometimes the only way to find out whether such tools work is to put them, and certain policies based on them, in practice and to evaluate whether they indeed are an adequate way to deal with the normative issues they are intended to deal with [cf. 38]. In the case of individual deliberation, it is the individual trying out the moral principles, values and norms who does this evaluation. However, in the tradition of democratic experimentation that has been inspired by Dewey's philosophy [39], it is usually the government that does the evaluation, although it can occasionally also be a judge or a company.

In this tradition, what exactly the criteria are by which to evaluate whether a certain policy works out well is not given beforehand but is itself open to learning. In this sense, normative learning is a form of what has been called second-order learning in which one learns not only with respect to the problem at hand but also about underlying theories and appreciation systems [cf. 2]. Such second order normative learning might thus result in the establishment of new norms and values.

4.5.3   Learning-by-Experimentation

The possibility to learn is a core feature of experiments. However, learning also takes place outside experiments. Below, I therefore position what I will call learning-by-experimentation in relation to two other broad modes of learning, which I will call learning-by-doing and learning-by-anticipation. The basic distinction between those two strategies is that learning-by-anticipation takes place *before* a technology is introduced in society while learning-by-doing takes place *after* a technology has been introduced in society.

I use the term 'learning-by-doing' here as a broad umbrella term that refers to all the kinds of learning that take place during the introduction of a technology in society. The term 'learning-by-doing' was initially used in innovation studies by Arrow to refer to the learning that takes place in manufacturing after the initial innovation and that may lead to considerable cost reductions [40]. Later authors introduced other types of learning like learning-by-using, learning-by-interacting and learning-by-failure [see 41]; the latter is the idea that (initial) failure of a technology may create lot of opportunities for learning and may indeed be the road to technological success [see 42]. My use of the term 'learning-by-doing' is broad and encompasses these various forms of learning during introduction and implementation of technology.

I use the term learning-by-anticipation to refer to the learning process that takes place due to the anticipation of the embedding of a technology in society and the possible impacts that brings. Learning-by-anticipation can happen in the form of attempts to predict certain risks and other consequences as in more traditional forms of risk assessment and technology assessment. Nowadays, learning-by-anticipation often takes more sophisticated forms by learning from anticipating possible futures. An example is the development of scenarios and the organisation of workshops in CTA [43]. The scenarios are not intended to predict the future, but they sketch possible futures and stakeholders are asked to anticipate the future with the help of scenarios. In this way learning might take place, for example, about expectations, strategies and the like. Similarly, in value sensitive design (VSD), value scenarios are used to explore what vales might be relevant in the design of a technology [see 44]. Another example are so-called serious games which mimic certain scenarios or possible development and strategic situations in which one can learn about how actors react to certain situations and each other's' behaviour [see 45].

For a proper embedding of a new technology in society, both learning-by-doing and learning-by-anticipation are probably indispensable, but both also have their limits and disadvantages.

A limit of learning-by-anticipation is that it cannot take into account things that have not been anticipated. In other words, it cannot deal with surprises or unknown unknowns. Apart from that, learning-by-anticipation runs a certain risk to be too much based on speculation; as in some form of nanoethics that are primarily based on ethically challenging but rather unlikely scenarios [46]. The dilemma in anticipation is that on the one hands one want to avoid neglecting certain possible scenarios while at the same time it is clear that some scenarios are much more likely than other and therefore might require more attention.

Learning-by-doing does not require anticipation because it takes place after a technology has been introduced in society. It might therefore be a possibility to discover unexpected things and to learn from actual developments. At the same time, learning-by-doing might be quite costly. It might be costly for two main reasons. First, although failure or the occurrence of undesirable effects may offer good learning opportunities they incur considerable social costs. A disaster like Fukushima might be epistemologically an excellent opportunity for learning; from a social and ethical point of view, it is primarily a disaster that is undesirable. Second, learning-by-doing might be costly because usually it takes place when a technology is already well embedded in society; changing the design of the technology or adapting certain institutions might at that stage have become quite costly. Or, if the introduction of a technology has failed, it might be impossible to revive the technology even if the failure of embedding might have led to learning that would allow a better embedding of the technology.

We might now position learning-by-experimentation between learning-by-doing and learning-by-anticipation. It is similar to learning-by-doing in that it takes place during the actual introduction of a technology in society, or at least in a part of society. Still, it is more anticipatory than regular learning-by-doing because it takes place in a research setting with at least the partial aim to learn something. Ideally then, learning-by-experimentation allows for learning things that cannot be learned by anticipation and at the same time is less costly than learning-by-doing. More realistically, learning-by-experimentation allows for new trade-offs between what can be learned and the costs of learning. Although learning-by-experimentation is probably somewhat more costly that anticipation because it takes place in society, it allows for learning things that cannot be learned by mere anticipation. On the other hand, learning-by-experimentation is likely to be less costly than learning-by-doing, although, arguably, some things will only be learned once a technology is introduced on a large scale in society. One way to deal with these trade-offs is to conceive of the introduction and embedding of a new technology as a gradual process of experimentation. We start experimenting on small scales, on which the costs of failure or risks are still limited and, if successful, we can start to experiment on larger scale on which more can be learned but also the potential costs of learning are higher.

4.6 An Example: Sunscreens with Titanium Dioxide Nanoparticles

I would like to illustrate the sketched approach with a case study on titanium dioxide nanoparticles in sunscreens. Sunscreens with TiO_2 nanoparticles are now on the market, and I will argue that their introduction and use in society amount to an unrecognised (de facto) social experiment because the potential risks of these nanoparticles were not fully known beforehand (and are still not completely known). I will also argue that this de facto experiment does not meet conditions of responsible experimentation. Still, it could be turned into a deliberate and responsible experiment if certain conditions are met. The analysis below is based on an earlier published article. It is necessary somewhat short and sketchy; for more details, the reader is referred to [5].

TiO_2 nanoparticles are now used in sunscreens for more than a decade. They are an alternative to chemical UV absorbers that degrade in the course of time. Consequently, they promise to offer better sun protection than traditional sunscreens. However, some have voiced the fear that these nanoparticles might cause cancer or may have otherwise adverse health effects. In 2006, the International Agency of Research on Cancer mentioned titanium dioxide as a possible cause of cancer [47]. However, this assessment applies to the bulk form of titanium dioxide and cannot be directly extrapolated to the nanoparticle size. Assessment of the safety and health risks of TiO_2 nanoparticles in sunscreens is further complicated by the fact that a range of other factors is potentially relevant for health effects like purity, morphology, crystal structure, the coating, the surrounding matrix, etc. Moreover various exposure routes should be considered like inhalation and skin contact, but also nose exposure (which may lead to penetration of nanoparticles in the brain). Moreover, a route like skin contact may depend on whether the skin is damaged (e.g. sun burn) which may lower the barrier to nanoparticles. In our 2010 article, we found that there are still important knowledge gaps in the risk assessments that had then been carried out.

Due to these knowledge gaps, there was (and probably still is) inconclusive evidence on the risks of TiO_2 nanoparticles. Consequently, the introduction of sunscreens with TiO_2 nanoparticles amounts to a de facto social experiment. Since some of these knowledge gaps could have been closed before the sunscreens were brought on the market, we concluded that this de facto experiment did not meet the first condition for responsible experimentation (cf. also Table 4.1), namely that experiments in society should only be done when other methods to gather knowledge of risks have been reasonably exhausted.

This conclusion is in line with an advice that the Health Council of the Netherlands voiced in 2006 concerning nanoparticles. On basis of the precautionary principle, they said: 'before nanoparticles are brought onto the market, their toxicological properties should be properly investigated' [48:15]. Later in their report they state that '[t]he goal of this investigation is to make nanoparticles a "simple" risk problem instead of an uncertain risk problem' [48:109].

Although I agree with this first part of the recommendation, the second part overestimates the possibility for assessing risks beforehand. Even if the knowledge gaps, which could have been closed before actual introduction, had been closed, risks would still to some extent have remained uncertain. This remaining uncertainty has three more specific reasons. First, the health effects of nanoparticles may be cumulative and depend on interaction with other substances or risks; so that long-term effects can hardly be tested in the laboratory. Second, the lab tests may not represent real-life circumstances. One issue, for example, is that tests are usually done on animals, so that the results have to be translated to humans. Third, some possible hazards may be overlooked due to ignorance. An interesting example of ignorance in this specific case is the appearance of unexpected defects on pre-painted steel roofs. These turned out to be caused by titanium dioxide from sunscreen used by workers during installation. While these defects do not necessarily indicate the presence of a health risk, it clearly shows that TiO_2 nanoparticles had unexpected effects.

Apart from the requirement that the social experiment through introduction of sunscreens with TiO_2 nanoparticles in society should not start before other sources of knowledge gathering about risk have been exhausted, we formulated three other more specific conditions for the acceptability of this social experiment.[2] First, the experiment should be controllable. More specifically, this includes the monitoring of possible effects, the feedback of such effects to the production process of sunscreens, the containment of possible hazards and a conscious up scaling of the experiment. A second condition we discussed is informed consent. In this case, this translates into two more specific conditions, namely (1) consumers should be informed about the use of nanoparticles in sunscreens and their potential risks and (2) consumers should be able to end their participation in the experiment, in particular they should be able to stop using such sunscreens might they wish so. The third condition related to the proportionality of risks and benefits. In this case, risks were still so uncertain that we reformulated this condition as requiring that there should be a continuous review of the risks. Moreover, decisions about continued use should be based on such a review.

None of these conditions is currently met. Possible risks are currently not actively monitored. Sunscreens with TiO_2 nanoparticles are not labelled nor is risk information provided so that consumers are not informed, and this also complicates the possibility to withdraw from the experiment. Also continuous review is missing.

---

[2]The conditions we applied for responsible experimentation in this case are this a bit different from Table 4.1, although the underlying ideas are rather similar.

We made the following recommendations. First, close the existing knowledge gap as far as possible. Second, monitor the possible effects of nanoparticles. Third, if necessary take action on basis of such monitoring. Fourth, ongoing design for safety. A possibility is to design nanoparticles in such a way that they can be traced. This would be helpful in monitoring. And the final recommendation is to change the law and legally require monitoring and labelling. If these recommendations are followed, the social experiment would largely become an acceptable social experiment in which learning takes place in our view.

4.7 Conclusions: Towards Responsible Experimentation

Society is increasing becoming the laboratory in which new technologies are tried out. In part, this is an inevitable development that is due to the experimental nature of new technologies and the uncertainties that are inherent in introducing and embedding new technology in society. Partly, it is also a deliberate development, or at least a phenomenon that can be put at deliberate use. I have argued that a proper experiment requires at least a deliberate attempt to learn as one of its partial aims. I have also argued that learning-by-experimentation may overcome some of the disadvantages of learning-by-doing and learning-by-anticipation.

This does not mean that I think that experimentation should completely replace other approaches for embedding new technology in society. But it is an important approach and one that has been relatively neglected. If we want to do more deliberate experiments with new technology, one of the main questions is that of responsible experimentation. Socials experiments with new technology need to be responsible, both in an epistemological sense of deliberating how to increase the possibilities of learning that is robust and reliable, and in an ethical sense, that arises from the potential impacts of such experiments on society.

Acknowledgement

This chapter was written as part of the research program 'New Technologies as Social Experiments', which is supported by the Netherlands Organization for Scientific Research (NWO) under grant number 277-20-003.

References

1. Robinson, D. K. R., P. Le Masson, and B. Weil (2012) Waiting games: Innovation impasses in situations of high uncertainty, *Technology Analysis & Strategic Management*, **24**(6), 543–547.
2. van de Poel, I., and S. D. Zwart (2010) Reflective equilibrium in R&D networks. *Science, Technology & Human Values*, **35**(2), 174–199.
3. Sollie, P., and M. Düwell (eds.) (2009) *Evaluating New Technologies. Methodological Problems for the Ethical Assessment of Technology Developments* (Springer, Dordrecht).
4. Keulartz, J., M. Korthals, M. Schermer, and T. Swierstra (eds.) (2002) *Pragmatist Ethics for A Technological Culture* (Kluwer Academic Publishers, Dordrecht).
5. Jacobs, J. F., I. Van de Poel, and P. Osseweijer (2010) Sunscreens with titanium dioxide (TiO_2) nano-particles. A societal experiment, *NanoEthics*, **4**, 103–113.
6. Van de Poel, I. (2009) The introduction of nanotechnology as a societal experiment. In: Arnaldi, S., A. Lorenzet, and F. Russo (eds.) *Technoscience in Progress. Managing the Uncertainty of Nanotechnology* (IOS Press, Amsterdam), pp. 129–142.
7. Brownsword, R. (2009) Regulating human enhancement: Things can only get better? *Law, Innovation and Technology*, **1**, 125–152.
8. Savulescu, J., and N. Bostrom (2009) *Human Enhancement* (Oxford University Press, Oxford).
9. Miller, P., and J. Wilsdon (eds.) (2006) *Better Humans? The Politics of Human Enhancement and Life Extensions* (DEMOS, London).
10. Schot, J., and A. Rip, (1997) The past and future of constructive technology assessment, *Technological Forecasting and Social Change*, **54**(2/3), 251–268.
11. Swierstra, T., D. Stemerding, and M. Boenink (2009) Exploring techno-moral change. The case of the obesity pill. In: Sollie, P., and M. Düwell (eds.) *Evaluating New Technologies* (Springer, Dordrecht), pp. 119–138.

12. Sandin, P. (1999) Dimensions of the precautionary principle, *Human and Ecological Risk Assessment*, **5**(5), 889–907.
13. Webster, M., and J. Sell (2007) Why do experiments? In: Webster, M., and J. Sell (eds.) *Laboratory Experiments in the Social Sciences* (Elsevier, Amsterdam), pp. 5–23.
14. Krohn, W., and J. Weyer (1994) Society as a laboratory. The social risks of experimental research, *Science and Public Policy*, **21**(3), 173–183.
15. Herbold, R. (1995) Technologies as social experiments. The construction and implementation of a high-tech waste disposal site. In: Rip, A., T. Misa, and J. Schot (eds.) *Managing Technology in Society. The Approach of Constructive Technology Assessment* (Pinter, London and New York), pp. 185–198.
16. Gieryn, T. F. (2006) City as truth-spot. Laboratories and field-sites in urban studies, *Social Studies of Science*, **36**(1), 5–38.
17. Levidow, L., and S. Carr (2007) GM crops on trial: Technological development as a real-world experiment, *Futures*, **39**(4), 408–431.
18. Fisher, E., and M. Lightner (2009) Entering the social experiment: A case for the informed consent of graduate engineering students, *Social Epistemology: A Journal of Knowledge, Culture and Policy*, **23**(3), 283–300.
19. Gross, M. (2010) *Ignorance and Surprise. Science, Society, and Ecological Design* (MIT Press, Cambridge).
20. Felt, U., B. Wynne, M. Callon, M. E. Gonçalves, S. Jasanoff, et al. (2007) *Taking European Knowledge Society Seriously* (Directorate-General for Research, Science, Economy and Society, Brussels).
21. *Charles in GM 'Disaster' Warning*. Available at http://news.bbc.co.uk/2/hi/uk_news/7557644.stm.
22. Martin, M. W., and R. Schinzinger (1996) *Ethics in Engineering* (McGraw-Hill, New York).
23. Hansson, S. O. (2004) Weighing risks and benefits, *Topoi*, **23**, 145–152.
24. Fisher, J. A. (2007) Governing human subjects research in the USA. Individualized ethics and structural inequalities, *Science and Public Policy*, **34**, 117–126.
25. Hoogma, R. (2002) *Experimenting for Sustainable Transport the Approach of Strategic Niche Management* (Spon, London).
26. Gross, M., and W. Krohn (2005) Society as experiment: Sociological foundations for a self-experimental society, *History of the Human Sciences*, **18**(2), 63–86.

27. Small, A. W. (1921) The future of sociology, *Publications of the American Sociological Society*, **15**, 174–193.

28. Oakley, A. (1998) Experimentation and social intervention: A forgotten but important history, *British Medical Journal*, **317**, 1239–1242.

29. Campbell, D. T., and M. J. Russo (1999) *Social Experimentation* (Sage Publications, Thousand Oaks, California).

30. Gonzales, W. J. (2010) Recent approaches on observation and experimentation: A philosophical-methodological viewpoint. In: Gonzales, W. J. (ed.) *New Methodological Perspectives on Observation and Experimentation in Science* (Netbiblo, La Coruna), pp. 9–48.

31. Hall, A. (2011) Experimental design: Design experimentation, *Design Issues*, **27**(2), 17–26.

32. Stoker, G., and P. John, (2009) Design experiments: Engaging policy makers in the search for evidence about what works, *Political Studies*, **57**(2), 356–373.

33. Ansell, C. (2012) What is a "democratic experiment"?, *Contemporary Pragmatism*, **9**(2), 159–180.

34. Belmont Report (1979) *Belmont Report*. Available at: http://ohsr.od.nih.gov/guidelines/belmont.html.

35. Beauchamp, T. L., and J. F. Childress (2001) *Principles of Biomedical Ethics*, 5th ed. (Oxford University Press, Oxford).

36. Council Directive 85/374/EEC of 25 July 1985 on the approximation of the laws, regulations and administrative provisions of the Member States concerning liability for defective products.

37. Dewey, J. (1957) [1920]) *Reconstruction in Philosophy* (Beacon Press, Boston).

38. VanderVeen, Z. (2011) John Dewey's experimental politics: Inquiry and legitimacy. *Transactions of the Charles S. Peirce Society*, **47**(2), 158–181.

39. Butler, B. E. (ed.) (2012) *Democratic Experimentalism* (Rodopi, Amsterdam/New York).

40. Arrow, K. J. (1962) The economic implications of learning by doing. *The Review of Economic Studies*, **29**(3), pp. 155–173.

41. Maidique, M. A., and B. J. Zirger (1985) The new product learning cycle, *Research Policy*, **14**(6), 299–313.

42. Petroski, H. (1982) *To Engineer is Human. The Role of Failure in Successful Design* (St. Martin's Press, New York).

43. Rip, A., and H. T. Kulve (2008) Constructive technology assessment and socio-technical scenarios. In: Fisher, E., C. Selin and J. M. Wetmore (eds.) *Yearbook of Nanotechnology in Society* (Springer, Amsterdam), pp. 49–70.

44. Czeskis, A., I. Dermendjieva, H. Yapit, A. Borning, B. Friedman, B. Gill, and T. Kohno (2010) Parenting from the pocket: Value tensions and technical directions for secure and private parent-teen mobile safety. In: *Proceedings of the Sixth Symposium on Usable Privacy and Security* (ACM, Redmond, Washington), pp. 1–15.

45. Altamirano, M., P. Herder, and M. de Jong (2008) Road roles Using gaming simulation as decision technique for future asset management practices. In: *IEEE International Conference on Systems, Man and Cybernetics* (IEEE, New York), pp. 2297–2302.

46. Nordmann, A. (2007) If and then. A critique of speculative nanoethics, *NanoEthics*, **1**(1), 36–46.

47. International Agency for Research on Cancer (2006) Carbon black, titanium dioxide, and talc. In: *IARC Monographs on the Evaluation of Carcinogenic Risk to Humans* (IARC, Lyon).

48. Health Council of the Netherlands (2006) *Health Significance of Nanotechnologies* (Health Council of the Netherlands, The Hague).

Chapter 5

Care and Techno-Science: Re-Embedding the Futures of Innovation

Christopher Groves

School of Social Sciences, Cardiff University, Cardiff, Wales CF10 3AT, UK
grovesc1@cardiff.ac.uk

5.1 Introduction

In recent years, the governance of new technologies has increasingly become framed as a re-negotiation of an implicit contract between science and technology (or techno-science), and society. The institutional inseparability of techno-science and public funding means that science has increasingly been framed as an inherently *public* endeavour that must account for its activities and their consequences to the societies in which it is embedded [1].

The direct motivation for re-negotiating the contract between public techno-science and society has been the uncertainties and ignorance [2] that often accompany the processes and products of technological innovation. These surround, on the one hand, the hazards and benefits associated with any emerging technology, and on the other, the potential contribution of such technologies to social change in relation to values, social structures, or material inequalities. If innovation produces novel artifacts, then the

---

Embedding New Technologies into Society: A Regulatory, Ethical and Societal Perspective
Edited by Diana M. Bowman, Elen Stokes, and Arie Rip
Copyright © 2017 Pan Stanford Publishing Pte. Ltd.
ISBN 978-981-4745-74-1 (Hardcover), 978-1-315-37959-3 (eBook)
www.panstanford.com

introduction of novelty into the world can also remake it in morally important yet unforeseeable ways [3]. If this is so, then a moral obligation devolves upon those involved in research and innovation to conduct these activities *responsibly*, even in cases where no human subject is directly involved in research.

David Koepsell [4] has argued that scientific research, the 'discovery of natural truths', must be conducted with a view to whether it can justify *ex ante* its researches to humanity as a whole, according to principles derived from bioethics. Such forward-facing calls for responsibility are reflected in concepts of 'responsible development' or 'responsible research and innovation' (RRI), governance approaches that increasingly focus on issues (such as uncertainty and ignorance about the future) inherent to innovation or emerging technologies as such [5], rather than on issues specific to individual technologies. However, applying principles familiar from fields like bioethics has been recognised as problematic in the face of limitations on foresight about potential consequences [6].

Outlining a possible framework for RRI, Stilgoe et al. [7], note that at the heart of RRI must be 'taking care of the future through collective stewardship of science and innovation in the present'. Care-full governance should be sensitive to novelty, and should not rely solely on past data and the assumption that the distribution of possible future events and their probabilities can simply be extrapolated from the past. Governance undertaken in the spirit of care for the future, Stilgoe et al. [7], argue, is governance characterised by what are essentially individual, collective and/or institutional 'virtues', such as *humility* in the face of incomplete information, practices that allow *adaptation* to a changing situation rather than creating 'lock-in' to particular technological options too quickly, and *responsiveness*, the capacity of governance institutions and innovation 'enactors' [8] to adjust their activities in response to (a) concerns about both the direct impacts of innovation and the priorities to which it is responding, and (b) wider social aspirations. 'For responsible innovation to be responsive, it must be situated in a political economy of science governance that considers both products and purposes' [7: 5]. The need for humility requires us to find 'disciplined ways to accommodate the partiality of scientific knowledge and to act under irredeemable uncertainty'.

In this chapter, I explore the meaning of what 'taking care' of the future through the governance of innovation might mean, if it is to extend beyond the realm of motherhood and apple pie. I flesh out an explicitly temporal concept of care that draws on both phenomenology and the ethics of care. This enables us to think about innovation and governance as a set of activities undertaken by particular institutions through which they 'perform' the future, constructing it as an object of knowledge and concern. Performing the future, I suggest, constitutes and is constituted by specific temporal relationships between past, present and future through which care for the future is exercised. These relationships can be lost, or in certain circumstances, in the reification of the performance itself, in the quest for socially valorised 'disembedded' knowledge of futures, as manifested in demand forecasts, cost–benefit analyses, profit projections and so on.

I explore how restoring an appreciation for the 'artisanal' performance of futures is essential to how innovation, and indeed governance of innovation, can be re-embedded in society as part of the broader goal of reconstructing the contract between techno-science and the societies that depend on it. In order to pursue this goal, we examine the meaning of 'care' itself, in relation to phenomenology and feminist ethics. In this way, I show that re-embedding innovation in society requires a reconsideration of ethical and moral subjectivity in order to make space for the 'virtues' and goals of RRI, one that is complementary [9] to more established concerns in science and technology studies (STS). To examine what this might mean in practice, I contrast specific performances of disembedded futures among innovation enactors with an example of how (and why) a particular nanotechnology company has practiced embedded innovation.

5.2 Care and Innovation

5.2.1 Needs, Attachments and Care

The discussion of care in this section moves us far beyond legal discussions of a 'duty of care'.[1] It links ethical and moral concerns

---

[1] In law a duty of care only arises where a special relationship is deemed to exist, e.g. between health care practitioners and patients, and is owed to individuals (rather

directly to how concern for the future more widely is central to the human condition (in Hannah Arendt's historicised, existential sense, distinct from a metaphysical concept of 'human nature'). The future is an 'issue for us' because of our vulnerability—vulnerability that is material (biologically, we have a range of survival needs) but also emotional (we need meaning and a sense of identity and self-efficacy). The ethical and moral significance of this existential concern with the future has been explored by care ethicists in a body of work recently distilled by Daniel Engster. Drawing on a range of thinkers from Sara Ruddick and Carol Gilligan to Eva Feder Kittay and Joan Tronto, Engster defines care as 'everything we do directly to help others to meet their vital biological needs, develop or maintain their innate capabilities and alleviate unnecessary suffering in an attentive, responsive and respectful manner' [10:31]. We are all dependent on others for the kind of care Engster describes either directly (as infants, sufferers from serious illness, or elderly) or indirectly (as reliant on capacities and resources provided by others that support our own autonomous projects) at points throughout our lives [10]. Consequently, Engster also follows thinkers like Joan Tronto in identifying care as a political as well as ethical principle.

Engster's account of care is thus directly relevant to RRI. Technological innovation is an excellent example of a collective activity that is dependent on particular institutions and policies, and which aims to meet vital needs, enhance capabilities, and alleviate suffering. RRI also foregrounds some of the defining qualities or virtues of care such as attentiveness and responsiveness.

However, Engster's definition of care remains somewhat externalist in its definition of 'what matters', and does not take into account the important of affect and emotion (highlighted by many care theorists) in constituting care. Nor does it take account of the temporality of care [9]. Taking care of needs is a process that engages particular ways of relating to the future, insofar as humans, as needy creatures, are needy in ways that implicate sense-making and emotion. Human beings are needy, not as particular instantiations of a universal species-being, but as individuals whose needs are met via 'satisfiers' that carry specific

---

than to society as a whole). Furthermore, the breach must be shown to have led to recoverable damage, which is within scope of the duty and/or not too remote a consequence of the failure to take care.

individual and shared meanings, and which can serve as the objects of attachment.

Practices of care are future-oriented, but in a specific way. The future they direct us to is not an abstract or generic one, a naturalistic space of probabilities in which outcomes are the more or less predictable product of natural laws, but instead to singular futures the specific futures of individual others who are the ultimate objects of care, futures that are part of biological narratives [11]. As such, care involves certain capabilities—imagination and compassion as well as reason—but also particular dispositions or attitudes towards specific others who are emotionally significant attachments for us.

In every case of care, we find instances of what Andrew Sayer argues are relationships of dependency that are not simply material in nature [12]. Instead, such relationships exhibit elements of emotional connection and meaning, elements that make things matter to people as *constitutive values*, objects that are affirmed by people as helping to 'make them up' as the specific agents they are, and to shape their sense of what they can and should do in the world.

Phenomenologically speaking, care is therefore a particular way of 'being in time' that participates in the futures of singular others. The singular individual cared-for is also a being-in-time, with a future that might go well or badly, whether the individual in question is a person, animal, place, institution, artifact or even a moral or political ideal [13]. All such objects of attachment have their own 'biographies' and their own internal measures of flourishing. It is not simply human individuals, *pace* Engster, who are objects of care, but all things that are enrolled as partners in constitutive relationships where singular futures are bound together to some degree in a common fate.

From this perspective, the caring individual is one who recognises in what they do and how they do it these temporal and more-than-material aspects of dependence. Engster's definition of care as looking after the biological needs and capabilities of others can thus be rewritten from the point of view of this explicitly temporalised definition of care that focuses, not on generic and disembedded needs, but on singular futures and needs that are embedded in particular networks of attachments.

To support the flourishing of others, including their own capacity to care (where this capacity can be possessed by the other who is cared for) a set of concentric circles of supporting conditions can be imagined in each case. Care for specific others necessarily extends into concern for the present and future durability of such supporting conditions. Both human individuals alive now and future people will require such conditions of care as food, shelter, places to inhabit, meaningful work, companionship, political participation, biodiverse ecosystems, and so on. Any responsibility to care for futures must necessarily encompass such conditions. An explicitly temporalised concept of care as rooted in attachment alerts us, however, to the fact that each of these conditions also has cultural, historical and perhaps biographical dimensions. Such conditions are always embedded in narratives about what matters [14]: appropriate foods, particular places, culturally important objects, resonant and seductive ideas, and so on. The attachments about which people care individually and collectively are thus more than accidents of biography, as they are also part of cultural histories.

This temporalised and expanded conception of care needs also to be expressed within what Engster identifies as a third level of care, that of policies and institutions which concern themselves with singular futures rather than generic ones. For example, policies that recognise indigenous people affected by infrastructure development as subjects holding their own, culturally specific, constitutive values [15] embody care for the (singular) future in this sense.

The account of care given here is necessarily both descriptive and normative. It includes a temporalised theory of what kinds of things matter to people and how (linking together, within particular narratives, the past and future through the present), and why this import of things is the source of moral obligations [12]. If care for the future is a key element of RRI, then these descriptive and normative elements can give us an insight into what kinds of objects care is and should be directed towards. I agree with Engster and others like Ruddick [16] and Walker [17] that the descriptive element of a theory of care focuses on practice—how subjects of care attend and respond to the needs or singular futures of others. Care is not a principle, but a form of social practice—indeed, insofar as it is also a way of making sense of the world as containing things that *matter*, it is an ingredient of *all* social practice. If we are moving from a descriptive theory of care to a normative theory that reflects the importance of care, then perhaps 'we should think of normativity more in terms of the ongoing flow of continual concrete evaluation, and less in terms of norms, rules, procedures, or indeed decisions and injunctions about what one ought to do' [12:97].

Understanding care in this way as practice can flesh out what 'care for the future' in RRI can mean, by injecting into it a focus on temporality and meaning as well as concrete needs. However, the contexts of RRI are typically interactions between institutions, and between individuals or groups of individuals and institutions, mediated by material (and more than material) objects. To what extent can these interactions be understood descriptively in terms of care, and what normative purchase can the concepts of care outlined above have on them? In the next section, we consider these questions in relation to how patterns of practice may perform the future in ways that either embody/support, or occlude, care-full innovation.

5.2.2 Expectations and Performing Futures

Meeting needs implies both material fulfilment and the creation of meaning through the production of expectations about the future that draw on particular pasts. These expectations are, in the interpersonal contexts discussed above, the substance of attachment and secure dependence. It is also in relation to the production of expectations that the nexus between care and innovation can be explored. Technological expectations contribute to the evolution of new technologies [18], and also shape regulatory responses [19]. Whether representational (e.g. speculations about the influence of emerging technologies on social change) or material in form (e.g. artifacts), technological expectations 'prehend' (the term is A. N. Whitehead's), or bring into relation, ongoing processes partially submerged in the past, and already completed products in which are congealed past processes. As a result, expectations can release latent potential, including here both potential for meaning in the form of new representations of promised futures, and material potential embodied in new affordances [20].

An example drawn from speculative representations of future nanotechnologies might be the production of concepts and images of 'nanogears' that, in combination with past innovations, such as the production of fullerenes [21], can become configured as precursors of speculative future developments (molecular manufacturing). Expectations thus create meanings around techno-scientific objects that connect to objects of attachment like moral values, cultural ideals, and the practices through which such things are cared for. For example, the value of new innovations may be articulated in relation to such goals and ideals as increased autonomy, the prevention of disease or promotion of particular conceptions of health, and/or the practices through which commitments to such important concerns are sustained or promoted.

Producing expectations is a way of answering the question 'what matters?' As a result, the effect of expectations on the content of debates about the desirability or otherwise of particular futures, and on the degree to which particular innovation pathways may represent examples of irresponsible or responsible innovation, is significant [22]. Through the spread of specific expectations, debates over the desirability of particular developments can be forestalled or brought forward [23], the 'obviousness' of some priorities established, and the politically dominant position of particular interests obscured [24].

The way expectations help frame such debates is dependent on their content and how specific knowledge practices are employed in producing them. They reflect particular ethical frameworks used in assessing what matters, exploiting in the process specific methodologies, and attendant assumptions used to make complex futures more manageable and tractable in decision-making. The techno-scientific institutional contexts in which decisions pertaining to innovation are made typically rely on recommendations that are justified in relation to disembedded, quantified, and depersonalised forms of knowledge [25]. The construction of expectations in these contexts thus often relies on forms of abstract modelling (cost–benefit analyses, profit projections, innovation scorecards and so on) which define 'what matters' in naturalistic terms as more or less quantifiable variables, and treat uncertainty as an equally quantifiable parameter of decisions.

The future performed on the basis of such resources tends to be equally disembedded and generic in nature, constructed as an extrapolation of past trends identified through observational data, tending towards the condition of a 'stationary' future [26]. This is the future as 'abstract' or 'empty' [27:57], produced through assemblages of knowledge practices, routinised forms of action and ethical frameworks that treat past and future as discrete domains connected mechanistically [28:15]. Grounded in routines of bureaucratic rationalisation, the performance of abstract and/or empty futures also express patterns of desire [29]. In particular, the desire 'for security and stability in the face of the world's hazards', 'a desire that may become a fear of what life may bring forth', creates a context where abstract futures, like abstract concepts, may appear as 'marks of contraction and withdrawal' (John Dewey, *Art as Experience* (1958), quoted in [30:22]) from singular futures. A desire for certainty may there cut against acting care-fully rather than encouraging actors to act in this way. A degree of certainty within the operational space of institutions is provided by disembedding, but this also entails disconnection and the implicit disavowal of the importance of 'messy' singular futures.

Knowledge practices and ethical frameworks that disembed the future thus 'prehend' past and future in an abstract way, creating the future as a reflection of a given dataset, a single slice taken through the past. In spatial contexts, it is well-documented how power acts through the gridding and standardising of territorial space [31], thus creating maps of bureaucratic certainties in order to govern an uncertain territory [32], activities through which 'an official landscape is forcibly imposed upon a vernacular one' [33:17]. In the temporal context of care and expectations, the abstract narratives that result from the disembedding of futures are often imposed on the multidimensional narratives about what matters that circle around attachments, their conditions of flourishing, and the expectations that connect them. Care-full practices that perform embedded futures recognise, by contrast, the interconnectedness of values that matter. They serve the need for careful practical judgment in constructing and acting out a narrative through which actors can strive to weave together as many of the singular narratives with which they are concerned as possible. Abstract narratives, on the other hand, tend to construct the future as a frame over and above these singular narratives, perhaps ultimately driven by a grand, top-down narrative that determines, once and for all, their essential meaning and role.

5.3 Disembedding: Generic Futures in Nano- and Biotechnology

We now examine two examples, one from promissory rhetoric around nanotechnology, and the other from formative debates about self-regulation in biotechnology, of ways in which the construction of expectations around innovation can construct such disembedded future narratives. Our second example also shows how generic futures, if care-less enough, can create problems for innovators.

Consider the following extract from a speech given in 2005 by the then-US Under-Secretary of Commerce for Technology, Phillip J. Bond, at a conference organised by the reinsurer Swiss Re:

> Nanotechnology offers the potential for improving people's standard of living, healthcare, and nutrition; reducing or even eliminating pollution through clean production technologies; repairing existing environmental damage; feeding the world's hungry; enabling the blind to see and the deaf to hear; eradicating diseases and offering protection against harmful bacteria and viruses; and even extending the length and the quality of life through the repair or replacement of failing organs. Given this fantastic potential, how can our attempt to harness nanotechnology's power at the earliest opportunity—to alleviate so many earthly ills—be anything other than ethical? Conversely, how can a choice to halt be anything other than unethical? (Bond, cited in [34]).

Bond performs here a strong consequentialist assessment of the future value of nanotechnology, one which draws on predictions made over the course of several years by nanotechnology enactors, ranging from materials scientists to technology strategists in government, and which were expressed emphatically in the strongly promissory document *Shaping the World Atom by Atom* [35]. He suggests that, based on expert consensus, the projected benefits of nanotechnological research are so plausibly great that not striving to realise them would be morally irresponsible. Uncertainty about consequences is often constructed asymmetrically by enactors, who tend to represent potential hazards as uncertain and benefits, by contrast, as more or less certain given the removal of any barriers to technical progress that may exist (such as regulation, or public aversion to risk). Connections are drawn between generic needs, technological potentials, and future benefits that extrapolate from past evidence of technical accomplishments to inevitable transformative futures in order to respond defensively to anticipated 'roadblocks' stemming from outsiders' concerns [36:359]. In this way, care for the future is translated into a single measure of success: given the *amount* of good that would be achieved by fulfilling all these needs, being care-full just means pushing ahead with whatever innovations emerge.

The audience is invited to occupy the same Archimedean point where the speaker, thanks to the expertise on which he claims to rely, stands to survey the coming landscape of innovation. This standpoint is cemented by the utilitarian assumptions behind the rhetoric: that what ultimately matters is the satisfaction of human preferences, in aggregate. This *particular* valorisation of 'what matters' (itself expressive of specific, culturally available ideals that serve as attachment objects which help to navigate uncertainty) obscures other questions: for example, how these benefits might, in fact, be realisable in concrete settings of care (for self and for others), whether they might, in fact, always be desirable, and how they might affect the narratives through which people, as members of particular social groups, understand themselves (we might ask which of these promised nanotechnology benefits—such as providing people without a given sense new sensory capacities— might be refused by individuals and why, and what effects might the sudden availability of such technologies have on the position of 'disabled' individuals in society?).

This definition of 'what matters' is also rhetorically powerful because it supports a deterministic view of the processes through which the future shall come about: 'nanotechnology is coming, and it won't be stopped' [37:7]: the train of new technology has always already left the station (see also [38]). Under such conditions it only remains to 'sort the legitimate concerns from imaginary ones', and to reduce uncertainty about risk (having assumed that benefits are certain). However, it has been pointed out that, for trains to run, tracks need to be laid first [34, 39]. As suggested by the sociology of expectations, the point of strong consequentialist arguments of the kind presented by Bond is to construct a generic future in which the start and destination of these tracks are already known, thanks to the informed extrapolation of future developments from past experience and data.

Bond's speech thus serves as a prime example of rhetorical labour that disembeds the future in order to generate authority for itself, authority that is buttressed by affective strategies and the evocation of a general moral framework in which care for the future is presented as being about maximising (and 'hastening') benefits while 'minimising any potential downside' [37:8]. The generic future thus performed becomes a kind of 'phantastic object' like 'the next Industrial Revolution' or 'the New Economy' [40] around which promissory narratives can circulate, narratives that then recruit actors by taming uncertainties.

Experiences from the evolution of other technologies indicate why such generic visions, constructed in order to 'think and act far ahead', are problematic. The process of producing a disembedded future may blind actors to the extent to which their future remains embedded in existing social expectations, and in broader networks of care that support the singular futures of other 'things' that matter. As a result, the performance in which they are engaged may be undermined, precisely because it loses the virtue of responsiveness articulated by Stilgoe et al. [7]. For a second example of how this can occur, we turn to Susan Wright's [41] account of how the National Institutes of Health in the USA developed their Guidelines on Recombinant DNA Technology.

The key condensation point for the process is generally represented as the Asilomar conference in 1975, an event generally cited as a 'landmark' [42:33] example of future-oriented self-regulation that brought about a successful precautionary 'solution' [43:14] to a regulatory dilemma by commentators on other emerging technologies like nanotechnology [44] and synthetic biology [45:48]. The image, common in performances of disembedded futures, of a society that has to 'prepare itself' for the arrival of an unstoppable emerging technology train, was central to the conference, yet undermined by its outcomes. As Wright outlines, the Asilomar meeting was represented in the invitations sent to participants as an opportunity to develop a position on regulation that would allow research and translation into applications to continue with minimal interference from regulators and other selectors [41:145]. This orientation has been represented by others as an attempt on the part of the nascent biotech community to 'give itself monopoly control of how the domain of biotechnology would be regulated' [46:69]. The rhetorical emphasis fell, as in Bond's speech, on the need to grant autonomy to enactors in order to achieve maximum aggregate future benefits.

The conference centred on a series of promissory, future-oriented presentations detailing new research approaches and potential future applications. Opening the meeting, MIT cancer researcher David Baltimore stated that the ultimate goal was 'a strategy to maximise the benefits and minimise the hazards in the future', but sketched a broad definition of possible benefits contrasted with a narrow definition of hazards, focused chiefly on lab hazards [41:148-149]. Three working panels then examined issues of risk identification, assessment and management, with their reports then forming the basis of collective discussion. Wider social concerns and ethical issues were ruled out of the working group terms of reference presented by Baltimore.

The discussion centred on what became the 'central dilemma' of the meeting: 'no-one had any real idea of what the risk might be or how to assess it' [41:151-152]. In the absence of a firm, consensual definition of hazards, the unquantifiability of risk was seen as a reason for preferring self-regulation, on the principle that expected benefits of continuing research trumped unquantifiable risks. Nonetheless, a central anxiety emerged. If no determinate response to the uncertainties surrounding even the limited risk factors admitted for discussion by the meeting was possible, then legislators might do the job for the scientists, and forcibly re-embed scientific research within society's normative frameworks through new statutory regulation.

A session of advice from legal scholars forced attendees to reconsider their commitments and, in particular, their belief in academic freedom beyond any reciprocal contract between science and society [41:153-154]. In particular, the legal experts stressed the need to consider the eventuality of being held liable for irreversible harm on grounds of professional negligence. This led the organising committee to draft a set of containment procedures based on the severity of potential hazards. Having proven contentious, this was then discussed by the full conference section by section, giving rise to debates conducted in terms that one observer suggested suited 'science fiction' more than 'hard science'. Later, the statement was published in *Science* with a problematic compromise at its heart. Huge uncertainties were acknowledged, but at the same time the adequacy of containment approaches for balancing benefits and risks was claimed.

The account Wright [41] provides, based on statements and testimony from participants and others present, dramatises how the performance of generic futures at Asilomar was undermined from within the performance itself. Optimistic techno-scientific expectations, which energised discussion of the need to certify continuing research through voluntary commitments, were eroded by the anxieties generated through the working groups' examinations of the uncertainties surrounding the singular futures of the specific institutional research programs with which they were concerned. The construction of a disembedded, generic future intended as a catalyst for consensus was eroded from within, as it became re-embedded in concrete research contexts hedged about with ignorance, and then within the social context of existing law regarding liability and the standards of care expected of professionals working in specific research contexts in relation to those whom may be harmed by their actions. Whether or not one interprets it as a signpost sited at a key point on a path of responsible self-regulation, the Asilomar process can be read as emblematic of difficulties that can result from the performance of a care-less, disembedded future as it rubs up against the singular contexts of care in which actors operate.

5.4 Care-Full Re-Embedding: Performing Concrete Futures

As well as providing plenty of instances of the rhetorical creation of disembedded futures, developments within innovation in nanotechnology also yield instances of how the care-full embedding of innovation performs more concrete futures, ones in which the meaning of technological artifacts for constitutive values are the focus of concern. For such an example, we turn now to a study of attitudes towards voluntary regulation among UK nanotechnology companies undertaken by the ESRC Centre for Business Relationships, Accountability, Sustainability and Society in which I was involved in 2008-10 [47]. From a series of interviews the authors conducted with representatives of companies of various sizes, from spin-outs to multinationals, various instances can be given of how generic futures may be re-embedded within relationships through which concerns with temporality and the meaning of objects enter into technological innovation.

I focus here on one example, a company (identified as Company N in the research) that occupied a place in the value chain typical of smaller firms included in the study. N develops enabling technologies (nanosensors designed to detect volatile organic compounds [VOCs]) using nanomaterials (carbon nanotubes) purchased from an upstream supplier. The uses of such sensors are several, but N focuses on finding ways to employ them into devices that can be used for medical diagnostic purposes by people with specific health conditions. On the company's website, the potential of the sensor is constructed using concepts that suggest disembedded, generic technological futures that may, in more radically promissory contexts, become phantastic objects—such as 'mobile health' and 'personalised preventive medicine'—with effects on innovation and also policy in healthcare hat have been studied by researchers in the sociology of expectations [48]. At the same time N's website depicts the specific content of these futures as continuous with present developments rather than being radically disruptive, as desirable and plausible outgrowths of current trends, and as evolving out of particular social needs rather than being driven by technology.

Although, at this level, the company's publicity constructs a disembedded future which its products are envisioned as helping to create, the innovation practices in which the company engages also focus on using technology as a means of caring for singular futures, and reflect an understanding that success depends on knowledge of the specific meanings that technological artifacts have for users of medical diagnostic technologies. The role of N in the value chain was identified in interview as 'like an R&D consultancy', with the core of business being product development intended to 'inspire' customers—consumer product manufacturers, in the main—to create new applications for nanosensors. Seeking collaborations with product manufacturers was, consequently, a key part of N's activities, identifying ways in which N's technology can 'meet a critical customer need'. In turn, this requires understanding end-user requirements. As a result, N identifies and consults with potential end-users of products based on nanosensors (such as individual sufferers of medical conditions and patient groups) to understand their needs.

Survival is dependent on a constructed generic future, measured by the company balance sheet: 'the bottom line for the company is to make money.' Nonetheless, attracting partners is the means through which this 'in the last instance' consideration can be taken care of. N is, like many micro businesses involved with nanotechnology, not engaged in attracting investment to e.g. scale up production processes as much as looking for ways to exploit existing intellectual property. It is perhaps for this reason that ways of demonstrating 'virtues' such as product reliability (such as a firm precautionary approach to risk minimisation, e.g. through removal of any possibility of end-user exposure to nanomaterials through nanosensor design), good management (as evidenced by being able to show one is aware of all relevant regulations relating both to one's own product and to the business activities of customer companies) and the positive contribution of N's innovations to customers' products ('the benefits to their solutions'), and to their customers' customers, are described as most vital to the company, rather than projected earnings or other measures of company health. The question constantly facing the company is how to project reliability and create trust in order to create and sustain commercial relationships based on the social value of products. The company's products, N suggests, must embody this concern for trust and the particular expectations of a range of stakeholders in order to be at all successful.

Answering this question involves, in addition, not just understanding need but understanding the different *meanings* of need. A product is not enough on its own to create expectations and build trust. N's engagement with potential end-users, such as asthma sufferers, takes place in focus groups, intended to establish whether 'they need the technology', establishing whether it 'would add value to their quality of life'. But this value, N suggests, will be derived not from simply fulfilling some abstract need, e.g. for maintaining the health of asthma suffers, where 'health' means the absence of asthma attacks. In this abstract sense, fulfilling the need just requires a technically reliably device, in this specific case, one able to monitor biomarkers such as the presence of volatile organic compounds in breath to spot, early on, signs of impending asthma attacks.

Fulfilling the need implies understanding it in the context of the meanings associated with the need. Through consultation with patient groups, N has decided to develop the technology (the nanosensor) in such as way that it can be easily be incorporated in artifacts like mobile phones. By doing this, the end product will enable people with asthma to overcome issues identified by focus groups such as the 'compliance of actually using those devices'. Compliance with monitoring to maintain health has been seen as compromised by resentment at the dependence one is forced into [49], through one's identity as an 'asthma sufferer'. N's nanosensors afford incorporation in near-future versions of a current technology, granting symbolic and emotional value ('liberation') to an artifact that supports an individual's identity and agency, making it more than just instrumentally valuable as a means of servicing a generic need. In this way, the technology incorporates en evaluation of the singular futures of its users. It has been reshaped in accordance with negotiation about the meaning of the specific ends it serves, rather than treating it as a disembedded tool through which an abstract outcome (absence of asthma attacks) may be achieved (cf. [9]).

While the future engaged with by N is, in the last instance, one that is abstract and empty in Adam and Groves' [27] terms—constructed through actuarial practices and consolidated in artifacts like balance sheets and profit projections—it is engaged with through a relationship with the lived future of others, the embedded expectations through which their lives are lived. This level of embedded engagement is established through imagination, emotional identification and dialogue—it is one that is performed between people, and mediated by objects that are not just commodities, but also emotionally meaningful, carriers of trust and care. As Steven Shapin has argued in relation to venture capitalists' investment decisions with regard to technological innovations, the projection of trustworthiness is, despite rhetoric of gambling and speculative betting being commonly seen as part of 'selling' innovation, the central factor in how these decisions are taken [50]. Negotiating highly uncertain futures that ride upon the backs of innovation depends on being able to elicit judgments from investors or front-end companies on the virtues of a person, company and/or product. If a company's balance sheet is, in the last instance, an anchor for judgments about its future health rooted in past performance, the repository for the organisation's latent potential is the concrete promise of products, how the uncertainties (including regulatory ones) surrounding them are managed by the company, and the capacity of representatives to make links, rhetorically and symbolically, with the activities of consumer-facing businesses and the needs of their users. For these reasons, N represents 'performing a credible technological and commercial future' [50: 289] as dependent on shaping singular futures of care through the re-embedding of technological artifacts in social contexts of use, as well as being dependent on commercial credibility produced by representations of a disembedded future, such as balance sheets and profit projections.

5.5 Conclusion

Disembedding the future, I have argued, is a way of performing the future in the present that relies on specific ways of representing the temporality of action. Specialist knowledge derived from past data informs present practice, symbolically taming an uncertain future through 'blueprints' and materially taming it by creating future realities. It sustains care, in the sense of an active concern for things that matter, in a limited form, minimising connection and the balancing of singular futures against each other in order to achieve a goal or goals. Such representations have a political significance, insofar as they create 'windows' through which certain futures may be imagined but not others, as well as rendering invisible the singular futures of certain objects of care, along with the concentric circles of supporting conditions that sustain them. Promoting a technological solution because of the aggregate balance of benefits over harms it is expected to bring tends to discount any concerns about the wider effects of this solution, which may drawn on concrete experiences of care and attachment. The performance of such disembedded futures crowds out the significance of other actors' ongoing performances of their own direct and indirect care for things that matter in favour of the autonomy of actors who are able to represent themselves as being the favoured enactors of an onrushing future heralded by emerging technologies.

Barbara Adam and I [27:133–138] make a similar distinction between two ways of 'acting' the future that ultimately derives from Deleuze and Guattari [51]. On the one hand, an 'architectonic' stance orients itself towards a future conceived of as the product of an imposition of a blueprint on inert matter, whose only significance is as 'stuff' fit to receive form. On the other, an 'artisanal' stance enacts the future through the exercise of *craft*, eliciting from the world its potential for becoming in ways that exhibit virtues similar to those that Engster [10] lists as essential to care (attentiveness, responsiveness, respect) or those called for by Stilgoe et al. [7] (humility, adaptation and responsiveness). 'Architects' map out generic, disembedded futures, while artisans strive for care-full re-embedding of the future within a multiplicity of irreducibly different concerns. In the one case, the performance is tied up with the symbolic reduction of uncertainty in order to act with an authority derived from representations of certainty, or at least of 'known unknowns', in order to secure the autonomy to take a decision and move forward. In the other, the performance is tied up with the need to care for the unfolding of singular futures while at the same time preserving their potential, constantly alive to the past of their bearers and revising estimations of what is desirable.

There are thus distinct (linear, non-linear) temporalities at work here, as Anne-Marie Mol has suggested in her work on the difference between 'logics of decision' and 'logics of care' [9]. Yet in the end they can only be kept apart rhetorically, as in the case of Bond's speech on the 'unethical' nature of deciding 'not to proceed' with nanotechnology. In practice, they mesh together as the future continues to be performed, as the case studies surveyed in this chapter show. Even where disembedded futures effectively disavow contexts of care, they still perform care for particular constitutive values (e.g. political ideals like efficiency, growth, utilitarian definitions of happiness and so on). It is through stories of how the dis- and re-embedding of futures intermesh with each other (as in the examples of Bond's speech, the Asilomar conference and Company N) that the ethical (and political) dimensions of the relationship between embedded and disembedded futures—the activities of subjects who are positioned by the practices in which they participate as architects and artisans—can be traced.

Bond's speech refers to a wide range of contexts of care in translating them into a single overarching context anchored by the concept of aggregate social good. The Asilomar conference shows how decisions on how to represent 'known unknowns' can unravel in confrontation with concrete contexts of direct care (e.g. for experiments and research programmes). Company N's activities show how representations of the future that perform certainty and the construction of known unknowns are relatively valueless without the exploration, through concrete relationships with others, of their singular futures and what matters to them.

Even the disembedded future expressed in Bond's speech is an expression of care, a rhetorical performance of a future that is designed to domesticate uncertainty, to build the relationships and marshal the resources needed to create secure expectations. But disembedded futures disavow wider connectedness to singular futures, and draw concern in tighter around a limited range of objects, emphasising above all else the ultimate value of autonomy to act and decide, and particularly its superiority over alternative ideals such as solidarity, i.e. the kinds of reciprocal commitments that are part of the performance of the future described in interviews with company N. If RRI is intended to re-embed techno-science within society, then an ethics of care for the future has indeed to be central to its efforts. But the meaning of care needs to be explored by questioning the ways in which techno-science relies on and promotes 'strategies of autonomy' and tends to exclude explorations of solidarity as an effective, but also a morally preferable ways of dealing with uncertainty [13].

Acknowledgements

Elements of the research discussed in this chapter were conducted with the support of the UK Department of Environment, Food and Rural Affairs through grant ref. CB0417. Thanks are due to two anonymous reviewers and also to Dr Elen Stokes for comments on an earlier draft—and particularly for her suggestion that 'performance', and especially its temporal aspects, is central to the account of care given here.

References

1. Ziman, J. M. (2000) *Real Science: What It Is, and What It Means* (Cambridge University Press, Cambridge).
2. Schummer, J. (2001) Ethics of chemical synthesis, *Hyle*, **7**(2), 103–124.
3. Grinbaum, A., and C. Groves (2013) What is 'responsible' about responsible innovation? Understanding the ethical issues. In: J. Beasant, M. Heintz and R. Owen (eds.) *Responsible Innovation: Managing the Responsible Emergence of Science and Innovation in Society* (Wiley, Chichester).
4. Koepsell, D. (2010) On genies and bottles: Scientists' moral responsibility and dangerous technology R&D, *Science and Engineering Ethics*, **16**(1), 119–133.
5. Felt, U., and B. Wynne (2007) *Taking European Knowledge Society Seriously: Report of the Expert Group on Science and Governance to the Science, Economy and Society Directorate, Directorate-General for Research, European Commission* (Office for Official Publications of the European Communities, Luxemburg).
6. Groves, C. (2009) Future ethics: Risk, care and non-reciprocal responsibility, *Journal of Global Ethics*, **5**(1), 17–31.
7. Stilgoe, J., R. Owen and P. Macnaghten (2013) Developing a framework for responsible innovation. *Research Policy*, **42**(9), 1568–1580.
8. Garud, R., and D. Ahlstrom (1997) Technology assessment: A socio-cognitive perspective, *Journal of Engineering and Technology Management*, **14**(1), 25–48.
9. Mol, A. (2008) *The Logic of Care: Health and the Problem of Patient Choice* (Routledge, New York).
10. Engster, D. (2007) *The Heart of Justice: Care Ethics and Political Theory* (Oxford University Press, Oxford).

11. Groves, C. (2011) The political imaginary of care: Generic versus singular futures, *Journal of International Political Theory*, **7**(2), 165–189.
12. Sayer, A. (2011) *Why Things Matter to People: Social Science, Values and Ethical Life* (Cambridge University Press, Cambridge).
13. Marris, P. (1996) *The Politics of Uncertainty: Attachment in Private and Public Life* (Routledge, London).
14. O'Neill, J., A. Holland A. Light (2008) *Environmental Values* (Routledge, London).
15. Schlosberg, D., and D. Carruthers (2010) Indigenous struggles, environmental justice, and community capabilities, *Global Environmental Politics*, **10**(4), 12–35.
16. Ruddick, S. (1989) *Maternal Thinking: Toward a Politics of Peace* (Beacon Press, New York).
17. Walker, M. U. (1998) *Moral Understandings: A Feminist Study in Ethics* (Routledge, New York).
18. Borup, M., N. Brown, K. Konrad and H. van Lente (2006) The sociology of expectations in science and technology, *Technology Analysis & Strategic Management*, **18**(3–4), 285–298.
19. Groves, C., and R. Tutton (2013) Walking the tightrope: Expectations and standards in personal genomics, *BioSocieties*, **8**(2), 181–204.
20. Michael, M. (2000) Futures of the present: From performativity to prehension. In: N. Brown, B. Rappert and A. Webster (eds.) *Contested Futures: A Sociology of Prospective Techno-Science* (Ashgate, London), pp. 21–42.
21. Robertson, D. H., B. I. Dunlap, D. W. Brenner, J. W. Mintmire and C. T. White (1994) Fullerene/tubule based hollow carbon nano-gears. *MRS Online Proceedings Library*, **349**, 283.
22. Simakova, E., and C. Coenen (2013) Vision, hype and expectations: A place for responsibility. In: R. Owen, J. Beasant and M. E. Heintz (eds.), *Responsible Innovation: Managing the Responsible Emergence of Science and Innovation in Society* (John Wiley & Sons, Chichester), pp. 241–268.
23. Hanson, V. L. (2010) Envisioning ethical nanotechnology: The rhetorical role of visions in postponing societal and ethical implications research, *Science as Culture*, **20**(1), 1–36.
24. Sparrow, R. (2008) Talkin' 'bout a (nanotechnological) revolution, *IEEE Technology and Society Magazine*, **27**(2), 37–43.

25. Giddens, A. (1990) *The Consequences of Modernity* (Polity Press, Cambridge).
26. Orléan, A. (2010) *The Impossible Evaluation of Risk* (Cournot Centre for Economic Studies, Paris).
27. Adam, B., and C. Groves (2007) *Future Matters: Action, Knowledge, Ethics* (Brill, Leiden).
28. Adam, B. (1995) *Timewatch: The Social Analysis of Time* (Polity Press, Cambridge).
29. Groves, C. (2013) Horizons of care: From future imaginaries to responsible research and innovation. In: K. Konrad, C. Coenen, A. Dijkstra, C. Milburn and H. van Lente (eds.) *Shaping Emerging Technologies: Governance, Innovation, Discourse* (IOS Press/AKA, Berlin), pp. 185–202.
30. Jackson, M. (1989) *Paths toward a Clearing: Radical Empiricism and Ethnographic Inquiry* (Indiana University Press, Bloomington).
31. Scott, J. C. (1998) *Seeing Like a State: How Certain Schemes to Improve the Human Condition Have Failed* (Yale University Press, New Haven).
32. Mitchell, T. (2002) *Rule of Experts: Egypt, Techno-Politics, Modernity* (University of California Press, Berkeley).
33. Nixon, R. (2011) *Slow Violence* (Harvard University Press, Cambridge).
34. Swierstra, T., and A. Rip (2007) Nano-ethics and NEST-ethics: Patterns of moral argumentation about new and emerging science and technology, *NanoEthics*, **1**, 3–20.
35. Amato, I. (1999) *Nanotechnology: Shaping the World Atom by Atom* (National Science and Technology Council, Washington D.C).
36. Rip, A. (2006) Folk theories of nanotechnologists, *Science as Culture*, **15**(4), 349–365.
37. Bond, P. J. (2005) Responsible nanotechnology development. In: Swiss Re (ed.), *Nanotechnology: Small Size–Large Impact?* (Swiss Re, Zurich).
38. Hsu, A. R., J. L. Mountain, A. Wojcicki and L. Avey (2009) A pragmatic consideration of ethical issues relating to personal genomics, *American Journal of Bioethics*, **9**(6–7), 1–2.
39. Hogarth, S. (2010) Myths, misconceptions and myopia: Searching for clarity in the debate about the regulation of consumer genetics, *Public Health Genomics*, **13**, 322–326.

40. Tuckett, D., and R. Taffler (2008) Phantastic objects and the financial market's sense of reality: A psychoanalytic contribution to the understanding of stock market instability, *The International Journal of Psychoanalysis*, **89**(2), 389–412.

41. Wright, S. (1994) *Molecular Politics: Developing American and British Regulatory Policy for Genetic Engineering, 1972–1982* (University of Chicago Press, Chicago).

42. Siune, K., E. Markus, M. Calloni, U. Felt, U. Gorski, A. Grunwald, et al. (2009) *Challenging Futures of Science in Society: Emerging Trends and Cutting-Edge Issues* (European Commission Directorate-General for Research, Brussels).

43. Healey, P., and H. Glimell (eds.) (2004) *European Workshop on Social and Economic Issues of Nanotechnologies and Nanosciences* (EC Research Directorate, Brusssels).

44. Jacobstein, N. (2006) *Foresight Guidelines for Responsible Nanotechnology Development* (Foresight Institute and Institute for Molecular Manufacturing, Palo Alto).

45. Capurro, R., J. Kinderlerer, P. Martinho da Silva and P. Puigdomenech Rosell (2009) *Ethics of Synthetic Biology: Opinion No. 25* (European Group on Ethics in Science and New Technologies to the European Commission. Brussels).

46. Dupuy, J.-P. (2004) Le problème théologico-scientifique et la responsabilité de la science. In: Health and Consumer Protection Directorate-General (ed.) *Nanotechnologies: A Preliminary Risk Analysis* (European Commission, Brussels), pp. 71–94.

47. Groves, C., L. Frater, R. Lee and E. Stokes (2011) Is there room at the bottom for CSR? Corporate social responsibility and nanotechnology in the UK, *Journal of Business Ethics*, **101**(4), 525–552.

48. Hedgecoe, A. (2004) *The Politics of Personalised Medicine Pharmacogenetics in the Clinic Cambridge Studies in Society and the Life Sciences*. Available at: https://http://www.dawsonera.com/guard/protected/dawson.jsp?name=https://idp.cardiff.ac.uk/shibboleth&dest=http://www.dawsonera.com/depp/reader/protected/external/AbstractView/S9780511317743.

49. Smelser, N. J. (1998) The rational and the ambivalent in the social sciences: 1997 Presidential Address, *American Sociological Review*, **63**(1), 1–16.

50. Shapin, S. (2009) *The Scientific Life: A Moral History of a Late Modern Vocation* (University of Chicago Press, Chicago).
51. Deleuze, G., and F. Guattari (1980). *A Thousand Plateaus* (Athlone, London).

Chapter 6

Division of Moral Labour as an Element in the Governance of Emerging Technologies

Arie Rip

Department of Science, Technology and Policy Studies,
School of Management and Governance, University of Twente,
PO Box 217, 7500 AE Enschede, The Netherlands

a.rip@utwente.nl

In the early nineties, and partly in response to David Collingridge's [1] *Control Dilemma* of new technology, Rip et al. [2] drew attention to a division of labour between power and control that had become institutionalised through the centuries, definitely since the late 19th century with the rise of new technologies together with the rise of the Regulatory State. A key indication is how government responsibility for new technology is now divided between Ministries of Trade and Industry, Economic Affairs and Innovation, which promote, and Ministries of Social Affairs, Health and Welfare, Environment, which attempt to control. There are struggles between the two sides, and compromises are achieved.

The notion of governance was not yet popular at that time, but the outlines of governance were there. Seen from a distance, it was not just a matter of (sometimes internecine) struggles about bureaucratic mandates. It was also a matter of how our societies are handling—managing would be too strong a word—new technologies. This division of labour between government ministries is a reflection of a larger pattern, where technology is developed first, and somewhat at a distance from society, and then inserted in society for others to respond to, to adapt to, to domesticate, sometimes to criticise it—but later, when investments have already been made. Control comes too late, by definition. That was the message of Collingridge's Control Dilemma [1], and part of his argument was that at an early stage, there was still a lot of uncertainty about a new technology: about what it might be, and might be able to do, how it would become embedded in society and what impacts it could have.

Since the 1980s there have been various attempts to critically assess new technology, also at an early stage, with Constructive Technology Assessment as one of the more elaborate and theorised approaches. One could speak of a reflexive turn in how our societies handle technology [3]. By now, it appears to be expected of the various 'responsables' (as it can be phrased in the French language). Actually, there is increasing interest in roles and responsibilities of various actors, and how they might do better—or be pushed to do better. The recent and fashionable discourse of Responsible Research and Innovation (RRI) [4] is an indication, and may push the interest and stimulate related practices. There could be a responsibility turn, and this would go further than regulatory approaches.

Be that as it may, there are good reasons to look at society and technology in terms of a division of moral labour: A division of roles and responsibilities, and linked to that (and sometimes preceding it) struggles about who and what is to be praised or blamed. The notion of a division of moral labour was first formulated by Rip and Shelley-Egan [5] and Shelley-Egan [6], who emphasised its historically evolving character as well as its often fragmented and divisive nature. In fact, divisions of moral labour have been there all the time. It is a powerful notion as it recognises how social order is shaped and maintained through roles and responsibilities, as well as it opens up ways to inquire into functioning and outcomes of divisions of moral labour—the stuff of governance.

I will explore division of moral labour in two steps. First, I develop a historical perspective, indicating some secular changes. This will also allow me to briefly present the existence of responsibility language, in which issues and possible changes can be discussed. Second, I discuss present-day aspects of moral division of labour, including narratives of praise and blame.[1] In conclusion, I return to the question of governance.

There is the classical trope of powerful, perhaps dangerous knowledge and how to handle it.[2] This may be seen as the responsibility of the owner of the knowledge, and the traditional response is secrecy, limiting the access to such knowledge to a small circle of the initiated. But who is the owner of knowledge in present day science with its collective character and its emphasis on its public character? And who can then, or should, decide about handling the potentially powerful knowledge?

Earlier on, there was no other division of moral labour than between the possessor of the knowledge and non-possessors, because of lack of access and/or lack of competence. The 16th century Italian mathematician and engineer Tartaglia had to make a difficult decision, whether he would make his ballistic equation (to be applied to predict the trajectory of a cannon ball) public or not.[3]

> Through these discoveries, I was going to give rules for the art of the bombardier. ... But then one day I fell to thinking it a blameworthy thing, to be condemned—cruel and deserving of no small punishment by God—to study and improve such a damnable exercise, destroyer of the human species, and especially of Christians in their continual wars. For which reasons, O Excellent Duke, not only did I wholly put off the study of such matters and turn to other studies, but also I destroyed and burned all my calculations and writings that bore on this subject. I much

---

[1]The body of this chapter is based on earlier and ongoing work of mine which I have also used in another publication [7]. Sometimes, I have used phrases and paragraphs from Rip [7] in the text of this chapter.

[2]As with the privileged knowledge of the priests and shamans, and picked up in the story of the sorcerer's apprentice. Scientists, and nanotechnologists explicitly, have been called intentional sorcerer's apprentice, experimenting with what they do not understand (yet).

[3]Niccolò Tartaglia, Nova scientia, the preface (1537), ed., trans. Drake & Drabkin 1969: 68ff. For the story, I base myself on a Dutch text, Bos [8], who refers to Charbonnier [9].

regretted and blushed over the time I had spent on this, and those details that remained in my memory (against my will) I wished never to reveal in writing to anyone, either in friendship or for profit (even though it has been requested by many), because such teaching seemed to me to mean disaster and great wrong. But now, seeing that the wolf [i.e., the Turkish Emperor Suleiman] is anxious to ravage our flock, while all our shepherds hasten to the defense, it no longer appears permissible to me at present to keep these things hidden.

Now move forward to a case from 2013. In the online version of the *Journal of Infectious Diseases*, October 7, 2013, Barash and Arnon [10] published their finding of the sequence of a newly discovered protein, but without divulging the actual sequence. The news item about this in *The Scientist Magazine* of 18 October 2013 says:

> [This] represents the first time that a DNA sequence has been omitted from such a paper. 'Because no antitoxins as yet have been developed to counteract the novel *C. Botulinum* toxin,' wrote editors at *The Journal of Infectious Diseases*, 'the authors had detailed consultations with representatives from numerous appropriate US government agencies.' These agencies, which included the Centers for Disease Control and Prevention and the Department of Homeland Security, approved publication of the papers as long as the gene sequence that codes for the new protein was left out. According to New Scientist, the sequence will be published as soon as antibodies are identified that effectively combat the toxin, which appears to be part of a whole new branch on the protein's family tree.

There are other cases where possible publication of sensitive details are prohibited, by the US National Science Advisory Board for Biosecurity, as in the case of the bird flu research by the Rotterdam team led by Fouchier (see also [11] for a discussion of the tensions involved).

My point here is about the similarities of the two cases, including the trope of powerful knowledge (at least, that is how the scientists and others see it), and how it can be used and misused. In the cases, the primary response to the possibility of misuse was to keep this knowledge hidden, but this will depend on the situation and the evolving balance of interests and visions. Whether to make such knowledge publicly available, and in fact, whether to invest in developing it at all, has to be evaluated again and again.

Thus, the structure of the considerations is the same, but the difference is that in the 21st century, the decisions are not individual but part of formal and informal arrangements and authoritative decisions by advisory boards and government agencies. What is also interesting is that there is no reference to responsibility of the researcher/scientist. In the 16th century this was because the word did not yet exist. In the 21st century, it was because the focus is now on what is permissible and expected, rather than an own responsibility of the researchers. The division of moral labour has changed.

In Tartaglia's time, the word 'responsibility' did not yet exist, even if the issues that we now address in terms of 'responsibility' were there. Elsewhere I have shown there is an evolving responsibility 'language', in general and for scientists and scientific research [12], where the words 'responsible' and 'responsibility' can be used to articulate roles and duties in an evolving social order. Such roles can then be part of longer-term 'settlements' of science in society (what is sometimes called a 'social contract' between science and society, cf. [13]).

It is instructive to trace the history of these key words. The big dictionaries of modern languages (Oxford English Dictionary, Grande Larousse etc.) offer historical data on the use of words. The adjective (sometimes used as a noun, as in the French 'responsable') has been in use for a long time, in French since the 13th century, in English since the 17th century, but in a variety of meanings.[4] It is in the late 18th century that some stabilisation occurs into the pattern of meanings that we see nowadays.

The noun 'responsibility' is only used since the late 18th century: since 1782 in French, since 1787 in English (those are the earliest quotes presented in the dictionaries). It is important to keep the relatively recent emergence of the term 'responsibility' in mind because the term is often used to refer to thoughts and analyses in texts of pre-19th century philosophers (e.g. Aristotle, Hume) who do not use the term. This suggests a continuity that is not there, and backgrounds societal developments through which the term 'responsibility' emerged and obtained its meanings.

---

[4]The quotes in the Oxford English Dictionary suggest the meaning of 'responsible' was not stabilised, different authors could use it in their own way. 'The Mouth large but not responsible (= correspondent) to so large a Body' (1698); 'This is a difficult Question, and yet by Astrologie responsible (= capable of being answered)'.

But the sociogenesis of the concept of responsibility is not visible in handbooks and studies of morality in the past, because almost all authors tend to project present-day language usage onto the past.[5]

What happened at the turn of the 19th century and stabilised in the course of that century is the emergence of bourgeois society and the idea of citizens (*citoyens*) with their rights and duties. To articulate those, an extension of language was necessary—the language of responsibility. Through that language, it became possible to discuss and fill in social order concretely. And some outcomes would find a place in the formal Constitution of the nation states as they organised themselves. This language of responsibility remains important to discuss evolving social orders, in the small and in the large. And it has become important for scientists (the term itself being an early 19th century invention, see Ross, 1962) and for science.

There is now a macro-narrative, similar in structure as for technology: the enactors of new science, the scientists, are only responsible for further development, unimpeded by broader considerations if they don't want to bother, while others may have to clean up eventual damage. There is the extension of the trope of powerful knowledge with the image of a sorcerer's apprentice who does not master the power he has unchained. And there is some gerrymandering, when scientists want to be praised for their achievements, but not blamed if things go wrong. This is captured in an aphorism by Ravetz [17]: 'Scientists take credit for penicillin, but Society takes the blame for the Bomb'.

---

[5]This tendency is frustrating in handbooks like the Dictionary of the History of Ideas [14] in which one would expect some sensitivity for historical developments. For example, in the Lemma on 'free will and determinism' [14:239–240] a brief sketch is offered of Hume's ideas, based on his Inquiry Concerning Human Understanding, Section VIII, using the terms 'responsible' and 'responsibility' all the time, while Hume himself speaks of 'blameable' and 'answerable' (and once of 'accountable') [15:107–109]. Somewhat of an exception is Adkins [16:4] who limits the anachronism to his title, and emphasises that moral responsibility is not an important concept for the Greeks (and does not occur as a term), because of their view of the world and society. It is only because of the Kantian turn, he claims, that a view of the world and society emerges in which 'For any man brought up in a western democratic society the related concepts of duty and responsibility are the central concepts of ethics' [16:2].

One can criticise such a position as gerrymandering, claiming the good and disavowing the bad (or at least, shifting the responsibility to others). But it is also a way of dividing labour (here, moral labour) to perhaps doing better overall. Think of the view that scientists have a moral obligation to work towards progress, and that is how they discharge their duty to society, while others (better qualified, or more responsible, or at risk) should look after social, ethical and political issues. What Ravetz's aphorism brings out is that one might want to query such a division of moral labour. This is a second-order ethics question: why would this be a 'good' division of labour? Second-order ethics discusses the ethical (and more broadly, normative, cf. [18]) aspects that become visible when one inquires into the justification of present overall social and institutional arrangements, rather than taking them for granted.

Developments since the 1970s, and particularly in the 2000s, are undermining the common justification of this 'settlement'. Responsibility language is used as a focus point to discuss the pros and cons of this division of moral labour and can include second-order ethics.[6] One indication is the emergence of the notions of 'responsible development' (around nanotechnology) and 'responsible research and innovation' (in the European Commission's R&D Framework program Horizon, 2020) [7, 20].

One can see evolving divisions of moral labour at work, at least being talked about. I will indicate some of the positions and their evolution. Let me start with the well-known industrialist's argument about the need to go for profit to survive, while other actors should take care of second-order, possibly negative effects (most often, government actors are assumed to have this

---

[6]This need not be a one-sided critique of closed science. One consideration is that it is important to have the scientific endeavour be protected from undue interference. This is quite clear for the micro-protected spaces of laboratories and other sites of scientific work, and the meso-level protected spaces of scientific communities and peer review, although there is also opening-up, ranging from citizen science to criticism of scientific practices and the knowledge that is being produced [19]. Seen from the side of society, the scientific endeavour is legitimate as long as scientists deliver, both in terms of their producing what is promised (progress, even if this can interpreted in different ways) and their adhering to the normative structure of science (cf. the issues of integrity of science). This is a mandate that justifies the relative autonomy of science–a sort of macro-protected space.

responsibility). While this argument continues to be heard, practices are different. The move towards corporate social responsibility is one example, and particularly important is the Responsible Care Program in the chemical sector [21]. From a sociological perspective, one can see the importance of the notions of 'good firms' and 'cowboy firms' (or 'rogue firms', cf. notion of 'rogue states' on the global scene). The 'good firms' behave well, and are to be praised for their efforts even if outcomes are not always ideal. While 'cowboy firms' transgress and must be condemned, particularly because they endanger the credibility of the 'good firms' in the sector.[7]

Analysis in terms of division of moral labour can also be used to understand the actual and possible role of lay people, citizens, and consumers. Consumers, for example, are projected as having a duty to buy, and be informed, and calculate rationally—if only to ensure that neo-classical economics remains applicable. But they can also go for political action through consumption decisions, including boycotts [22]. And there are evolving liability regimes which shift the responsibilities between producers and consumers.[8]

The present interest in public engagement often remains within traditional divisions of moral labour by positioning members of the public as articulating preferences that may then be taken up in decision making as additional strategic intelligence. But one could have joint inquiry into the issues that are at stake [24].

In Codes of Conduct (as for nanotechnology) and broader accountability of scientists and industrialists generally, there is an assumption that there will be civil society actors willing and able to call them into account. That may not be the case: civil society actors may not be able, or not be willing, to spend the necessary time and effort. This is already visible in so-called 'engagement fatigue'.

For newly emerging science and technology, there is rhetoric of progress. The promise of progress, somehow, is assumed, and when there is criticism, or just queries, rhetorics kick in. At

---

[7]Interestingly, discussions about integrity of science and the occurrence of fraud have the same structure. Fraud is positioned as deviation from a general good practice, and done by 'rogue scientists'.
[8]Cf. Lee and Petts [23:153].

the height of the recombinant DNA debate, in the second half of the 1970s, the medical possibilities were emphasised: 'Each day we lose (because of a moratorium) means that thousands of people will die unnecessarily'. The justificatory argument about genetically modified organisms, in the contestation about its use in agriculture, refers to hunger in developing countries (which need biotechnical fixes, it appears). When such promises are contested, a subsidiary argument kicks in: people don't understand the promise of the technology so we have to explain the wonders of the technology to them.[9]

One sees here how narratives of praise and blame become short-circuited: only praise for the new technology is allowed. While this can be seen as a reduction, some such short-circuiting is inevitable, though. For example, hype about new technologies may be necessary to draw attention to them and mobilise resources for their further development—otherwise their promise would never materialise. Even while hype may lead to disappointment later on, something that actors realise might happen but apparently can't do much about [18, 25]. At the same time, there might be concerns about the nature and impacts of the new developments, which again can be exaggerated in order to get a hearing. In a sense, there is shadow boxing all around.

There are further implications for a division of moral labour. For example, the importance of early warning is now widely recognised (cf. [26]) on late lessons of early warning), but who can tell, at an early stage, whether a warning is significant? Mandates might become articulated for who may legitimately warn. Critical NGOs might be candidates for the task of voicing concerns and pushing them on the agenda; but this can also be viewed as unnecessarily harassing the promoters of technology. There is some institutionalisation now, in the sense that proponents of a new technology actually expect there will be opponents [27], and look for signals (and sometimes try to appease possible opponents beforehand).

Such anticipation can lead to further evolutions of the division of moral labour. An interesting example are the meetings of the International Dialogue on Responsible Research and

---

[9]This is the equivalent of the well-known deficit model shaping exercises of public understanding of science.

Development of Nanotechnology, positioned as opening up a space for broad and informal interactions [20, 28] but hopefully, having consequences. In the first meeting in 2004, there was a proposal to develop a Code of Conduct, which was eventually taken up by the European Union [29]. Interestingly, the Code is much broader than the consequentialist ethics visible in the review of the US National Nanotechnology Initiative; see in particular the reference to a culture of responsibility (N&N here stands for Nanoscience and Nanotechnologies).

> Good governance of N&N research should take into account the need and desire of all stakeholders to be aware of the specific challenges and opportunities raised by N&N. A general culture of responsibility should be created in view of challenges and opportunities that may be raised in the future and that we cannot at present foresee (Section 4.1, first guideline).

This perspective, often found in recent documents about RRI and responsible development, assumes too much harmony. There are not only tensions and conflicting interests, but there is also a basic political problem in the whole notion of 'responsible'.

Because of the variety of values and societal convictions, there will be no consensus about who/what is going to count as 'responsible'. This constitutes a structural problem, not only because of essential contestations in our societies, but also because 'responsible' is not well specified, and can be used as a 'feel good' term (similar to terms like 'sustainability').

An illustrative example is the Round Table on Responsible Soy, which offers certification of soy production practices to those firms who apply and satisfy the criteria.[10] Members of the Round Table are Monsanto, Syngenta, Cargill, Nidera, BP, Shell (which are all interested parties), but also Wereldnatuurfonds and Solidaridad. One could applaud this as an action of 'good firms', but at least Monsanto's record is not without blemishes. Critical societal groups have looked into the criteria for certification, and labelled them, at least in part, as 'greenwash'. And they have been alerting various actors (organisations) to

---

[10]For this example, I draw on Hanssen and De Vriend [30], who did an interesting study (in Dutch) on the role of social media around biotechnology.

what they see as problematic with the certification.[11] All this is possible because the term 'responsible' allows the various interpretations. What remains is that there is a debate from which the parties cannot extract themselves because of the shared use of the term 'responsible'.[12]

It will be clear that divisions of moral labour occur, emerging and sometimes contested, that they can stabilise and become master narratives. But they can also become unsettled again, open up and change. As such, they are an integral element of governance. Recognising them for what they are is then a step towards considering meta-governance.

Concrete governance arrangements and approaches are embedded in historically evolving divisions of moral labour. For example, regulation assumes compliance at the side of the regulatees, and while the problem of compliance is recognised, it tends to be addressed with more-of-the-same approaches, attempting to force the regulatee into compliance. Instead, compliance could (should) be reformulated as part a pattern of moral labour, as with 'good' firms in a sector complying to environmental regulation in contrast to the 'cowboy' firms [32].

For new technologies, the additional complication is their uncertainty in terms of shape, performance, and possible embedding in society. Division of moral labour is still there, but it is more difficult to pin down. To some extent, it is a matter

---

[11]The Round Table appears to be active and relatively successful (http://www.responsiblesoy.org/en/). 'Responsible Soy' appears to have become a commonly used term, even if it stretches the meaning of 'responsible'. The critical assessments continue as well, e.g. http://www.toxicsoy.org/toxicsoy/RTRS_files/RTRS%20backgrounder%20v2.pdf. The dynamics involved are brought out in a quote on http://www.isealalliance.org/tag/round-table-for-responsible-soy-rtrs: 'Driven by a growing private sector preference to use standards to build sustainability and economic security into supply chains, the demand for certified products is outstripping supply in many sectors. This supply gap is opening new challenges for a standards movement whose prior concerns were mainly on building demand. The closing session of the ISEAL Conference 2012 brought together business, NGOs and standards-setters to look at the innovations needed to reach the next wave of producers.' The ISEAL Conference is providing a space for interactions, including negotiations about mandates and scope of responsibilities.

[12]Thus, the term 'responsible' is a so-called boundary object, open in content, but in each of the worlds it functions in having a specific meaning and role [31].

of learning by anticipating, learning by doing (including trial and error), and learning by negotiating. This can be improved by realising the nature of these multi-actor learning processes.

What does this mean concretely? First, the importance of what has been called dynamic governance ('supporting dynamics rather than constraining them'), the willingness to accept changes, further evolutions, in spite of the need to be effective here and now.[13] Second, to be willing to accept concrete approaches, even if they reduce complexity, but evaluate them in terms of what they mean, and imply, for evolving divisions of moral labour. Such a possibility has been discussed for division of carrying risk and vulnerability in our risk societies [34]. For divisions of moral labour this is more diffuse and intangible, but the difficulty should not lead us to give up on it.

References

1. Collingridge, D. (1980) *The Social Control of Technology* (Frances Pinter, London).
2. Rip, A., T. J. Misa, and J. Schot (eds.) (1995) *Managing Technology in Society. The Approach of Constructive Technology Assessment* (Frances Pinter, London).
3. van Lente, H., and A. Rip (2017) Reflexive co-evolution and governance patterns. In: Bowman, D. M., E. Stokes, and A. Rip (eds.) *Embedding and Governing New Technologies* (Pan Stanford, New York).
4. Owen, R., J. Bessant, and M. Heintz (eds.) (2013) *Responsible Innovation. Managing the Responsible Emergence of Science and Innovation in Society* (John Wiley & Sons, New Jersey).
5. Rip, A., and C. Shelley-Egan (2010) Positions and responsibilities in the 'real' world of nanotechnology. In: von Schomberg, R., and S. Davies (eds.) *Understanding Public Debate on Nanotechnologies: Options for Framing Public Policies: A Working Document by the Services of the European Commission* (European Commission, Brussels), pp. 31–38.
6. Shelley-Egan, C. (2011) *Ethics in Practice: Responding to an Evolving Problematic Situation of Nanotechnology in Society* (PhD Thesis, University of Twente, Enschede, defended on 13 May 2011).

---

[13]This is a general point, increasingly recognised but difficult to realise in practice. The reference to 'dynamic governance' is from the conclusion of the MASIS Expert Group Report [33].

7. Rip, A. (2014) The Past and Future of RRI, *Life Sciences, Society and Policy*, **10**(17). DOI: 10.1186/s40504-014-0017-4.
8. Bos, H. J. M. (1975) *Mathematisering en maatschappij, of Hoe loopt een success-story af?* (Mathematisch Instituut, Amsterdam, Lecture, September).
9. Charbonnier, P. (1928) *Essais sur l'histoire de la ballistique* (Société d'Editions Géographiques, Maritimes et Coloniales, Paris).
10. Barash, J. R., and S. S. Arnon (2014) A novel strain of clostridium botulinum that produces type b and type h botulinum toxins, *Journal of Infectious Diseases*, **209**(2), 183–191.
11. Evans, S. A. W., and W. D. Valdivia (2012) Export controls and the tensions between academic freedom and national security, *Minerva*, **50**, 169–190.
12. Rip, A. (1981) *Maatschappelijke verantwoordelijkheid van chemici* (PhD Thesis, University of Leiden, Leiden).
13. Guston, D. H., and K. Kenniston (1994) Introduction: The social contract for science. In: Guston, D. H., and K. Kenniston (eds.) *The Fragile Contract: University Science and the Federal Government* (MIT Press, Cambridge), pp. 1–41.
14. Wiener, P. P. (ed.) (1973) *Dictionary of the History of Ideas; Studies of Selected Pivotal Ideas* (Charles Scribner's Sons, New York).
15. Hume, D. (1955) *An Inquiry Concerning Human Understanding*. Hendel, C. W. (ed.) Indianapolis: The Liberal Arts Press (Bobbs-Merrill). Originally published in 1748.
16. Adkins, A. H. (1975) *Merit and Responsibility. A Study in Greek Values* (Chicago University Press, Chicago).
17. Ravetz, J. (1975) … et augebitur scientia. In: Harré, R. (ed.) *Problems of Scientific Revolution: Progress and Obstacles to Progress in the Sciences* (Clarendon Press, Oxford), pp. 42–57.
18. Rip, A. (2013) Pervasive normativity and emerging technologies. In: van der Burg, S., and T. Swierstra (eds.) *Ethics on the Laboratory Floor* (Palgrave Macmillan, Basingstoke), pp. 191–212.
19. Rip, A. (2011) Protected spaces of science: Their emergence and further evolution in a changing world. In: Carrier, M., and A. Nordmann (eds.) *Science in the Context of Application: Methodological Change, Conceptual Transformation, Cultural Reorientation* (Springer, Dordrecht), pp. 197–220.
20. Fisher, E., and A. Rip (2013) Responsible innovation: Multi-level dynamics and soft intervention practices. In: Owen, R., Bessant, J.,

and Heintz, M. (eds.) *Responsible Innovation: Managing the Responsible Emergence of Science and Innovation in Society* (John Wiley & Sons, Chichester), pp. 165–183.

21. King, A. A., and M. J. Lenox (2000) Industry self-regulation without sanctions: The chemical industry's responsible care program, *Academy of Management Journal*, **43**(4), 698–716.

22. Throne-Holst, H. (2012) *Consumers, Nanotechnology and Responsibilities. Operationalizing the Risk Society* (PhD thesis, University of Twente, Enschede).

23. Lee, R. G., and J. Petts (2013) Adaptive governance for responsible innovation. In: Owen, R., Bessant, J., and Heintz, M. (eds.) *Responsible Innovation: Managing the Responsible Emergence of Science and Innovation in Society* (John Wiley & Sons, Chichester), pp. 143–164.

24. Krabbenborg, L. (2013) *Involvement of Civil Society Actors in Nanotechnology: Creating Productive Spaces for Interaction* (PhD Thesis RU Groningen, Groningen).

25. Rip, A. (2006) Folk theories of nanotechnologists, *Science as Culture*, **15**, 349–365.

26. Harremoës, P. (ed.) (2001) *Late Lessons from Early Warnings: The Precautionary Principle 1896–2000* (European Environmental Agency, Copenhagen).

27. Rip, A., and S. Talma (1998), Antagonistic patterns and new technologies. In: Disco, C., and B. J. R. van der Meulen (eds.) *Getting New Technologies Together* (Walter de Gruyter, Berlin), pp. 285–306.

28. Tomellini, R., and J. Giordani (2008) *Report: Third International Dialogue on Responsible Research and Development of Nanotechnology* (European Commission, Brussels).

29. European Commission (2008) *Commission Recommendation of 07/02/2008 on a Code of Conduct for Responsible Nanosciences and Nanotechnologies Research*. Brussels, 07/02/2008, C(2008) 424 final.

30. Hanssen, L., and H. De Vriend (2011) *De komst van sociale media. Een Nieuwe Dynamiek in het Debat over Biotechnologie?* (COGEM onderzoeksrapport CGM, Nijmegen, Driebergen).

31. Star, S. L., and J. Griesemer (1989) Institutional ecology, 'translations', and boundary objects, *Social Studies of Science*, **19**, 387–420.

32. Rip, A. (2011) Interlocking socio-technical worlds. In: Rip, A. (ed.) *Futures of Science and Technology in Society* (Institute for Innovation and Governance Studies, University of Twente, Enschede), pp. 149–170.

33. Markus, E., K. Siune, M. Calloni, U. Felt, A. Gorski, et al. (2009) *Challenging Futures of Science in Society. Emerging Trends And Cutting-Edge Issues* (European Commission, Brussels).
34. Rip, A. (1991) The danger culture of industrial society. In: Kasperson, R. E., P. J. M. Stallen (eds.) *Communicating Risks to the Public. International Perspectives* (Kluwer Academic, Dordrecht), pp. 345–365.

Chapter 7

Ethical Reflexivity as Capacity Building: Tools and Approaches

Clare Shelley-Egan[a] and Federica Lucivero[b]

[a]*Oslo and Akershus University College of Applied Sciences, Postboks 4 St. Olavs plass, 0130 Oslo, Norway*
[b]*School of Social Science, Health & Medicine, King's College London, The Strand, London, UK*

Clare.Shelley-Egan@afi.hioa.no, federica.lucivero@kcl.ac.uk

7.1 Introduction

The notion of 'reflexivity' has emerged as a key dimension of responsible research and innovation (RRI), a policy discourse particularly visible at European Union (EU) level since around 2010 [1–3]. The interest in RRI seeks to extend scientific responsibility so as to include future societal impacts of technological development [4]. As Stilgoe et al. note, this call for responsibility 'demands reflexivity on the part of actors and institutions (...)' [5:1571]. Reflexivity 'at the level of institutional practice, means holding a mirror up to one's own activities, commitments and assumptions, being aware of the limits of knowledge and being mindful that a particular framing of an issue may not be universally held' [5:1570].

This is second-order reflexivity [6], in which the value-based socio-ethical premises and background theories of the research system are interrogated. 'Reflexive' scientists think about their own ethical, social and political assumptions, consider and expand their role responsibilities in research, and may even change their practices over time [4]. The increased pressure for such ethical reflexivity makes untenable scientists' use of Mertonian norms to argue that it is their duty to produce scientific knowledge by upholding the norms of objectivity and disinterestedness and in so doing closing off space for responsibility for broader social and ethical issues [7]. Their use of the norms to make this argument is particularly inappropriate in the case of applied sciences such as nanotechnology in which ethical and societal considerations play an important role in the motivation, practices and outcomes of the research [6]. Indeed, an emphasis on the need for consideration of broader aspects of nanoscience and nanotechnology research is visible in nanotechnology policies in many countries, in particular in the Netherlands [8], the United States (US) [9–11] and the United Kingdom (UK) [12].

Within this call for reflexivity, there is an assumption that, by enhancing the reflexivity of scientists in relation to broader issues and the (possible) need to do something about them, scientists will be able to make decisions that can contribute to research that better meets society's needs [13]. To that end, a number of reflexivity-building initiatives have been introduced over the last decade to some areas of new and emerging science and technology. These initiatives are very diverse, ranging from codes of conduct for natural scientists [6] to research funders' requirements for 'societal relevance' or 'broader impacts' descriptions in project proposals [12], the inclusion of ethics and social science subjects in scientists' training and education [14], and the encouragement of experimental 'embedment' of ethicists and social scientists within laboratories [10]. Efforts to stimulate the reflexivity of natural scientists initially arose in research programmes aimed at studying the ethical, legal and social implications (ELSI) of new technologies, were advanced in different approaches including technology assessment [15, 16], anticipatory governance [17], upstream engagement [18] and midstream modulation [9], and are currently flagged as being important in the academic discourse on RRI [19, 20]. These studies have facilitated the development of a range of tools and approaches to stimulate scientists' reflexivity, and have contributed to a critical second-order discussion on the meaning, effectiveness, and legitimacy of these initiatives, including critical perspectives on ethical, legal and social aspects (ELSA) programmes and productive divisions of moral labour [21]; criticism of the role of proceduralism in ethics committees [26]; and critical analysis of the functionality of codes of conduct [6].

This chapter delineates some institutional initiatives and supportive tools that aim to enhance scientists' ethical reflexivity and offers a discussion of their objectives and challenges. While the chapter does not offer an exhaustive overview of existing approaches, it sets forth a proposal to shift the framing of the often-vague call for reflexivity from an emphasis on 'enhancing' reflexivity to a vocabulary that stresses the importance of 'building capacity' for reflexivity. To that end, the approaches and tools that are described are discussed as attempts to build capacity for ethical reflexivity that may be employed to address some of the limitations of current institutional requirements.

The chapter proceeds as follows. In Section 7.2, we describe the institutional requirements that aim to cultivate ethical reflexivity and discuss some of their limitations. In the following section, we reframe the endeavour to 'enhance' reflexivity in terms of capacity building, namely working with the capacities for ethical reflexivity that scientists already possess and facilitating further building of capacities, specifically with regard to the articulation and performance of ethical reflexivity and the broadening of perspectives. In Section 7.4, we describe and discuss tools and approaches with regard to how they contribute to enabling the articulation and performance of reflexivity. In the concluding section, we highlight issues and challenges that still need to be addressed.

7.2 Institutional Requirements and Opportunities to Engage in Ethical Reflexivity

Institutional context enables and constrains researchers in their response to the pressure to be reflexive [22, 23]. In the following, we briefly describe and discuss three types of initiative that aim to stimulate ethical reflexivity through requiring it as a part of institutional requirements: funding institutions' requirements for a description of the societal relevance of scientific research, codes of conduct, and intensive collaborations between natural and social scientists.

One means of stimulating and enhancing competencies in ethical reflexivity comes from the side of funding agencies. These bodies increasingly require consideration of ethical and social aspects in all research proposals, over and above the usual requirements concerning ethical screening of experiments on humans and animals. The European Commission (EC) already has a requirement for extended impact assessments in research proposals. This requirement has the same weight as the regular requirements of scientific quality and management of the proposed project in evaluations. Since 1997, the National Science Foundation has introduced a second criterion with which to evaluate research proposals, in addition to 'intellectual merit' [24]. This second 'broader impact' criterion requires authors of research proposals to indicate the 'societal benefit' of the proposed research. Similar requirements have been set out by national funding institutions that ask applicants to describe the 'societal relevance' of their proposed research (e.g. in the Dutch context, see [25]). The provision for broader scientific responsibility requires scientists to demonstrate that they are actively taking broader issues into account and feeding them into their research agendas. Such institutional requirements force scientists to think in broader terms than the ethics of experiments, for which they can have 'recourse to the procedural'. However, the meaning of these requirements is often unclear to the scientific community which has expressed resistance towards them [24]. Moreover, the formulation of these criteria suggests that activities to increase societal relevance may be attached to, rather than integral to, the content of the research [26]. Furthermore, over time, practical considerations may gain the upper hand, for example, when scientists comply with the requirement by 'copying and pasting' from earlier proposals. While researchers may pursue this broader work themselves, it is more likely that they will take a pragmatic stance and outsource these new tasks to social scientists, ethicists and humanists who work to support and stimulate scientists to enhance their reflexivity with respect to their research [21, 27].

Codes of conduct are widely used as detailed guidelines for the behaviour of various professional communities and come in a variety of types, ranging from aspirational to enforceable. Examples include the *Code of Conduct for Scientific Practice* developed by the Association of Universities in the Netherlands (VSNU) and the *Code of Conduct and Guidance on Professional Practice* issued by the Royal Society of Chemistry. Codes of conduct may be used as a point of reference, i.e. this is how one must behave. Codes of conduct can also be important for enhancing ethical reflexivity if one looks not only at behaviour but also at the discussion and reflection that might follow. Indeed Schuurbiers [6] argues that the implementation phase of codes of conduct is just as important as their establishment as it addresses questions relating to how the code is adopted and how it can be implemented in ongoing practices.

Such questions may lead to an enhancement of the ethical reflexivity of actors, including scientists. Numerous voluntary codes of conduct for nanoscience and nanotechnology have been developed [28], one of which is the European Commission's (EC) *Code of Conduct for Responsible Nanoscience and Nanotechnologies Research*. The Code implicitly communicates a notion of responsible nanoresearch [29] that depends on adherence to seven principles by nanoscience and nanotechnologies stakeholders, including researchers and research organisations. The EC Code aims to 'responsibilise' stakeholders—including researchers—to play their part in contributing to responsible nanoscience and nanotechnology development [30].

Responsibilisation demands a degree of reflexivity, as researchers should not only be aware of their responsibilities but also interrogate and even broaden their responsibilities so as to contribute to a broader, collective responsibility [29, 31]. This ambition has not been realised for a number of reasons however, including insufficient dissemination, amongst a host of other issues [32] and crucially, an absence of links with the rationality of researchers regarding rewards, benefits and incentives [30]. Perhaps most striking is Kjølberg and Strand's finding—in their study of the responses of 17 University of Bergen nanoresearchers to the EC Code [29]—that while the code extends an invitation to stakeholders to participate in an open forum with the EC regarding the code and its implementation and monitoring, the nanoresearchers did not see it as an invitation to participate; they saw the invitation as deriving from distant bureaucrats with a vague understanding (even misunderstanding) of the science and with unclear consequences for their practices.

Another initiative involves intensive collaboration between researchers in the humanities, natural sciences and social sciences in order to contribute to the appropriate embedding of technological and scientific advances in society through the incorporation of research on ethical and societal issues in large-scale research programmes. For example, the UK's Engineering and Physical Sciences Research Council (EPSRC) has just recently funded a multi-million-pound project on synthetic biology applications for water supply and remediation, which includes a cross-cutting work package on 'responsible innovation'. The aim of this work package is to understand the ethical, legal and social issues of this particular application of synthetic biology, and to ensure that the scientific work programme is aware of and shaped by these issues. While this major initiative aims to embed reflexivity as an important dimension in the research and innovation process, earlier experiences with similar programmes show that there are limitations to such an approach.

Indeed, Owen and Goldberg [12] highlight some limitations in their study of an initiative taken by the EPSRC which, with the introduction of risk registers, required applicants to reflect on the wider implications of their proposed research, to identify potential impacts and qualitatively assess their associated risks. The risk registers—which were externally peer reviewed—were all completed 'conservatively', with the overwhelming majority of identified impacts narrowly focused on health impacts associated with exposure to nanomaterials. Few potential impacts on the wider environment and no future societal impacts were identified [12]. This focus on—or 'recourse' to—technical issues further reinforced *de facto* responsibilities and back-grounded broader consideration about research agendas.

These institutional initiatives introduce some requirements to scientific practice that compel scientists to reflect on their responsibilities and the impacts of their research outcomes on society. Although they aim to reconcile the divide between science and society, these initiatives often fail to achieve the objective of integrating scientific research with societal needs and relevance. Indeed, as it has emerged in several commentaries reported above, institutional requirements for ethical reflexivity, such as 'societal relevance' criteria in funding schemes, codes of conduct and interdisciplinary collaboration, remain external requirements for individual scientists who only 'fill in the forms' and 'tick the boxes'. The pursuit of ethical reflexivity, however, as expressed in RRI documents, seems to entail higher-level engagement of scientists, moved by a genuine awareness of their responsibility. For this reason, it is crucial to move away from the institutional enforcement of reflexivity towards approaches that foster the development of capacities for reflexivity within a specific social context.

7.3 Ethical Reflexivity and Building Capacity

The programmes discussed above seem to, in our view, miss their target of engaging scientists to be reflexive about their responsibilities and the societal implications of their research. There are, we argue, a number of reasons for why this is the case.

Enhancement of their ethical reflexivity is not a priority for scientists. First, scientists are more interested in advice about what should be done and what should not be done than in being reflexive [33]. Second, as 'enactors' of science and technology, scientists bring particular views and perspectives in their response to the pressure for 'reflexivity' [34, 35]. They work in 'enactment' cycles by constructing scenarios of progress and identifying obstacles to be overcome. This stance tends to sidestep reflexivity; a potential 'lack of acceptance' by publics, for example, becomes a problem, while the real issue should concern the appraisal of new and emerging science and technologies by all relevant actors, including publics [21].

When enactors are willing to be reflexive, they may still feel helpless as to how to go about it [36]. This suggests a need to build capacity in terms of furnishing scientists with the necessary skills and competencies to identify issues, to be productive about them and, at times, to be proactive about them.

In our view, enhancing the ethical reflexivity of actors is to encourage them to articulate their reflexivity. Although enhancing ethical reflexivity can have a training component such as in the inclusion of ELSA (and insights of ELSA studies) in training at the master's and PhD levels and in training courses in ethics for scientists, we argue that actors do not need help to be ethically reflexive. In so doing, we argue against what we call an 'ethical deficit model', which is explicit in the latter activities (inclusion of ELSA in training at master's and PhD levels and training courses in ethics) and implicit in many others. The ethical deficit model serves to perpetuate the notion that scientists' ethical reflexivity requires 'enhancing', independent of a diagnosis regarding the actual deficiencies of scientists. In addition, social scientists, humanists and ethicists are positioned as knowing 'better'—these experts can use their expertise to help scientists overcome their shortcomings by enhancing their ethical reflexivity. We argue that scientists have the capacity to be reflexive but may require help in order to perform their reflexivity.

We argue that there is no need to delegate ethical reflexivity because of a supposed lack or basic deficit. Enactors already have competence for ethical reflexivity—it is the performance that may be lacking. This competence/performance distinction can be understood as similar to the distinction between linguistic competence and performance. There is the basic competence to speak a language. A first round of performance involves learning to speak a particular language; this ability is a competence in its own right. In linguistics, performance of a language relates to actually speaking a (particular) language. Our argument is that, with regard to reflexivity of scientists, they are not lacking competence in reflexivity but are often lacking the skills to perform the articulation of reflexivity. This particular limitation derives from scientists limiting their reflexivity by recourse to roles and mandates. Consequently, justifications are made, not in terms of ethics but in terms of their role, work, institutional context and so on. While these are legitimate justifications for scientists, they are offered without reflective inquiry and thus allow short-circuiting of challenging or novel situations [23].

Enhancing the ethical reflexivity of scientists from a non-deficit perspective is to encourage them to articulate their reflexivity, in addition to helping them to perform their reflexivity, i.e. translating their reflexivity into concrete actions. In addition, articulation alone is not sufficient as reflexivity is also a moral activity. In other words, reflexivity is not merely an activity that warrants rational arguments and logical justifications; it is an activity that requires the capacity to 'feel' a moral problem. In the terminology of Swierstra and Rip, reflexivity necessitates approaches that 'heat up' the perspectives of scientists such that their morality or 'cold' ethics—moral routines that are unproblematically accepted—are 'heated up' by new problems and dilemmas which are made visible by the new technology [33]. To that end, scientists can profit from tools and incentives to enable them to see that they are making shortcuts by falling back on old routines and modes of justification.

The notion of 'moral imagination', inspired by John Dewey, plays an important role in the articulation and performance of reflexivity. Moral imagination is a cognitive human competence important for making decisions; specifically, it is significant when we make decisions according to particular epistemic and moral criteria. We decide (not) to do something on the basis of what we (think we) want, what we know about the world, and what we think is right to do. In this sense, deliberation is a process of moral reasoning, and therefore making the moral content of this reasoning explicit should improve the process. Moral imagination plays an important role in this process of deliberation, because it allows people to project themselves and their actions into the future and to 'feel' it. This ability to 'put yourself in someone else's shoes'—imagining how you would feel and act in a potential future situation—is a key reflexive skill for moral reasoning.

Moral imagination is therefore an important skill for scientists to develop in order to gain awareness of the societal application of their laboratory research [37, 38]. Technology developers have to make daily decisions that concern future technologies, and which determine their nature and application. For example, researchers developing biomedical technologies explore different possibilities for the application of scientific and technological innovations in a clinical context; they have to discuss these possibilities and make a decision as to the best direction to take. In such a decision-making process, different values and considerations play a role, for example, questions of feasibility or profit. In this instance, the ability to imagine how you would feel and act in the future is a key capacity for the reflexive scientist.

7.4 Building Capacities: Supportive Tools and Approaches

There appears to be a need to rethink ethical reflexivity and approaches to facilitating reflexivity if the desired change, i.e. changes in roles and responsibilities and practices, is to be achieved. In the previous section, we highlighted three items that require particular attention:

(1) the need to develop capacities to articulate ethical arguments and issues
(2) the need to develop capacities with which to imagine and discuss societal implications or to 'heat up' long-held views and routines
(3) the need to develop capacities with which to translate reflexivity into concrete actions, or, in other words, the capacity to perform reflexivity

In the following sections, we will describe some approaches that offer tools for the development of these capacities.

7.4.1 Articulating Reflexivity

As scientists tend to fall back on roles and mandates, or uncritically follow required procedures, the articulation of their ethical concerns can be a demanding task. Their educational background and past experience do not necessarily provide them with the toolbox to recognise the ethical concerns related to their work and to communicate their arguments. Some approaches have focused on filling this gap, as, for example, the 'Socio-Technical Integration Research' (STIR) decision protocol and argumentation scenarios.

The STIR decision protocol is a reiterative process that highlights scientists' ethical decisions in laboratory daily practice. Initially developed in a pilot study by Erik Fisher [10], this tool for capacity building has been further implemented, tested and expanded in the US National Science Foundation's funded project, STIR. In 20 case studies, introducing an 'embedded humanist' in a laboratory in different scientific and geographical areas, the project assesses scientists and engineers' capacities to integrate socio-technical considerations in their daily practice and engages laboratory research practitioners in ongoing reflections and deliberation.

The embedded humanist in the laboratory has the task of observing the *de facto* social, physical, and normative factors that influence research and communicating these observations to researchers in order to make them aware (reflexive) of the role of these factors in their research practices and decisions, in addition to their position in socio-technical systems. Embedded scholars also document any directed decision, in which laboratory practitioners consciously act in light of specific goals, values and concerns. The decision protocol is a research tool used by the embedded scholar in semi-structured interviews with laboratory practitioners. It is based on a model of decision components and it aims to explore the perceived state of affairs that elicit a response (opportunity), perceived available course of actions (alternatives), criteria selected to respond (considerations) and effects of selected alternatives (outcomes).

Entering the laboratory and using this protocol to engage with research practitioners and discuss their daily decisions, embedded humanists induce a process of awareness in lab practitioners by feeding back to them their observations regarding goals, values and social structures. This process is characterised by the expansion of the range of considerations and the alternatives that are taken into account in scientists' deliberations. The manner in which such awareness leads to changes in daily practices is still a matter of exploration within the STIR project.

Being aware of the social goals, norms and systems that influence their decision is only one step towards the development of scientists' capability to articulate their reflexivity. Another important step involves offering arguments to communicate their ethical standpoints and positions. Argumentation scenarios have been developed as one tool with which to foster ethical reflexivity of scientists and to provide them with capacities with which to articulate their arguments [23]. Based on Stephen Toulmin's Model of Argumentation and on interviews with scientists, these scenarios show how a claim is justified by a warrant (based on an ethical theory), and is counter-argued with a reservation. These tools were introduced to a small focus group with scientists in order to develop their understanding of the argumentative structure of moral dilemmas around scientists' responsibility and accountability. Argumentation scenarios share with socio-technical scenarios the goal of stimulating scientists to discuss their positions, specifically by focusing on ethical arguments. As schematic representations of the functions of and relations between claims, warrants and reservations, these tools can enhance actor reflexivity regarding the discursive structure of ethical arguments. However, they do not trigger actors' moral imaginations. Moreover, there is a danger that scientists could use such argumentation scenarios in a calculative manner, such that the emphasis would be on solving the 'problem' of the argument, rather than on providing justifications for the arguments. For this reason, articulation of arguments in action and interaction with heterogeneous actors such as representatives from industry, civil society organisations and research funding councils (in a workshop environment, for example) is important.

7.4.2 Imagining and Discussing Societal Relevance

The articulation of existing ethical considerations, positions and social goals is critical to scientists' reflexivity and responsible decision making in R&D and communication. Another set of capacities of utmost importance concerns the anticipation of agency roles and responsibilities as well as social and ethical implications of their research activity. Several tools have been developed to respond to this need.

The fictive script analysis offers a tool to anticipate actors' agency, roles and responsibilities in R&D. Building on the concepts of script, according to which the design of existing artifacts 'embodies' some assumptions about use and users and 'prescribes' how they should be used [13], de Laat notes that this concept can guide a prospective analysis of emerging objects [39, 40]. In the case of emerging technologies, designers assume that users and other actors will relate in a certain way. However, there might be some gaps between the designers' assumptions and the present situation with respect to expectations regarding emerging technologies. Since the objects of expectations are still under construction, these assumptions cannot be described by analysing material design choices.

Emerging objects often exist as texts in funding proposals, public oral communications, interviews or patents; for this reason their 'description' can only have a 'fictive' character. De Laat's 'fictive script' analysis provides some methodological tools with which to address this issue. It devises three exercises to make fictive scripts explicit and to describe the users, the actors, and the roles that are assumed in emerging technologies. Because they are based on expectations rather than on material objects, these descriptions have a fictional character and have to be considered as 'thought experiments' [40:194].

The first exercise is a '100% thought-experiment'. It involves describing how the world would be if the emerging object would deliver exactly as expected. By employing this exercise, 'the general characteristics of the object in question' can be quickly grasped and described together with the 'barriers' that exist in today's world and that constrain the realisation of this object; for example, if the object is the battery for an electric vehicle, this experiment would lead to the statement that gas stations should deliver electricity, which is not the case today.

The second exercise is a 'black boxing' experiment. It involves black-boxing the object as it is conceived by its designers and analysing it as an 'input-output device'. In this way, the actors connected to this device can be identified; for example, Renault engineers building the battery in electric vehicles or the users of this vehicle. The third thought experiment involves analysing the 'relationships between actors incorporated into the object'. In this way, a description can be offered of what is expected from future users. Taking again the electric vehicle as an example, de Laat explains that

> If the battery is designed as a single module, it may be rapidly exchanged for a new, fully charged one. If however it is divided over different places in a car (which for instance was the case with the first electric Peugeot's 106) it necessarily will have to be recharged at home or at an electricity station [40:194].

These three exercises are complementary and 'provide a preliminary description of the future world the object defines' [40:195]. Such descriptions enable a comparison between different anticipated futures as well as an eventual deliberation on emerging technologies. 'Fictive script' [39, 40] can be used as a tool to create greater agency for actors. The notion refers to the conditions necessary for an envisaged innovation to take place and to be successful. Actors can be confronted with a fictive script and be asked to fill in the gaps required for the implementation of a successful innovation; such an exercise stimulates greater agency for actors because they can now see what has to be done and how.

Den Boer et al. [41] have used the idea of 'fictive scripts' in a mapping methodology aimed at turning open-ended promises of nanoresearchers into concrete challenges. The authors specify 'future scripts' as a possible future innovation chain that could take up and realise the potential of some nanoscience research. By comparing current innovation value chains (including external contexts and structures or 'framework conditions') with future innovation chains, gaps and barriers in present networks can be identified. This kind of scripting exercise can be carried out in considerable detail, with checks and counterchecks with various relevant and knowledgeable actors.

The primary aim of the study of Den Boer et al. was to develop a methodology to enhance the ethical reflexivity of (nano) scientists, in particular junior scientists, in relation to present and future societal contexts of their work, by exploring potential research connections even when they are not (yet) linked to applications [41]. Stimulating agency is an important aspect of this 'reflexivity-in-action'.

While these fictive script exercises are often conducted as one-to-one conversations between the social scientist and the scientist, it is also important for scientists to imagine social and ethical implications of their research in a setting in which different perspectives are confronted and discussed. Constructive Technology Assessment (CTA) workshops aim to create such a space. CTA aims to improve the conditions for technology development at an early stage of the innovation process by providing stakeholders with tools with which to incorporate knowledge about socio-technical dynamics into decision-making processes [15]. CTA is an approach that can be operationalised in workshops which function as a kind of forum for the performance of the articulation of reflexivity. CTA workshops are interactive, heterogeneous settings that facilitate a 'protected space' for scientists to reflect on technological development, position themselves with respect to other stakeholders, probe each other's worlds, articulate their perspectives, and listen to other points of view. Stakeholders therefore have to be able to articulate their position, defend their arguments, and criticise and learn from others. CTA workshops 'act as a microcosm of the real world and allow the mingling of actor-roles and experiences of those participating' [42:206]. Moreover, such workshops introduce elements that would normally be quite distant or invisible to participants' traditional activities in the real world (ibid). These particular features of CTA workshops facilitate the performance of the articulation of reflexivity of actors.

Socio-technical scenarios can be used as inputs to these interactive workshops. They are fictional narratives that, beginning with an existing socio-technical configuration, describe potential future developments and dynamics triggered by decisions by different stakeholders [43]. Building on previous analytic research on socio-technical dynamics, these scenarios describe how technological innovation breaks up the existing socio-technical order and requires actors to 're-align' in different combinations [42]. In the field of nanotechnology, Robinson has used socio-technical scenarios in CTA workshops as a platform for the exploration of issues around emerging nanotechnologies and to facilitate the 'probing' of the worlds of the various participants [42]. Scenarios are used as tools to broaden stakeholder understanding of socio-technical dynamics and to develop reflexivity regarding one's own role in shaping future configurations, in addition to stakeholder interdependence. In this sense, socio-technical scenarios offer both a window of opportunity for modulating decision-making and a strategy for moving in a desired direction [42]: 'broadening perspectives and providing insights in[to] socio-technical dynamics enables actors to do better in their normal working environment and can eventually contribute towards more desirable paths' [44:329]. Although a discussion of the desirability of an emerging technology can be triggered by this interactive exercise, this is not its primary aim; scenarios offer entrance points for ethical discussion but as Robinson observed in his workshops, these were rarely taken up, with other issues such as start-up firms and regulation dominating [42].

Inspired by socio-technical scenarios, techno-ethical scenarios have been developed. Van Rijswoud et al. have used an interactive scenario study in the field of community genetics in order to articulate the 'endogenous futures' of community genetics and to stimulate debate among various stakeholders [45]. This method has been taken further by Swierstra et al. [46] and Stemerding et al. [47] in their techno-ethical scenario approach. The construction of techno-ethical scenarios facilitates the exploration of potential controversies around new and emerging science and technology beforehand by applying the 'grammar' of ethics of new and emerging science and technology [33].

These scenarios are built according to a three step approach in which

(1) the current moral landscape is outlined;
(2) the technology is introduced and some potential ethical controversies are played out; and
(3) some closure to these controversies is provided through (fictive, but plausible) technical, regulative or organisational solutions.

These three steps can be repeated several times so as to articulate further future controversies and closures. Techno-ethical scenarios provide a tool for the anticipation of ethical controversies around a new technology and for the exploration of the dynamics of interaction between current morality and new technologies.

This approach can help stakeholders to recognise, understand and manage potential controversies and to contribute to public debate about the desirability of the new and emerging technology in question. Techno-ethical scenarios facilitate opening up of moral perspectives but not in a straightforward way. The development of such scenarios replaces actors' own anticipation of potential ethical issues and arguments. However, it is the use of scenarios in interaction with other stakeholders that may lead to opening up, for example, through the occasion of justifying their perspectives to others. Techno-ethical scenarios have been used in CTA workshops to initiate a discussion on the roles and responsibilities of different stakeholders in the development of a specific technology [48, 49]. They have also been used in workshops with scientists in order to feed results of ethical assessments back to scientists and to trigger a discussion on the desirability and acceptability of emerging technologies [50].

Techno-moral vignettes have been used as tools to initiate discussion regarding potential 'soft impacts' of emerging technologies [50]; that is, the impacts of the technology on forms of life, concepts, and morality. These impacts are considered important by the lay public and should therefore be included in democratic deliberation on emerging technologies. However, some social actors consider these impacts 'soft' because they are not clearly quantifiable, objective risks [51]. For this reason, such impacts are often neglected in public discussions. Like scenarios, vignettes are narratives that focus on potential impacts of new technologies and on ways of life and which offer accounts of techno-moral change. However, unlike scenarios, they do not offer a line of action and course of conduct. In fact, scenarios, like movies, are a narrative form that describe a temporal unravelling of events. In this way, they draw attention to alternative pathways and dilemmas that can emerge as a consequence of particular decisions. In contrast, vignettes are more similar to photographs: they are 'snapshots' of the future [52]. Unlike scenarios, vignettes do not demonstrate causal connections between events or provide explanations of how a certain situation occurs. While scenarios take stakeholders by the hand to show the potential co-evolution of technology and morality step-by-step, demonstrating causal pathways between events, vignettes describe a future state of affairs.

7.4.3 Performing Reflexivity

The articulation of ethical and social considerations, in addition to the anticipation of techno-social implications, ethical debates and moral changes are crucial capacities to build in order to foster scientists' ethical reflexivity. The tools presented above constitute some attempts to develop these capabilities by several means including helping laboratory practitioners to become aware of the values and social considerations implicit in their deliberative processes and to expand the range of alternatives; offering them tools to clarify their claims, warrants and reservations; fostering scientists' capabilities to identify relevant stakeholders, together with their roles and responsibilities; offering them tools to foster their imagination of social implications and dynamics as well as ethical debates; and creating spaces for the discussion of such scenarios.

Articulating and imagining social and ethical implications and considerations is valuable because it enhances scientists' capability to act autonomously [26]. It is also important in order to perform this reflexivity in their practices and make them more ethically and socially responsible. In the 'midstream modulation' vocabulary, we could ask how the 'reflexive' modulation of scientists is conducive to a 'directed' modulation through their conscious choices.

All of these tools aim indeed to create capacities that allow scientists to gain awareness of ethical and societal issues and direct their actions and choices accordingly. Although some changes in behaviour and performance have been documented [10], it is difficult to evaluate how the tools have contributed to these changes in practices and how pervasive these changes are. Building capacities for performing reflexivity therefore remains open challenge.

7.5 Conclusions and Discussion

We have argued for the need to stimulate the articulation and performance of reflexivity through developing supportive tools and approaches. While these different aspects do allow for the 'heating up' of the perspectives of scientists, by pushing scientists to offer justifications for their perspectives in interaction not only with social scientists and ethicists but with other stakeholders, and enabling scientists to see how and where they are making shortcuts, challenges remain.

A first challenge concerns the translation of reflexivity into improved action. In the short term, marginal changes may be the most that can be expected. For example, in a 2006 midstream modulation study, Fisher reported four changes in research practices, namely, modification of an experimental setup, a change in disposal methods, the introduction of an alternative chemical catalyst and the formulation of safety rules [11]. In his description of a laboratory engagement study a few years later, Schuurbiers [53] reported that his presence as an outsider facilitated a change in laboratory practice, namely the wearing of lab coats. While the midstream modulation did succeed in affecting change in research practices—even succeeding where regulations had failed—the fact that change focused on environmental health and safety issues rather than broader questions about the socio-ethical context of their research, is striking. Nonetheless, such changes may accumulate over time and lead to changing role expectations.

A second challenge involves the institutional embedding of reflexivity. Some of the tools described in the previous section are designed in order to ameliorate the danger that the institutional requirement to take into account broader issues remains an 'add-on' activity and, indeed, is viewed as such. However, these tools cannot be successful without institutional support. For instance, it may be necessary to have incentives in the research system so as to facilitate the consideration of broader aspects of research in the core activities of researchers [54]. Incentives to work more closely with social scientists and ethicists may be strengthening with the inclusion of ELSA aspects/responsible innovation in large scientific research programmes; however, whether such interactions will allow for the articulation of reflexivity or exacerbate a division of moral labour (with an emphasis on 'enhancing'), remains to be seen.

In such a division of moral labour, ethical reflection is delegated to 'socio-humanistic consultants' [21] or 'convergence workers' [27]. This division of moral labour may not be ideal, given that it would allow researchers to 'follow the rules' without having to be sure that what they are doing is 'responsible' [21]. There is another scenario, however, in which social scientists and ethicists would work to actively contribute to the articulation of directions in which to proceed; in so doing, they would work as 'collaborators' alongside natural scientists [55]. While their expertise would continue to be 'a form of reflexive mediation' [21:668], it would require active involvement of scientists and other actors and at various levels/institutions. Moreover, it would be part of a larger attempt to modulate ongoing co-evolution of technology and society and make it more reflexive. Whichever form future arrangements may take, the adequacy of the division of moral labour pursued should be put up for discussion in relation to whether it delivers in terms of accepted goals and whether it is a 'good' division of moral labour (however 'good' may be defined) [20].

A third challenge centres on a fundamental tension in the drive for reflexivity, which we have shown here. On the one hand, there are institutional initiatives and requirements that fail to take into account the capacities for reflexivity that have to be fostered within a specific social context. On the other hand, specific tools developed with the aim of stimulating the capacity for ethical reflexivity have been developed but are not taken up because they lack institutional reinforcement. This tension, we argue, requires reconciling institutional requirements with supportive tools. This challenge is further exacerbated by the fact that approaches focus mainly on individual scientists and research groups as a means of modulating responses at the individual and group levels. However, 'organised irresponsibility' in relation to the newly emerging science and technology of which individual scientists are a part cannot be addressed by such individual reflexivity alone. It is necessary to broaden the perspective to reflexiveness in the overall process at the meso- and macro-levels. There are already indications of reflexiveness at the meso-level, visible in mechanisms such as codes of conduct and in initiatives taken by funding agencies. Initiatives at this level can influence dynamics at the level of actors' practices; the requirements of funding agencies at the meso-level can be leveraged as a means of influencing reflexivity of scientists at the micro-level. As Fisher and Rip observe,

> It is this multi-level constellation ... that carries the moves in the direction of responsible innovation, more so than any particular type of initiative, some of which would by themselves not seem very effective [4:167].

The question is whether these initiatives can be operationalised with the help of the tools/practical approaches described already in order to affect changes in the practice of doing science. How might we go about doing this? We focus on initiatives taken by funding agencies because of their pivotal role in providing means for doing science. Funding agencies' interest in extended social impact assessment can be operationalised by requiring 'fictive script' to be part of the proposal. The requirement of a 'fictive script' would imply that scientists would no longer be able to 'pimp' or overextend the promises made in their proposals in a sort of reality vacuum but would be required to translate open-ended promises into potential research connections and reflect on what would be required in order to make these successful. Moreover, scientists would be required to think about the current and future societal contexts of their work.

Acknowledgements

The authors are grateful to Professor Arie Rip for his earlier collaboration on this topic, in addition to comments and feedback on this chapter.

References

1. Jacob, K., and J. van den Hoven (2013) *Options for Strengthening Responsible Research and Innovation: Report of the Expert Group on the State of the Art in Europe on Responsible Research and Innovation* (European Commission, Brussels).
2. Owen, R., P. Macnaghten, and J. Stilgoe (2012) Responsible research and innovation: From science in society, to science for society, with society, *Science and Public Policy*, **39**, 751–760.
3. Rip, A., and C. Shelley-Egan (2010) Positions and responsibilities in the 'real' world of nanotechnology. In: von Schomberg, R., and S. R. Davies (eds.) *Understanding Public Debate on Nanotechnologies: Options for Framing Public Policy* (European Commission, Brussels), pp. 31–38.
4. Fisher, E., and A. Rip (2013) Responsible innovation: Multi-level dynamics and soft intervention practices. In: Owen, R., M. Heintz, and J. Bessant (eds.) *Responsible Innovation* (John Wiley, London), pp. 165–183.
5. Stilgoe, J., R. Owen, and P. Macnaghten (2013) Developing a framework for responsible innovation, *Research Policy,* **42,** 1568–1580.
6. Schuurbiers, D. (2010) Social Responsibility in Research Practice-Engaging applied scientists with the socio-ethical context of their work (PhD Dissertation, Delft University of Technology, Delft).
7. Ziman, J. M. (1998) Why must scientists be more ethically sensitive than they used to be?, *Science*, **282**(5395), 1813–1814.

8. Rip, A., and H. van Lente (2013) Bridging the gap between innovation and ELSA: The TA program in the Dutch nano-R&D program NanoNed, *NanoEthics*, **7**, 7–16.
9. Fisher, E., R. L. Mahajan, and C. Mitcham (2006) Midstream modulation of technology: Governance from within, *Bulletin of Science Technology Society*, **26**(6), 485–496.
10. Fisher, E. (2007) Ethnographic invention: Probing the capacity of laboratory decisions, *NanoEthics*, **1**(2), 155–165.
11. Schuurbiers, D., and E. Fisher (2009) Lab-scale intervention, *EMBO Reports*, **10**(5), 424–427.
12. Owen, R., and N. Goldberg (2010) Responsible innovation: A pilot study with the UK engineering and physical sciences research council, *Risk Analysis*, **30**(11), 1699–1707.
13. Akrich, M. (1992) The description of technological objects. In: Bijker, W., and Law, J. (eds.) *Shaping Technology Building Society: Studies In Sociotechnical Change* (MIT Press, Cambridge).
14. Schuurbiers, D., S. Sleenhoff, F. J. Jacobs, and P. Osseweijer (2009) Multidisciplinary engagement with nanoethics through education–the Nanobio-RAISe advanced courses as a case study and model, *NanoEthics*, **3**(3), 197–211.
15. Schot, J., and A. Rip (1997) The past and future of constructive technology assessment, *Technological Forecasting and Social Change*, **54**(2–3), 251–268.
16. Van Eijndhoven, J. C. M. (1997) Technology assessment: Product or process?, *Technological Forecasting and Social Change*, **54**(2–3), 269–286.
17. Guston, D. H., and D. Sarewitz (2002) Real-time technology assessment, *Technology in Society*, **24**(1–2), 93–109.
18. Wilsdon, J., and R. Willis (2004) *See-Through Science* (Demos, London).
19. Owen, R., J. Stilgoe, P. Macnaghten, M. Gorman, and D. Guston (2013) A framework for responsible innovation. In: Owen, R., J. Bessant, and M. Heintz (eds.) *Responsible Innovation: Managing the Responsible Emergence of Science and Innovation in Society* (John Wiley, London) pp. 27–50.
20. Beck, U., W. Bonss, and C. Lau (2003) The theory of reflexive modernization: Problematic, hypotheses and research programme, *Theory, Culture & Society*, **20**(1), 1–33.
21. Rip, A. (2009) Futures of ELSA, *EMBO Reports*, **10**(7), 666–670.

22. Wynne, B. (2011) Lab work goes social, and vice-versa: Strategising public engagement processes, *Science and Engineering Ethics*, **17**, 791–800.
23. Shelley-Egan, C. (2011) *Ethics in Practice: Responding to An Evolving Problematic Situation of Nanotechnology in Society* (Phd Dissertation, Universiteit Twente, Enschede).
24. Holbrook, J. B. (2005) Assessing the science–society relation: The case of the US National Science Foundation's second merit review criterion, *Technology in Society*, **27**(4), 437–451.
25. Royal Netherlands Academy of Arts and Science, Association of Universities in the Netherlands and the Netherlands Organization for Scientific Research (2009) *Standard Evaluation Protocol 2009–2015. Protocol for research assessment in the Netherlands.* Available at www.knaw.nl/sep.
26. Fisher, E., and M. Lightner (2009) Entering the social experiment: A case for the informed consent of graduate engineering students, *Social Epistemology*, **23**(3–4), 283–300.
27. Guillemin, M., and L. Gillam (2004) Ethics, reflexivity, and 'ethically important moments' in research, *Qualitative inquiry*, **10**(2), 261–280.
28. Bowman, D. M., and G. A. Hodge (2009) Counting on codes: An examination of transnational codes as a regulatory governance mechanism for nanotechnologies, *Regulation& Governance*, **3**, 145–164.
29. Kjølberg, K. L., and R. Strand (2010) Conversations about responsible nanoresearch, *NanoEthics*, **5**, 99–113.
30. Dorbeck-Jung, B., and C. Shelley-Egan (2013) Meta-regulation and Nanotechnologies: The challenge of responsibilisation within the European commission's code of conduct for responsible nanosciences and nanotechnologies research, *NanoEthics*, **7**(1), 55–68.
31. von Schomberg, R. (2010) Organising collective responsibility: On precaution, codes of conduct and understanding public debate. In: Fiedeler, U., C. Coenen, S. R. Davies, and A. Ferrari (eds.) *Understanding Nanotechnology: Philosophy, Policy and Publics* (IOS Press, Amsterdam), pp. 61–70.
32. Meili, C., W. Widmer, S. Schwarzkopf, E. Mantovani, and A. Porcari (2011) *NanoCode MasterPlan: Issues and Options on the Path Forward with the European Commission Code of Conduct on Responsible N&N Research.* Available at: http://www.nanocode.eu/files/NanoCode-MasterPlan.pdf.

33. Swierstra, T., and A. Rip (2007) Nano-ethics as NEST-ethics: Patterns of moral argumentation about new and emerging science and technology, *NanoEthics,* **1**(1), 3–20.

34. Garud, R., and D. Ahlstrom (1997) Technology assessment: A socio-cognitive perspective, *Journal of Engineering and Technology Management,* **14,** 25–48.

35. Rip, A. (2006) Folk theories of nanotechnologists, *Science as Culture,* **15,** 349–365.

36. Rip, A. (2007) Research choices and directions-in changing contexts. In: Deblonde, M., L. Goorden, A. Nordmann, and A. Rip (eds.) *Nano Researchers Facing Choices. The Dialogue Series #10* (Universitair Centrum Sint-Ignatius, Antwerpen), pp. 33–48.

37. van der Burg, S. (2009) Imagining the future of photoacoustic mammography, *Science and Engineering Ethics,* **15**(1), 97–110.

38. van der Burg, S. (2010) Ethical imagination: Broadening laboratory deliberations. In Roeser, S. (ed.) *Emotions and Risky Technologies* vol. 5 (Springer, Amsterdam), pp. 139–155.

39. De Laat, B. (1996) *Scripts for the Future. Technology Foresight, Strategic Evaluation and Socio-Technical Networks: The Confrontation of Script-Based Scenarios* (PhD Dissertation, University of Amsterdam, Amsterdam).

40. De Laat, B. (2000) Scripts for the future: Using innovation studies to design foresight tools. In: Brown, N. B. Rappert, and A. Webster (eds.) *Contested Futures. A Sociology of Prospective Techno-Science* (Ashgate, Aldershot), pp. 175–208.

41. Den Boer, D., A. Rip, and S. Speller (2009) Scripting possible futures of nanotechnologies: A methodology that enhances reflexivity, *Technology in Society,* **31,** 295–304.

42. Robinson, D. K. R. (2010) *Constructive Technology Assessment of Emerging Nanotechnologies: Experiments in Interactions* (PhD Dissertation, University of Twente, Enschede).

43. Rip, A., and H. Te Kulve (2008) Constructive technology assessment and socio-technical scenarios. In: Fisher, E., C. Selin, and J. M. Wetmore (eds.), *The Yearbook of Nanotechnology in Society, Volume 1: Presenting Futures* (Springer, Amsterdam), pp. 49–70.

44. Van Merkerk, R. O., and R. E. H. M. Smits (2008) Tailoring CTA for emerging technologies, *Technological Forecasting and Social Change,* **75**(3), 312–333.

45. van Rijswoud, E., D. Stemerding, and T. Swierstra (2008) *Genetica, genomics en gezondheidszorg: een toekomstverkenning* (Centre for Society and Genomics, Radboud University, Nijmegen).
46. Swierstra, T., D. Stemerding, and M. Boenink (2009) Exploring techno-moral change: The case of the obesity pill. In: Sollie, P., and M. Düwell (eds.) *Evaluating New Technologies* (Springer, Amsterdam), pp. 119–138.
47. Stemerding, D., T. Swierstra, and M. Boenink (2010) Exploring the interaction between technology and morality in the field of genetic susceptibility testing: A scenario study, *Futures*, **42**(10), 1133–1145.
48. Krabbenborg, L. (2013) Dramatic rehearsal on the societal embedding of the lithium chip. In: van der Burg, S., and T. Swierstra (eds.) *Ethics on the Laboratory Floor* (Palgrave Macmillan, Basingstoke), p. 21.
49. Krabbenborg, L. (2013) *Involvement of Civil Society Actors in Nanotechnology: Creating Productive Spaces for Interaction* (Proefschrift Rijksuniversiteit, Groningen).
50. Lucivero, F. (2012) *Too Good to Be True? Appraising Expectations for Ethical Technology Assessment* (Universiteit Twente, Enschede).
51. Swierstra, T., and H. Molder (2012) Risk and soft impacts. In: Roeser, S., R. Hillerbrand, P. Sandin, and M. Peterson (eds.) *Handbook of Risk Theory SE-42* (Springer, Amsterdam), pp. 1049–1066.
52. Van Notten, P. W., J. Rotmans, M. B. van Asselt, and D. S. Rothman (2003) An updated scenario typology, *Futures*, **35**(5), 423–443.
53. Schuurbiers, D. (2011) What happens in the lab: Applying midstream modulation to enhance critical reflection in the laboratory, *Science and Engineering Ethics*, **17**, 769–788.
54. Shelley-Egan, C., and D. R. K. Robinson (2013) *Can Devices of Responsibility Shape the Core Activities of Nanotechnology Research Organisations and Firms? A Pilot Investigation into RRI Approaches and Incentives for Change. Shaping Political Spaces for Research, Innovation and Markets*, Paris, Ecole des Mines, 12–13 September.
55. Calvert, J., and P. Martin (2009) The role of social scientists in synthetic biology, *Science & Society Series on Convergence Research*, *EMBO Reports*, **10**, 201–204.

Part 2

Promises, Politics and Particularities of Nanotechnologies

PART 2

Chapter 8

The Demand Side of Innovation Governance: Demand Articulation Processes in the Case of Nano-Based Sensor Technologies

Haico te Kulve and Kornelia Konrad

*Department of Science, Technology and Policy Studies,
School of Management and Governance, University of Twente,
7500 AE Enschede, The Netherlands*

h.tekulve@utwente.nl, k.e.konrad@utwente.nl

8.1 Introduction

Governance of innovation can be understood as the coordination of interdependent actors' actions and interactions regarding the development and 'societal embedding' [1] of new and emerging technologies. Much of the coordination occurs within, and through, ongoing processes along the innovation journey of new technologies. Among these processes, demand articulation processes, i.e. the unfolding and specification of user requirements and preferences regarding new technologies [3, 4], play an important role. The emergence and specification of preferences and requirements may generate new innovation trajectories, guide actors' pursuit of innovations and support market introduction [5].

Embedding New Technologies into Society: A Regulatory, Ethical and Societal Perspective
Edited by Diana M. Bowman, Elen Stokes, and Arie Rip
Copyright © 2017 Pan Stanford Publishing Pte. Ltd.
ISBN 978-981-4745-74-1 (Hardcover), 978-1-315-37959-3 (eBook)
www.panstanford.com

However, particularly in early phases of innovation when technologies are still under development, their shape and performance is not yet fully known, and there is little or no user experience, user preferences and requirements may not be clear and therefore cannot be easily 'captured' and translated into specifications guiding further development. In such situations, demand articulation processes can be considered to be little effective as coordinating mechanisms. Especially in cases of promising technologies such as novelties which are expected to contribute to sustainability or health and safety of consumers, governments, firms and other stakeholders may then want to intervene to stimulate their exploration and exploitation. In policy circles, demand-based policies are currently drawing increasing interest [6]. Such policies consist primarily of procurement, but can also include stimulation of demand articulation to support users and producers in assessing and making sense of novel technologies in order to identify added value and requirements [7]. As we will show later, there are further possibilities to stimulate and support demand articulation processes.

In this chapter, we argue that demand articulation processes constitute an important and relevant part of innovation governance when understood in a broader sense, and that this opens up new opportunities to intervene in the demand-side of innovation processes. We will show that demand articulation occurs at more places and involves a broader set of dynamics than user-producer interactions in the context of new product development, which are usually discussed in the literature, see for example [8-10]. This implies that actors aiming to intervene in the demand side can benefit from a broader perspective as it may contribute to a wider range of opportunities for intervention, i.e. open up engagements with a broader set of actors, while at the same time it implies that they have to take into account such broader processes and their dynamics in order to be productive. Ideally they also have tools to capture what is happening in demand articulation processes and identify possibilities for intervention. We will show that Constructive Technology Assessment (CTA) approaches offer such tools.

To illustrate our argument we will discuss findings of our empirical studies into demand articulation processes of newly emerging sensor applications for business users in the drinking water, and food and beverages sectors and interactive workshops where actors were supported in articulating and reflecting on demands. Both sectors are envisaged as application fields for a similar set of micro- and nanotechnology-enabled platforms for sensor applications, but differ in terms of governance structures. In both domains, actors inquire into requirements regarding future sensor applications and occasionally find themselves in impasses. They differ in terms of how envisioned users take into account other actors in these domains during demand articulation processes. In the concluding section, we will reflect on the findings and draw general lessons regarding governance of innovation by demand articulation.

The data in this chapter draw upon a study into demand articulation processes in the Dutch drinking water, and food and beverages sectors. For this study, we conducted and transcribed 29 semi-structured interviews with stakeholders from April to November 2013, and gathered data through desktop research and participatory observation at conferences held in the Netherlands. In Fall 2013 we organised a one-day interactive event with a plenary discussion and two parallel workshops, which gathered 35 participants from knowledge institutes, nationally operating water companies, internationally operating food companies, and representatives of governmental agencies, standardisation and certification organisations. Discussions in the workshops were recorded and transcribed to facilitate data analysis.

8.2 Distributed Processes of Demand Articulation

8.2.1 Demand Articulation as Part of Innovation Governance

Demand articulation processes, i.e. the unfolding and specification of user requirements and preferences regarding technologies, play an important role in the innovation journey of new technologies by acting as selection mechanisms, offering incentives and reducing uncertainties [5, 8, 11, 12]. As processes, they have dynamics of their own, while at the same time they fulfil a steering or governance role in the overall innovation process.

Demand articulation processes contribute to how actors such as innovating firms and potential users make sense of the meanings of new technologies and the action strategies they employ accordingly. We suggest that a further governance effect of demand articulation works via attempts of actors to actually initiate and pursue demand articulation processes in interaction with other actors which sets in motion or alters existing technology development trajectories. Illustrations of these effects can be found in the emerging field of organic and large area electronics where firms struggled with uncertainties regarding future demands and engaged in dedicated attempts to articulate expectations via roadmapping activities [13].

Demand articulation can thus be characterised as 'de facto' governance patterns which direct and coordinate actors' actions and interactions with some authority (see also Chapter 2 of this volume).[1] They are partly intentional and partly unintentional and emerge from interactions between actors (in contrast to patterns initiated by authorities) [14, 15].

The scope of this 'governance by demand articulation' then is dependent on the understanding of actors and interactions included in demand articulation. Commonly, demand articulation is associated with marketing research, the elicitation of customer preferences and requirements, which is then fed back to product differentiation or improvement [11, 16, 17]. Mariampolski, in his guide on market research, positions demand articulation as a predominantly local affair, and as something that needs to be managed by the firm[2]:

> 'User needs' and 'user friendliness' have become the principle injunctions of corporate development laboratories. What better way is there to determine the match between products and their consumers than by soliciting their advice beforehand and running concepts and prototypes past them afterwards [17:2].

The process of demand articulation and its relation with technology development is however more complex and involves

---

[1]While the overall argument is applicable more generally, analysing de facto governance patterns may be relatively easier in cases of professional (business) users than in the case of consumers which may have a broader (even) less clearly defined set of interests and goals. See also Lundvall [10].

[2]See also Jiao and Chen [18] on the management of capturing and understanding customer requirements.

more dynamics and actors. Rather than a static, one-way affair, demand articulation is a dynamic co-evolutionary process where suppliers need to learn about relevant product characteristics and customers learn to handle and evaluate new technologies [8, 19–21]. When actors are formulating demands, they also do more than envisioning desired characteristics of a new technology and their implementation in an organisation.[3] The introduction of new products also requires societal embedding [1] which generates additional articulation processes regarding directions and requirements. Novel products and their practices need to be

(1) integrated in a specific product value chain;
(2) be accepted in the regulatory environment, including regulations and standards in the sector where it is applied;
(3) be accepted by society more generally.[4]

From this perspective, demand articulation adds up to the formulation of requirements and preferences for not just a novel product, but for a future socio-technical world including the application of novel products and their embedding in the broader environment.

Articulation of societal embedding is particularly important, if demand articulation follows a so-called 'stretch-strategy' [24:18–22]. The formulation of preferences and requirements regarding the embedding of new technologies and products can be compatible, and 'fit' with existing social structures and governance arrangements in the domain of application. Alternatively, formulations may not be compatible and 'stretch' the existing structures. In principle, the fit and stretch strategies can refer to various analytic dimensions of demand articulation, such as the technology and use environment characteristics discussed by Hoogma, [24:1], but also to the broader business, regulatory and societal environments highlighted in this chapter. Fit strategies guide technological innovations along existing trajectories, whereas stretch strategies initiate new, possibly disruptive, trajectories [25, 26].

---

[3]Cf. the notion of 'innofusion' [22], which refers to the ongoing development of products and user requirements at the point of application.
[4]See also McMeekin [23:114], who characterises demand as selection criteria emerging from an evolving 'selection environment fabric'.

If demand articulation follows a fit-strategy, the various dimensions of the envisaged socio-technical world can by and large remain as they are. In this case, active embedding into the use context is still required, but such a strategy does not pose major challenges to existing value chains, the regulatory environment and/or public acceptance. In case of stretch strategies, however, changes are required and articulation of societal embedding becomes an important part of demand articulation. Moreover, in situations where stretch strategies are followed in demand articulation processes, they can be accompanied by dedicated attempts to alter the institutional environment in order to ensure a future 'fit' between the novel product and its broader environment. Firms can act as institutional entrepreneurs [27] by introducing new evaluation criteria and use practices through advertising [28, 29] or the creation of new standards [30] in order to shape customer preferences. Users and interest groups may enact new practices of consultation and information exchange between end-users and technology developers [31], or introduce novel risk assessment approaches and codes of conduct [32]. When novel institutional patterns are introduced and become part of governance arrangements affecting innovation processes, this changes the governance of innovation in a specific domain.

Seen from this broader perspective, demand articulation processes become especially relevant for the governance of the whole innovation process, by including not only technology development processes but also their embedding in society.

8.2.2 Dynamics in Distributed Demand Articulation Processes

Demand articulation processes have dynamics of their own, the understanding of which is relevant if one aims to anticipate their governance effects or intervene in the processes themselves. Our previous discussion suggested that the social locus of demand articulation and its governance effects are not limited to users and producers and their spaces of interaction. Demand articulation processes are thus not confined to, and determined by, market interactions.[5] Instead, processes of demand articulation are distributed across a variety of actors and spaces, as suggested by the notion of 'distributed innovation' [34]. This perspective on the distributed nature of demand articulation processes is also suggested by other studies. Schot and Albert de la Bruhéze [35] have pointed out the importance of various 'mediation actors' and 'mediation junctions' involved in the alignment of product characteristics and user requirements. Others have highlighted the importance of intermediaries and stakeholders in demand articulation processes [36–38].

How interactions in demand articulation unfold will differ due to different interaction patterns between, among others, users and producers, which depend on the type of technology [39] and characteristics of sectors regarding the roles of users and producers in innovation processes [40]. Demand articulation is also political in the sense that actors formulating demands are influenced by their interactions with other actors and may debate and disagree about preferred directions and the assumptions underlying such directions [3, 38]. In cases where demand articulation processes include priority-setting of a variety of preferences and requirements, differences in the power and influence of actors on such processes and institutionalised patterns in demand articulation processes can influence the unfolding of these processes and which 'demands' eventually are foregrounded [37]. Demand articulation processes thus evolve within a broader governance setting.

Benz [33] offers an analytical approach that is helpful to study dynamics in demand articulation processes and their governance effects. He proposed that governance occurs through interconnected arenas where actors define and struggle with different problems. Arenas, such as parliaments, dedicated committees, advisory councils and deliberative bodies, are defined by functions and rules and may be loosely or strongly connected to each other. We suggest that this approach is also fruitful in the case of demand articulation, as the various dimensions of technology development and embedding (industry, regulation,

---

[5]In general, market modes of governance are considered to be inappropriate to capture what is happening as actors 'coordinate their actions in the shadow of the law (hierarchy) and in the shadow of competition among consumers and producers of goods in more or less explicit negotiations over contracts' [33:7].

society) are often articulated in different arenas which nevertheless will influence what is happening elsewhere, such as in meetings organised by branch organisations, intermediaries, governmental policy makers or standardisation committees. Regulation, in terms of legislation and policies, but also standardisation, is a clear example, as it will affect interactions in the business environment. Indeed, regulation and standards are part of the tool-kit of demand-based policies, and recognised as supporting demand articulation processes [7:15].

Actors formulating and discussing preferences and requirements (within different arenas) will be guided by interactions and rule sets in the different arenas—which can be called 'governance of demand articulation', complementing the 'governance by demand articulation' discussed so far. Actors' actions and interactions regarding demand articulation will also be influenced through the emerging socio-technical paths [34] they co-create.[6] Especially in early stages, where technological developments are relatively open-ended, demand articulation processes will be shaped by regimes in a domain where new technologies are applied and embedded. Hoogma argued that

> the introduction of new technologies takes place against the backdrop of existing technological regimes. The existence of these regimes sets criteria for a new technology, which translate to preferences of producers, users, and regulators [24:17].

These regimes can be characterised as rule sets that can be specific for an arena of technology development [43] but may also include a variety of arenas (see also Geels [44] on socio-technical regimes as linked regimes). In early phases of technology developments, expectations will definitely be important as well—in general [45, 46] and for demand articulation in particular [47, 48].

The unfolding of demand articulation processes then will be influenced by fit or stretch strategies of actors involved, the use possibilities afforded by technologies [49] and the governance arrangements (rules and arenas) in the domain of application.

---

[6]See also Teubal [41] and Clark [42] on hierarchical patterns in the specification of designs.

8.2.3 Interventions in Demand Articulation and Tools to Support Them

While demand articulation processes in principle can play an important role in innovation governance, they may not easily unfold. In such situations, actors, such as firms and governments, may want to intervene. For interventions, the interactive nature of demand articulation processes is important to take into account. To some extent this is already visible in market research projects that work with focus groups to elicit preferences and user requirements [16, 17]. Interactive elements are even more prominent in the case of 'Living Lab' methodologies which aim to develop more user-centred products through involving users interacting with novel technologies in everyday life settings (or simulations thereof) [50]. Such approaches assume though that the performance of novel products and user requirements to some extent already exist or emerge within these confined, short-term settings and can be captured accordingly. Secondly, they conduct less analysis of what is and might be happening (including, for example, across various arenas relevant to societal embedding of new technologies) and therefore do not structure discussions accordingly. This structuring is especially important in the case of new and emerging technologies, because their performance and demand is not only uncertain as to whether they will materialise, but directions and characteristics are also indeterminate, meaning that they may unfold in various ways which may not yet be clear. Then, analysis and diagnosis of ongoing socio-technical dynamics is useful to make sense of the situation and to reduce some of the complexities in order to have productive discussions [51, 52]. This allows participants to connect present and future developments and support their reflexivity in articulation of preferences and requirements as well as formulation of strategies to further such articulations.

CTA is an approach which combines interactive discussions and anticipation on future developments, and it pays explicit attention to demand articulation [53]. In CTA, innovation processes and their dynamics play a key role and outcomes are seen as being constructed, or co-produced during socio-technical changes. As an intervention, CTA aims at supporting processes, stimulating open-ended learning and actors' reflexivity rather than steering toward specific outcomes. In that respect it does not take an explicit normative position other than the objective of stimulating actors' reflexivity, their awareness of other actors' perspectives and their knowledge of innovation processes and dynamics. The stimulation of open-ended learning of and between stakeholders is especially relevant for demand articulation processes which struggle to emerge or unfold and which are 'inherently creative process[es]' [3].

CTA has recently been further developed for new and emerging technologies, such as micro- and nanotechnologies where uncertainties regarding demand and supply, and integration in society play an important role [13, 54, 55]. Specific elements in this approach include the organisation of workshops which aim to stimulate learning of and between actors with different perspectives towards new technologies and the use of scenarios exploring the innovation journey of new technologies including their embedding [51, 52]. Designing the workshop as a micro cosmos, with key players involved in the technologies and sector studied, participants and the intervention agent can and do learn about perspectives of actors in the sectors, ongoing dynamics and possible effects of the innovation process they are often involved in, but usually only with regard to particular aspects of it, and they can articulate and discuss strategies [56, 57].

To what extent these learning experiences and identified strategies have broader impacts will depend on the set of involved participants, but also on circumstances in the context of the participants which make it more easy or difficult to take up the insights gained in the workshops. Credibility pressures such as pressures to move the field forward may create incentives (for firms, researchers, but also governmental actors) to take up acquired insights in the workshops and thus enable CTA workshops to have broader impacts [56]. We applied the CTA approach for our case studies into demand articulation of sensors in the drinking water and food & beverages sectors, which will be discussed in Section 8.4.

8.3 Demand Articulation of New Sensor Applications for the Drinking Water and Food and Beverages Sectors

In this section, we discuss findings from our empirical studies into demand articulation processes in the drinking water and food and beverages sectors. The findings apply to the Dutch context, unless otherwise indicated. We will especially pay attention to fit and stretch demand articulation strategies, as stretch strategies are particularly challenging in terms of societal embedding articulation and may require, or contribute to, changes in innovation governance.

8.3.1 Sensors and Characteristics of the Drinking Water and Food and Beverages Sectors

Sensor technologies such as those enabled by micro- and nanotechnologies are expected to contribute to quality and safety issues in the drinking water, and food and beverage sectors through offering innovative solutions via miniaturisation, high sensitivity and specificity, and speed of measurement [54, 58, 59]. Many of these technologies are still in early phases of their 'innovation journey' as few are introduced on the market. The performance of sensors is not always clear and there are significant technological challenges involved in sampling and sampling preparation, and the maintenance of sensor technologies. For the prospective users of sensors in our case studies, the performance and added value of such novel technologies is not necessarily evident, as indicated in our interviews and the workshop discussions. While in both domains potential users can formulate requirements regarding measurements of specific parameters, they find it more difficult to determine which of the variety of possible parameters to be measured are preferred and to specify their requirements.

Comparing the institutional characteristics of the two sectors suggests that different dynamics will be at play in the demand articulation processes. The food and beverages sector in the Netherlands (and globally) has a relatively complex value chain consisting of more players and a larger variety in supplied products compared to the drinking water sector. It is highly competitive, which is not the case in the publicly owned drinking water sector in the Netherlands.[7] Customers play a more prominent role in the food value chain (see Te Kulve et al. [60]) on differences in the role of end-users in expectations around novel nanotechnology-enabled products in these domains). While sensors as such are unlikely to be questioned as a legitimate technology by the public, possible effects on the quality and safety of food or drinking water may be, as well as possible changes in the distribution of responsibilities regarding safety and quality [61]. Both domains have explicit regulation regarding end-products and this becomes an issue when sensors are considered as alternatives for existing approaches to control quality and safety of products.[8]

Thus, we expect that demand articulation processes in the case of food and beverages will be more influenced by rules and discussions in arenas where professional users of novel sensors interact with their value chain partners. In the drinking water sector rules and discussions in policy arenas will be more prominent in the dynamics of demand articulation processes. What will likely be shared among both domains are efforts towards efficiency improvement which act as a 'guiding principle' [62, 63] in their production environments.

8.3.2 Fit and Stretch Strategies in Demand Articulation Processes

The extent to which demand articulation is open-ended will differ between application areas within the domains. In the case of efforts for efficiency improvements, requirements regarding products and their embedding in production environments are

---

[7]In other countries, the drinking water sector is more fragmented, consisting of more players, which can be either publicly or privately owned.

[8]In the case of drinking water: the European Drinking Water Directive (98/83/EG) applies, as well as the Dutch law on drinking water (Drinkwaterwet) and regulations (Drinkwaterregeling) specifying monitoring programs and measurement methodologies according to ISO standards. In the case of food & beverages there exist a range of regulations, including the General Food Law (178/2002/EC), the regulation on microbiological criteria (2073/2005/EC) which also specifies monitoring programs and measurement methodologies according to ISO standards, and the Dutch commodities law (Warenwet).

relatively clear—assuming that users have already identified room for improvement in their processes. Efficiency improvements will mostly be incremental and often not challenge the prevailing evaluation criteria in the existing environment, thus leading to fit strategies in societal embedding articulation. In the case of water, examples of sensor applications following fit strategies can be found in purification plants for drinking water to enable reduction of chemicals during water treatment, or automation of the treatment plant in order to save costs or due to a foreseen lack of skilled personnel in the future. In the case of food, automation is an objective, as well as the prevention or reduction of contaminated batches that currently are often only detected at the end of the process. These type of monitoring efforts are not required by authorities and firms are interested out of economic motivations.

Efficiency improvement is an important and taken for granted driving and guiding force for exploring and exploiting novel sensor technologies in production environments. In our case studies, actors were more ambivalent regarding improving quality assurance and control through sensors where determining added value is less straightforward. We will highlight a number of salient issues.

8.3.3 Sensors for Monitoring Water Quality in the Distribution Network: Fit or Stretch the Monitoring Regime?

To assure and control water quality, sensors measuring water quality parameters may be used after water purification or in the distribution network. Especially the water distribution network with its many kilometres of piping is considered to be of great interest (in the Netherlands, and globally) as little is known of what happens within the network.

Less clear is which parameters should be measured, how many sensors are needed, and at which positions in the network. These are important questions, not the least because they affect investment and maintenance costs.[9] Large numbers would increase

---

[9]Clearly, requirements and preferences regarding purchase and maintenance costs are also important in demand articulation processes. Interviewees highlighted uncertainties regarding these economic dimensions and how they are intertwined with discussions regarding anticipated value of the application of sensors.

costs and operational costs would rise when sensors are difficult to reach for data transfer, and (if necessary) for inspection or service. Further questions are which organisational responses should be taken if anomalies are detected, and how these novel monitoring practices affect the relationships between water companies and water laboratories (who usually perform the analysis). Finally, it is not clear if authorities would accept measurements based on new sensors. All these specific questions add up to the more general question how novel sensors should be embedded in a monitoring regime and what this means for concrete requirements for sensors.

The current monitoring regime can be characterised as consisting of

(1) a prescribed set of parameters to be measured according to specific analysis techniques, and
(2) additional monitoring activities by drinking water companies to assure water quality.

Ideally, novel sensors have the same sensitivity and specificity characteristics as analysis done within the labs, but then quicker: real-time or near real-time. If the sensors would have similar performance as the analyses of laboratories, which are currently used for compliance with legal obligations, they would likely be accepted as alternative options by authorities. Ideally, such sensors should also be able to detect all contaminants and would be very cheap. Except for the relationship between drinking water companies and water laboratories, which would be likely to change, the emphasis on equal performance follows a fit strategy in terms of societal embedding articulation.

For many years, research and development efforts regarding sensors for measuring water quality have, unsurprisingly, focused on increasing the number of parameters to be measured and higher levels of sensitivity. Recently however, there is a shift in direction towards more generic measurements [64]. In the Netherlands, this shift is promoted by a number of researchers and water companies who expect that novel sensor technologies will not be able to deliver equal or better performance (compared to laboratory equipment) in the near future. In addition, there is the ongoing inquiry into which parameters should be measured at all, also given the impossibility to measure all known (and newly emerging) contaminants in drinking water. Instead of sensitive and specific measurements, a generic approach is advocated which is referred to as 'water footprinting' in order to detect anomalies.

This approach is not broadly shared in the water sector and is at odds with the established monitoring strategies and approaches. Some water companies are sceptic regarding monitoring water quality in the network at all or the approach of measuring deviation without knowing for sure whether there are indeed deviations and whether they are for the better or worse. The footprinting approach would also require re-thinking the implications of such approaches for response protocols in case of actual incidents and their embedding in a monitoring regime. Actors advocating this approach can thus be considered as institutional entrepreneurs, promoting new practices and institutions. Thus, such a generic monitoring strategy would challenge and thus stretch the current monitoring regime due to its fundamentally different measurement approach compared with currently accepted and legally prescribed approaches and its (presently unclear) consequences for follow-up action. This implies that more actors than only (local) users and producers of sensors have to be involved in articulating requirements and institutional changes may be required.

8.3.4 Sensors for Monitoring Food Quality: Stretching Commercial Relationships along the Value Chain

Issues regarding integration in the business environment are very visible in discussions on preferences and requirements for novel sensors. Given the highly competitive nature of the food sector and its high pressure on margins, an important question is who in the (extensive) product value chain is willing to pay for sensors, and who benefits from such sensors, if at all. In situations where the introduction of novel sensors may enable food producers better to characterise quality of their end products, this may lead to re-structuration of the supply relationship. A food producer may be able to harvest products with a high quality, but if the timing of this harvesting does not fit with the production planning of their customers or with the pre-set arrangements with their customers, they may not be able to sell their high quality products. To be of added value, food producers then may need to re-negotiate institutionalised trade relationships with other customers (see also Van de Poel [63] for discussion on battery cages and 'scratching eggs').

The competitive element is also visible in the application of sensors on packaging or containers that are visible for both the supplier and user, for instance a producer and a trader, or a retailer and a consumer. Customers may value such applications as it contributes to transparency of product quality. Suppliers do not always value such transparency as they are concerned that customers may not buy these products even if the product is still safe to consume, or that such improvement in transparency weakens their position in commercial negotiations. This is visible in, for instance, the reluctance of retailers to use sensor technologies on packaging. In their role as gatekeeper in the food chain, large retail chains have a powerful voice in formulating requirements regarding the application of such technologies [32]. The introduction of sensors at interfaces between actors in the supply chain then will likely include stretch strategies in societal embedding articulation as it would be accompanied with new trade practices.

What underlies both discussions is disagreement between supply chain partners about the roles of sensor technologies and power struggles. That is, the preference for quality improvement of products by one actor is not automatically a legitimate requirement according to other actors. The further question then is who will be able to win over the other actor. Hence, here the politics involved in demand articulation processes become apparent.

8.3.5 Certification and Standardisation in Societal Embedding Articulation: Fit or Stretch

In both sectors, trust in sensors by prospective users is an important issue, but is driven by different considerations. The water sector has had negative experiences with sensors in the past that has led to a cautious attitude towards sensors [64]. Validation contributes to trust in sensors and is considered to be very important in the drinking water sector. Certification is not yet a prominent theme (see also Van den Broeke et al. [64]), but might be in the future when sensors are considered to replace existing monitoring and measurement practices that are currently accepted by regulatory authorities.

In the food and beverages sector, certification is already a prominent theme, definitely when novel measurement technologies are applied for quality control of end products, but also for process control purposes. Food companies considering purchasing novel measurement technologies require their certification according to ISO standards. This is to show to and convince authorities, but in particular also their customers, that they are reliable partners. Buyers of produced food such as retailers can require food suppliers to comply with specific standards regarding food safety, in addition to regulations prescribed by law.[10] These customers may audit their suppliers and discuss quality assurance methods, including monitoring technologies, whereas this normally does not happen in the drinking water sector.

A general challenge for both domains is that certification will follow particular standards. Novel sensor technologies may however not easily be compared with existing standards because of different performance characteristics, requiring separate certification trajectories or the creation of new standards. Introducing novel standards implies stretch strategies in societal embedding articulation.

This discussion of demand articulation processes in our cases offers some background to the challenges actors face, and formed the starting point for our interactive CTA workshops.

8.4 Supporting Demand Articulation Processes via CTA Workshops

For the preparation of the workshops we worked with the diagnosis that demand articulation processes regarding sensor applications in the two sectors did not, or not easily, unfold and that this was, at least partially, due to lack of clarity regarding

---

[10]Examples of sector regulations include: the International Food Standard (IFS), the British Retail Consortium standard (BRC), the EN-ISO 22000. These regulations, as well as legislative requirements worldwide, make use of the Codex Alimentarius, a framework developed by the World Health Organization in co-operation with governments and branch organisations.

their future societal embedding. The objective of the workshops then was to explore requirements and preferences regarding societal embedding of sensors, including strategies to stimulate articulation of such requirements. To do so, we organised the workshops in three separate sessions. After a plenary introduction and keynote, we held two parallel sessions on the application of sensors for either drinking water or food and beverages applications. The first sessions focused on deepening understanding of user needs and underlying dynamics, whereas the second sessions explored challenges and routes to specify and realise these needs. In a final plenary session, participants ranked assessments and strategies regarding the application of sensors. To support the discussions, scenarios were developed which sketched different routes along which sensor applications could be explored and developed, anticipating developments in technologies, requirements of potential users and their interactions with stakeholders such as customers, consumers and regulatory authorities. The scenarios explored how the involvement of other actors than users or producers might support the further articulation and specification of preferences and requirements for novel sensor technologies.

During the workshops, participants discussed present day challenges and assessments of areas where novel sensors were expected to be applied. Participants discussed their considerations commented on and inquired about each other's considerations. This is a general feature of interactive sessions, but especially relevant here as the participants represent different actor positions in a sector and thus—as a group—are able to discuss demand articulation including societal embedding aspects thereof.

Participants in both sessions observed that possible users of sensor technologies do not always formulate clear needs. This was attributed to a lack of decisions regarding directions of applications or to their perceived lack of urgency regarding application of novel sensors. In both sector-specific workshops, participants expected that novel sensors were most likely to be introduced to optimise and automate production processes. This suggests, not surprisingly, that processes where actors follow fit strategies are likely to unfold more easily and are therefore more prominent as a governance pattern. This assumption was confirmed by participants' comments that sensors for measuring quality were more of a challenge. According to participants in the food sector this was due to ambivalences of companies regarding increased transparence of product quality enabled by such sensors, the importance of new commercial relationships for the diffusion of such sensors, and challenges in validation and certification. Participants in the drinking water session noted that the interpretation of results and added value of novel sensors such as in the distribution network was not clear, and that novel sensors would not be readily accepted by authorities as valid for the measurements prescribed by regulations.

The discussions in the workshops gave explicit attention to future developments and routes to advance sensor developments and demand articulation. Governance aspects such as regulation and standardisation, and the suggestion of the scenarios to involve external parties, were taken up in both workshops. Participants in the food session discussed in detail challenges of measuring in fluid and solid matrices and the difficulties of certifying novel sensors against existing reference norms. According to participants, the formulation of new standards could pave the way for the introduction of new sensors. They anticipated that this might lead to conflicts with laboratories who use the reference standards to their advantage. Interestingly, participants appreciatively noted that the formulation of standards would force the involved actors to become more specific about performance and requirements of such sensors. At this moment it is however unclear which norms exactly need to be developed. A possible solution, proposed by one of the participants, would be to develop different norms in parallel (cf. classic example of VHS versus Betamax).

The drinking water workshops discussed the involvement of governmental actors responsible for policies regarding regulation and inspection of drinking water quality. Existing regulations specifying monitoring programs and measurement methods were, to some extent, seen as a barrier to the introduction of sensors in cases were these sensors were to replace existing measurement approaches. According to participants, the drinking water sector needed to be pro-active and demonstrate the added value to authorities. Also here participants considered that the development of novel standards and certification schemes could play a positive role.

The workshop discussions highlighted that to further demand articulation processes, actions at the collective level were required. The introduction or modification of rules and standards was specifically mentioned as important as they would set conditions for demand articulation processes. During the ranking exercise of identified actions to stimulate demand articulation, participants considered this as more relevant than platforms for deliberation convening users, producers and other stakeholders. We see this as an affirmation of the importance of viewing demand articulation processes from a broader perspective by including aspects of societal embedding.

8.5 Conclusions

In this chapter, we suggested that demand articulation processes constitute an important element of the governance of an innovation journey. We understand demand articulation as a distributed process which is not confined to specifying requirements of immediate user groups in a more or less independent way. Rather, articulation of demands by users and other stakeholders is linked to further requirements for societal embedding of an innovation, such as the need to integrate an innovation into value chains or to adjust to or create regulation and standards. Our case studies showed that articulation of preferences and requirements regarding sensor technologies was not a simple task and that issues related to societal embedding were indeed part of the considerations and articulation processes of potential users and other stakeholders.

The case studies further showed that the types of actors and arenas which were relevant and prominent in demand articulation processes varied between sectors and led to different emphases. Whereas in the case of drinking water discussions regarding monitoring practices were prominent, commercial relationships were highlighted in the world of food and beverages. The food case showed how competitive dynamics made the formulation of user requirements regarding sensor applications more complex compared with the water case, due to different appreciations by actors along the value chain of the possible advantages and disadvantages of sensor applications.[11] These specific discussions are likely to guide the innovation process in these sectors in particular directions. The workshops corroborated these analyses and added further items. They especially contributed to articulating the stretch strategies in these domains. In the case of water, participants concluded that the formulation of requirements for novel sensor technologies required rethinking current paradigms in the sector regarding the quality and safety of drinking water. In the case of food, participants noted the importance of standards and certification during formulation of sensor specifications.

What does this tell us about governance of innovation by demand articulation processes and for possible interventions in such processes? A first insight is that governance by demand articulation processes reaches beyond guidance of technology design. It may also relate to the shaping of standardisation and regulation processes or value chains. As a consequence, uncertainties regarding preferences and requirements cannot be simply reduced by designer-user interaction, as their formulation is partially beyond the realm of envisioned users and producers. Secondly, the governance structures in particular domains influence demand articulation processes, by structuring who is involved, if and how they interact with other actors during such processes, and whose requirements may prevail in case of conflict. We speculate that in cases of stretch strategies, where co-ordination between actors will be important, co-ordination may be relatively easier in domains characterised by co-operative relationships than those where competitive relationships are dominant.

This discussion makes clear that actors aiming to intervene in the demand side benefit from identifying and assessing stretch strategies, and taking into account developments across different arenas of interaction, as well as domain specific dynamics. This broadened perspective on demand articulation processes then suggests considering interventions including other arenas in addition to users and designers. CTA offers tools to do so, as we demonstrated for our case studies. Even if demand articulation

---

[11]We expect similar discussions to occur within large integrated organisations such as the drinking water companies.

constitutes only a subset of processes in the innovation journey, due to its being interwoven with other processes it plays a significant role in the overall governance of innovation and therefore is worthy of closer attention.

Acknowledgements

The authors thank the editors for helpful suggestions and comments and the participants in the empirical studies for their contributions. This work is supported by NanoNextNL, a micro and nanotechnology consortium including 130 partners which is supported by the Dutch Government.

References

1. Deuten, J. J., A. Rip, and J. Jelsma (1997) Societal embedding and product creation management, *Technology Analysis & Strategic Management,* **9**(2), 131–148.
2. Van de Ven, A. H., D. E. Polley, R. Garud, and S. Venkataraman (1999) *The Innovation Journey* (Oxford University Press, Oxford).
3. Boon, W. P. C., E. H. Moors, S. Kuhlmann, and R. Smits (2011) Demand articulation in emerging technologies: Intermediary user organisations as co-producers, *Research Policy,* **40**, 242–252.
4. Lee, J. J., K. Gemba, and F. Kodama (2006) Analyzing the innovation process for environmental performance improvement, *Technological Forecasting & Social Change,* **73**, 290–301.
5. Di Stefano, G., A. Gambardella, and G. Verona (2012) Technology push and demand pull perspectives in innovation studies: Current findings and future research directions, *Research Policy,* **41**, 1283–1295.
6. Organisation for Economic Co-operation and Development (2011) *Demand-Side Innovation Policies* (OECD, Paris).
7. Edler, J. (2013) *Review of Policy Measures to Stimulate Private Demand for Innovation. Concepts and Effects* (University of Manchester, Manchester).
8. Bohlmann, J. D., J. Spanjol, W. J. Qualls, and J. A. Rosa (2013) The interplay of customer and product innovation dynamics: An exploratory study, *Journal of Product Innovation Management,* **30**(2), 228–244.

9. Joshi, A. W., and S. Sharma (2004) Customer knowledge development: Antecedents and impact on new product performance, *Journal of Marketing,* **68**(October), 47–59.
10. Lundvall, B.-A. (1988) Innovation as an interactive process: From user-producer interaction to the national system of innovation. In: Dosi, G., C. Freeman, R. Nelson, G. Silverberg, and L. Soete (eds.) *Technical Change and Economic Theory* (Pinter Publishers, London), pp. 349–369.
11. Bharadwaj, N., and Y. Dong (2014) Toward further understanding the market-sensing capability–value creation relationship, *Journal of Product Innovation Management,* **31**(4), 799–813.
12. Fontana, R., and M. Guerzoni (2008) Incentives and uncertainty: An empirical analysis of the impact of demand on innovation, *Cambridge Journal of Economics,* **32**(6), 927–946.
13. Parandian, A., and A. Rip (2013) Scenarios to explore the futures of the emerging technology of organic and large area electronics, *European Journal of Futures Research,* **1**(1), 1–18.
14. Voss, J.-P. (2007) Innovation processes in governance: The development of 'emissions trading' as a new policy instrument, *Science and Public Policy,* **34**(5), 329–343.
15. Rip, A. (2010) De facto governance of nanotechnologies. In: Goodwin, M., B.-J. Koops, and R. Leenes (eds.) *Dimensions of Technology Regulation* (Wolf Legal Publishers, Nijmegen), pp. 285–308.
16. Hauser, J. R., S. Dong, and M. Ding (2014) Self-reflection and articulated consumer preferences, *Journal of Product Innovation Management,* **31**(1), 17–32.
17. Mariampolski, H. (2001) *Qualitative Market Research: A Comprehensive Guide* (Sage, London).
18. Jiao, J., and C. H. Chen (2006) Customer requirement management in product development: A review of research issues, *Concurrent Engineering: Research and Applications,* **14**(3), 173–185.
19. Leonard-Barton, D. (1988) Implementation as mutual adaptation of technology and organization, *Research Policy,* **17**, 251–267.
20. Lundvall, B.-A., and A. L. Vinding (2004) Product innovation and economic theory: User-producer interaction in the learning economy. In: Christensen, J. L., and B.-A. Lundvall (eds.) *Product Innovation, Interactive Learning and Economic Performance* (Elsevier, Amsterdam), pp. 101–128.

21. Rip, A. (1995) Introduction of new technology: Making use of recent insights from sociology and economics of technology, *Technology Analysis & Strategic Management,* **7**(4), 417–431.
22. Fleck, J. (1988) *Innofusion or Diffusation? The Nature of Technological Development in Robotics* (vol. 4) (University of Edinburgh, Research Centre for Social Sciences, Edinburgh).
23. McMeekin, A. (2001) Shaping the selection environment: 'Chlorine in the dock'. In: Coombs, R., K. Green, A. Richards, and V. Walsh (eds.) *Technology and the Market: Demand, Users and Innovation* (Edward Elgar, Cheltenham), pp. 112–135.
24. Hoogma, R. (2000) *Exploiting Technological Niches: Strategies for Experimental Introduction of Electric Vehicles* (PhD Thesis, Twente University Press, Enschede).
25. Bower, J. L., and C. M. Christensen (1995) Disruptive technologies: Catching the wave, *Harvard Business Review,* **73**(1), 43–53.
26. Markides, C. (2006) Disruptive innovation: In need of a better theory, *Journal of Product Innovation Management,* **23**, 19–25.
27. Garud, R., C. Hardy, and S. Maguire (2007) Institutional entrepreneurship as embedded agency: An introduction to the Special Issue, *Organization Studies,* **28**(7), 957–969.
28. Munir, K. A., and N. Philips (2005) The birth of the 'Kodak moment': Institutional entrepreneurship and the adoption of new technologies, *Organization Studies,* **26**(11), 1665–1687.
29. Tripsas, M. (2008) Customer preference discontinuities: A trigger for radical technological change, *Managerial and Decisions Economics,* **29**, 79–97.
30. Garud, R., S. Jain A. Kumaraswamy (2002) Institutional entrepreneurship in the sponsorship of common technological standards: The case of Sun Microsystems and Java, *Academy of Management Journal,* **45**(1), 196–214.
31. Maguire, S., C. Hardy, and T. B. Lawrence (2004) Institutional entrepreneurship in emerging fields: HIV/AIDS treatment advocacy in Canada, *Academy of Management Journal,* **47**(5), 657–679.
32. Te Kulve, H. (2010) Emerging technologies and waiting games: Institutional entrepreneurs around nanotechnology in the food packaging sector, *Science, Technology & Innovation Studies,* **6**(1), 7–31.
33. Benz, A. (2007) Governance in connected arenas: Political science analysis of coordination and control in complex rule systems. In:

Jansen, D. (ed.) *New Forms of Governance in Research Organizations: Disciplinary Approaches, Interfaces and Integration* (Springer, London), pp. 3–22.

34. Garud, R., and P. Karnøe (2003) Bricolage versus breakthrough: Distributed and embedded agency in technology entrepreneurship, *Research Policy*, **32**, 277–300.

35. Schot, J., and A. Albert de la Bruhèze (2003) The mediated design of products, consumption, and consumers in the twentieth century. In: Oudshoorn, N., and T. Pinch (eds.) *How Users Matter: The Co-Construction of Users and Technologies* (The MIT Press, Cambridge), pp. 229–245.

36. Boon, W. P. C., E. H. M. Moors, S. Kuhlmann, and R. Smits (2008) Demand articulation in intermediary organisations: The case of orphan drugs in the Netherlands, *Technological Forecasting & Social Change*, **75**, 644–671.

37. Klerkx, L., and C. Leeuwis (2008) Institutionalizing end-user demand steering in agricultural R&D: Farmer levy funding of R&D in the Netherlands, *Research Policy*, **37**, 460–472.

38. Roelofsen, A., W. P. C. Boon, R. R. Kloet, and J. E. W. Broerse (2011) Stakeholder interaction within research consortia on emerging technologies: Learning how and what?, *Research Policy*, **40**(3), 341–354.

39. Nahuis, R., E. H. Moors, and R. E. Smits (2012) User producer interaction in context, *Technological Forecasting and Social Change*, **79**(6), 1121–1134.

40. Van de Poel, I. (2003) The transformation of technological regimes, *Research Policy*, **32**, 49–68.

41. Teubal, M. (1979) On user needs and need determination: Aspects of the theory of technological innovation. In: Baker, M. J. (ed.) *Industrial Innovation: Technology, Policy, Diffusion* (MacMillan Press, London), pp. 266–289.

42. Clark, K. B. (1985) The interaction of design hierarchies and market concepts in technological evolution, *Research Policy*, **14**, 235–251.

43. Rip, A., and R. Kemp (1998) Technological change. In: Rayner, S., and E. L. Malone (eds.) *Human Choice and Climate Change: Resources and Technology* (Batelle Press, Columbus).

44. Geels, F. (2004) From sectoral systems of innovation to socio-technical systems Insights about dynamics and change from sociology and institutional theory, *Research Policy*, **33**, 897–920.

45. Borup, M., N. Brown, K. Konrad, and H. van Lente (2006) The sociology of expectations in science and technology, *Technology Analysis & Strategic Management,* **18**(3/4), 285–298.

46. van Lente, H., and A. Rip (1998) The rise of membrane technology: From rhetorics to social reality, *Social Studies of Science,* **28**(2), 221–254.

47. Konrad, K. (2008) Dynamics of type-based scenarios of use: Opening processes in early phases of interactive television and electronic marketplaces, *Science Studies,* **2**(2), 3–26.

48. Martin, P. (2001) Great expectations: The construction of markets, products and user needs during the early development of gene therapy in the USA. In: Coombs, R., K. Green, A. Richards, and V. Walsh (eds.) *Technology and the Market: Demand, Users and Innovation* (Edward Elgar, Cheltenham), pp. 38–67.

49. Hutchby, I. (2001) Technologies, texts and affordances, *Sociology,* **35**(2), 441–456.

50. Dell'Era, C., and P. Landoni (2014) Living lab: A methodology between user-centred design and participatory design, *Creativity and Innovation Management,* **23**(2), 137–154.

51. Rip, A., and H. Te Kulve (2008) Constructive technology assessment and socio-technical scenarios. In: Fisher, E., C. Selin, and J. M. Wetmore (eds.) *The Yearbook of Nanotechnology in Society, Volume 1: Presenting Futures* (Springer, New York), pp. 49–70.

52. Te Kulve, H., and A. Rip (2011) Constructing productive engagement: Pre-engagement tools for emerging technologies, *Science and Engineering Ethics,* **17**(4), 699–714.

53. Schot, J., and A. Rip (1997) The past and future of constructive technology assessment, *Technological Forecasting and Social Change,* **54**(2/3), 251–268.

54. Robinson, D. K. R., and M. Morrison (2011) Nanotechnologies for improving food quality, safety, and security. In: Frewer, L. J., W. Norde, A. Fischer, and F. Kampers (eds.) *Nanotechnology in the Agri-Food Sector: Implications for the Future* (Wiley-VCH Verlag GmbH & Co. KGaA, Weinheim), pp. 107–126.

55. Van Merkerk, R. O. (2007) *Intervening in Emerging Nanotechnologies. A CTA of Lab-on-a-Chip Technology* (PhD thesis, Utrecht University, Utrecht).

56. Parandian, A. (2012) *Constructive TA of Newly Emerging Technologies: Stimulating Learning by Anticipation through Bridging Events* (PhD thesis, Delft University of Technology, Delft).

57. Krabbenborg, L. (2013) Dramatic rehearsal on the societal embedding of the lithium chip. In: Van der Burg, S., and T. Swierstra (eds.) *Ethics on the Lab Floor* (Palgrave Macmillan, New York), pp. 168–187.

58. Organisation for Economic Co-operation and Development (2011) *Fostering Nanotechnology to Address Global Challenges: Water* (OECD, Paris).

59. Posthuma-Trumpie, G. A., and A. Van Amerongen (2011) Using nanoparticles in agricultural and food diagnostics. In: Frewer, L. J., W. Norde, A. Fischer, and F. Kampers (eds.) *Nanotechnology in the Agri-Food Sector: Implications for the Future* (Wiley-VCH Verlag GmbH & Co. KGaA, Weinheim), pp. 75–87.

60. Te Kulve, H., K. Konrad, C. Alvial Palavicino, and B. Walhout (2013) Context matters: Promises and concerns regarding nanotechnologies for water and food applications, *NanoEthics*, **7**(1), 17–27.

61. ETC Group (2004) *Down on the Farm: The Impact of Nano-scale Technologies on Food and Agriculture* (ETC Group, Ottawa).

62. Smit, W. A., B. Elzen, and B. Enserink (1998) Coordination in military socio-technical networks: Military needs, requirements and guiding principles. In C. Disco, and B. Van der Meulen (eds.) *Getting New Technologies Together: Studies in Making Sociotechnical Order* (Walter de Gruyter, Berlin), pp. 71–105.

63. Van de Poel, I. (1998) Why are Chickens housed in battery cages? In: Disco, C., and B. Van der Meulen (eds.) *Getting New Technologies Together: Studies in Making Sociotechnical order* (Walter de Gruyter, Berlin), pp. 143–177.

64. Van den Broeke, J., C. Carpentier, C. Moore, L. Carswell, J. Jonsson et al. (2014) *Compendium of Sensors and Monitors and Their Use in the Global Water Industry* (Water Environment Research Foundation, Alexandria).

Chapter 9

Evolving Patterns of Governance of, and by, Expectations: The Graphene Hype Wave

Kornelia Konrad[a] and Carla Alvial Palavicino[b]

[a]*Department of Science, Technology and Policy Studies,*
School of Management and Governance, University of Twente,
7500 AE Enschede, The Netherlands
[b]*NUMIES, Alameda 1845, dpto 301, Santiago, Chile*

k.e.konrad@utwente.nl, carla.alvial@gmail.com

9.1 Introduction

Anticipation in the form of expectations, visions, scenarios, market forecasts, foresight and impact assessments is a pervasive element in the governance of new and emerging science and technologies [1–3]. Broadly shared expectations on promising technology fields and applications mobilise and guide research and innovation actors, visions and scenarios sketch how a future world may look and feel, public funders draw on foresight processes to prioritise the most promising fields, anticipatory impact assessments inform regulatory action, roadmaps specify tasks and distribute roles in order to coordinate further action, and industry actors refer to market forecasts to decide if and when a field may become strategically relevant to them.

There is increasing recognition of the importance of anticipatory governance in a wide range of academic fields, in particular around newly emerging science and technology, but also beyond [4–8]. In fact, it has been claimed that anticipation and anticipatory governance are becoming institutionalised as a facet of governance [2]. In parallel, there has emerged a large strand of research, strongly related to science and technology studies, which investigates the strategic and performative role of expectations in science and technology [9–12].

Taking the general importance of future-orientation for the governance of (new and emerging) science and technology as a backdrop, this chapter aims at understanding concrete modalities of future-orientation in a wide range of forms as different modes of governance by expectations. We would like to emphasise that we do not see such modalities purely instrumentally, as a governance tool applicable at any time and place. Rather, they are forms of anticipatory practices that are part of an evolving governance structure in a particular field, time and place. As such, anticipation in the form of expectations, visions, etc. themselves may be governed, or at least shaped by the evolving governance structures. Thus, there is also governance of expectations, so to say, which is a key part of the dynamics.[1]

Expectations, just as many of the other forms of future representations, partly overlapping with expectations, are not static; they evolve over time. By expectations we refer to specific assumptions—as statements or embodied in other forms visible to the outside world—about future conditions or developments assigned a certain likelihood—in our context about and related to science and technology [14]. This means that some of the above mentioned forms of anticipation as visions or roadmaps often include expectations, though not necessarily: for instance, a vision may come without further specification of its likelihood. A common and well-known phenomenon related to expectations are hypes: promises, that is optimistic expectations on the potential of particular technologies which nevertheless require work to be done, surge quickly, but often turn out as overly optimistic, resulting in some form of disappointment [15]. The notion of hype cycles has become a common term in the academic

---

[1] The concept of governance of and by expectations was first presented by one of the authors in a paper presented at the *EASST 2010 conference* [13].

world [16, 17], the world of consultants [18] and industry [19, 20] and among scientists [21–24]. Expecting hype cycles to occur has turned into a sort of second-order expectation or folk theory [25].

In this chapter, we further suggest that thinking of hypes as a hype wave rather than a cycle can be a fruitful shift in perspective and we will illustrate this for the case of graphene, a new and—as it has often been called—'miracle' material. Hype may not occur for all actor groups at the same time, but rather evolves, as a wave moves through time and spaces. The wave may follow different dynamics in different communities—just as a physical wave may do when moving through different spaces and materials. Related to this, the forms of anticipations and anticipatory practices, the particular modes of governance and governance effects may evolve as well. More specifically, we will argue that governance of and governance by expectation reflexively relate to and build on each other; a dynamic which leads to governance patterns evolving over time.

In the remainder of the chapter, we will first present the conceptual frame of governance of and by expectations. We will then follow how both governance of and by expectations have evolved in the field of graphene, a field that emerged about a decade ago in the broader realm of nanotechnologies. We show how the rise of graphene, as a techno-scientific field, has been accompanied by promises and expectations leading to a hype dynamic. This 'hype' moves as a wave through different spaces: originating in science, then linking and moving to policy, to the market, and ultimately creating repercussions in the regulatory space. Each 'space' is characterised by a set of actors, institutions, and anticipatory practices, by specific rules and governance arrangements [26, 27]. These spaces are not static, but they evolve over time, expanding the boundaries, incorporating new actors, etc.

Methodologically, this chapter is based on an in-depth case study in the development of expectations on graphene, ways of producing expectations[2] and their interaction with the development of the field, carried out from the end of 2012 to the first half of 2014. In total, 29 semi-structured interviews

---

[2]The particular anticipatory practices involved in expectation building have been investigated in more detail elsewhere [28].

with graphene actors (EU and US) were conducted following a snowball sampling, covering: graphene scientists ($n = 9$), science journalists ($n = 2$) research program coordinators ($n = 3$), policy actors ($n = 2$), consultants ($n = 3$), venture capitalists ($n = 2$), graphene company CEOs ($n = 6$), website managers ($n = 2$), and a standards organisation ($n = 1$). Interviews ranged from 40 min to 1 hour. One of us attended three graphene conferences (two scientific conferences and one business meeting). This was complemented with an extensive document analysis (scientific publications, press releases, interviews and videos, roadmaps, research proposals), selected on the basis of themes raised in the interviews and in relevant social media. It is important to note that this account of the graphene field reflects mostly the developments in Europe and United States, excluding important developments in Asia.

9.2 Governance of, and by, Expectations

9.2.1 Modes of Governance

In general, the concept of governance draws attention to the different modes or institutional rules of coordination among individuals, organisations, societal subsystems and states, ranging from hierarchical steering to networks, communities, associations, and market-like forms of coordination organised by both formal and informal rules [29, 30]. Accordingly, the concept of governance of and by expectations is meant to capture the different modes how expectations contribute to the coordination of innovation processes (governance by expectations) and the different modes how expectations themselves are coordinated among individuals, organisations, communities and arenas (governance of expectations). Furthermore, related to the modes or ways of governance, attention is raised for the sort of actors and institutional arrangements contributing to or being influenced by expectations.

Taking up a common distinction of modes of governance, we can say that expectations may be coordinated in a 'market-like' manner, when various actors voice expectations that others may accept, modify or contest, up to a competition for the most spectacular promises. These 'markets' of expectations [31] are ordered by particular structures and institutions, for instance the specific constellation of actors, communication platforms (e.g. journals, conferences), formats (e.g. types of articles, reports, presentations) and rules of communication within specific discourse arenas, such as scientific, public or policy discourses. Furthermore, over the last decades very concrete expectation markets have emerged with expectations being commercially produced, and packaged into particular products, such as market reports, and traded by consultancies and forecasting institutes [32].

Networks as modes of governance become important, when expectations are negotiated among the participants of networks and communities following more or less codified rules. For instance, foresight and roadmapping processes may show this characteristic, when following rather clearly delimited processes with regard to participating actors and duration over time, and according to specified procedures. While a fully hierarchical 'prescription' of expectations seems to be a somewhat unlikely case, some actors and organisations may hold a prominent or even dominant position in shaping collective expectations (cf. [32]) for the role of the Gartner Group in the ICT world. As with governance arrangements in general, these different modes of governance are not mutually distinctive, but may interrelate [29].

Governance *by* expectations encompasses different modes and different forms of coordination as well, exhibiting different degrees of bindingness. There is the comparatively 'soft' mode of coordination of actor strategies and actor constellations at times when expectations are rather fluid, e.g. in emerging technological fields, when actors reciprocally position themselves by way of discursively exchanging and mutually adapting expectations, which may result in the emergence of patterns and paths [33].

Expectations become more binding, when certain expectations turn into institutionalised, collective expectations, which are part of a debated or taken-for-granted social repertoire, which constitutes a reference point for actors who feel a certain pressure to respond to them [34]. Expectations may become further institutionalised and solidified, if they are taken up in formal institutional arrangements, for instance when integrated as requirements and yardsticks into funding schemes [11] or as highly organised roadmaps in industry. A well-known example for this process is Moore's law—the expectation of a continuous increase of performance of chips—which turned from a retrospective observation, to a broadly shared expectation and requirement [35] to the well-organised industry-wide roadmapping process it is now [36, 37].

Governance of, and by, expectation are not independent, but are more or less directly linked. A number of anticipatory governance tools, such as foresight, aim to coordinate and shape expectations of diverse actors and at the same time create governance effects, e.g. for an innovation field.

In a less direct way, we may think of both aspects as reflexively linked. As implied by the concept of governance by expectations, expectations contribute to coordinating actors, and shaping strategies and institutional arrangements, and at the same time the form and content of expectations depends on particular actors, strategies and institutional arrangements, as implied by the notion of governance of expectations. We can, furthermore, assume that there is some, though only partial, overlap between those actors and communities who contribute to expectation building and those actors who are influenced by the resulting, more or less shared expectations and sometimes hype dynamics. For instance, some innovation actors will be quite active in expectation work, while others less, and some actors, as journalists and consultants, are involved in expectation work, yet usually not in innovation activities [11].

As an implication, expectations that emerged within a societal domain may reflexively feedback on the actors and structure that shaped them, and thus actor constellations, actor strategies and structures may change affecting the further process of expectation building and coordination.[3] In addition, more actors may get involved, or new types and communities of actors may become interested. As a result, expectations may grow, up to hype, spread to new communities, shift in focus, and thus lead to the phenomenon of a hype wave, introduced above.

---

[3]For discourse in general, Hajer [38] has described similar dynamics as institutionalisation of discourses, that is, the translation of a discourse into institutional arrangements.

9.2.2 Intentional and De Facto Governance

It is important to recognise that in addition to the more common understanding of governance as intentional attempts at governance, there is de facto governance. De facto governance refers to the patterns and structures of coordination that emerge largely non-intentionally from the interaction of many actors, through mutual dependencies of perspectives and actions [39, 40]. Intentional governance and de facto governance are not independent of each other. In fact, intentional governance is one element in the overall emerging and stabilising de facto governance. Actors can be reflexive about this and take into account the dynamics of evolving de facto governance when considering intentional interventions in order to be more effective. For expectations, both are visible. Social dynamics of expectations are the result of strategic 'discourse activities' of many actors [11, 41, 42], yet the actual outcome can hardly be controlled by anyone, and actors can adapt their strategies if the aggregated outcome of their own and others activities, e.g. hype dynamics, evolves in unforeseen ways [11].

Moreover, increasingly attempts at creating more structured and formalised forms of expectation building aimed at supporting policy processes and strategy formation in firms have emerged over the last decades, indicating that the more expectations are recognised as part of the de facto governance structure, the more they are integrated into intentional governance. This tendency is obvious in the vast array of foresight processes that has emerged in the last decades as a dedicated governance tool at various policy levels [43] and also in the corporate world [44]. Which collective expectations will eventually emerge, become dominant or contested, depends on the interplay of intentional strategies, formal processes and aggregated social dynamics [45].

Different ways how expectations coordinate and 'govern' innovation processes have been highlighted in a strand of research within science and technology studies, the sociology of expectations, which considers specifically the performative role of expectations in science, technology and innovation, that is, governance *by* expectations. Expectations mobilise innovation actors to enter new fields, legitimate the allocation of resources, give definition to roles for various actors, structure the societal debate on new technologies [12, 47], guide the interpretation of novel technologies [14, 20], and support the formation of new techno-scientific fields [48–50]. As a result, expectations channel efforts into certain directions and contribute to the emergence and stabilisation of socio-technical structures, institutional arrangements and paths—the actual directions taken may well deviate from early expectations though, thus pointing to the importance of de facto governance [51–53].

An explicit consideration of different governance modes *of* expectations and their specific role, such as competitive voicing of expectations or negotiation and adjustment within networks and communities [11], or more or less formalised forms of expectation building has been less of a focus in this literature. What has emerged in recent years is a number of studies which investigate a variety of anticipatory practices contributing to the shaping and spreading of expectations in different ways [28, 54]. These practices and their specific performative effects have been studied for instance for the work of consultants [32], online news providers [55], or venture capitalists [56]. More formal 'tools' for expectation building have been extensively addressed in the literature on foresight [57–60].

In recent years, roadmaps are becoming more and more popular, not only as a strategic instrument at the level of firms, but as governance tools deployed by policy actors for addressing the development of industries or fostering socio-technical transitions [61], or vice versa as a tool deployed by a set of actors in a field to persuade policy [62]. In this chapter, we will pay attention to the broad variety of anticipatory practices and their products, either explicitly aimed at expectation building, such as roadmaps or market forecasts, or practices which embody expectations in a more implicit form, such as the creation of spin-off companies or patents. The latter are not seldom based on expectations and promises as well, but exactly for this reason are perceived by others as supporting the very promises they are based on. We would like to stress that foresight processes and other anticipatory practices should not be considered as more or less isolated processes, but rather be seen as local sites of expectation building within broader societal discourses, or—put differently—as embedded in the larger 'sea of expectations' [12, 45, 63] various interactions taking place between these different forms of expectation building, which will mitigate the type of governance effects they may create.

Therefore, in the following study of expectations and their relation to emerging governance patterns in graphene, we will consider both formal and informal expectation building processes. We will trace how expectations related to graphene have been produced and governed by an evolving set of actors and anticipatory practices and if and how this interacts with the evolvement of governance patterns in the field more generally. We will, furthermore, show how in this process graphene expectations and eventually hype move and expand through different spaces.

9.3 The Graphene Hype Wave

Graphene, the so called 'miracle material' [65], has generated high expectations since its first isolation in 2004. Graphene, which consists of a single layer of carbon atoms, presents exceptional thermal, electrical and strength properties, and promises to enable disruptive innovation via a wide range of applications, from the next wave of electronics, to energy storage and water purification [66].

The expectations about the possibilities opened by graphene have played an essential role in shaping the field, e.g. the actor constellations, the resources available, and the research directions taken. In this section, we show how governance patterns evolve in interaction with expectation dynamics, including hype. We follow this process across different spaces [26], which are each characterised by specific institutions, rules and anticipatory practices. Graphene expectations largely originated in the science space, but soon linked up with and moved to other spaces as well. While there is a temporal order in the way expectations and hype moved across these spaces, which is also reflected in the structure of this section, spaces and expectation dynamics are interlinked and feedback on each other.

9.3.1 A Graphene Hype Emerging in the Science Space

Graphene as a material and as a research field emerged about a decade ago (2004) in the 'science space'. Already the beginnings of graphene were characterised by a great enthusiasm from the scientific community, led by the charismatic figure of Andre Geim. Geim, a Russian scientist working at Manchester University (UK) became the spokesperson of the field. He was one of the discoverers of graphene, and well known in the physics community before. Geim and his collaborator Konstantin Novoselov isolated graphene in 2004, and they eventually also succeeded in publishing their results in one of the most prestigious journals 'Science' and later on in various *Nature* publications.[4] Graphene was quickly framed under the overarching promise of the 'next big thing' in electronics, providing a response to the expected end of Moore's law, which predicted that silicon-based electronics would soon reach its limitations in achieving continuous miniaturisation and performance increase [65, 67, 68].

The enthusiasm grew fast in the scientific community, creating a whole new field of research. This was facilitated by a relatively easy and cheap method of isolation of graphene (mechanical exfoliation or scotch tape method) (see Novoselov et al. [45]) (Interview 4, standards organization, January 2014) but also by the circulation of expectations mediated by high impact journals.

In a few years, the number of publications and patents increased exponentially both in quantity and diversity [69] (see Fig. 9.1), and the number of people interested in the material grew exponentially as well (Interview 3, graphene flagship coordinator, May 2013). Graphene became the centre of attention of many scientific conferences, up to the point of being considered a hype [71] and becoming a buzzword for publications in high impact journals (Interview 17, graphene scientist, February 2013).

This emergent community was composed of scientists—mostly physicists—and some industrial actors, loosely organised around conferences and some funding schemes (e.g., Eurographene). Expectations about graphene were circulated through scientific media such as scientific publications, some scientific blogs and in increasingly popular scientific conferences dedicated to graphene. As we have described elsewhere [28], high impact journals, in particular *Nature*, engaged actively in shaping expectations and building a community around graphene

---

[4]An extensive discussion of the anecdote around this publication can be found in Geim's Nobel Lecture. Available at: http://www.nobelprize.org/nobel_prizes/physics/laureates/2010/geim_lecture.pdf.

by editorials and other highly visible publication formats, which played an important role in defining the promises, expectations, challenges and requirements for graphene.

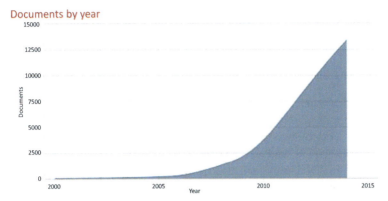

Figure 9.1 Steep increase of scientific publications using the word graphene (2000–2014). *Source*: Scopus, generated by the author.

As explained by a *Nature* editor, graphene was from its beginnings compared to carbon nanotubes in terms of its properties and potential applications (Interview, journal editor, July 2013). One of the biggest problems associated with carbon nanotubes was their production. For this reason, producing graphene at a reasonable scale and price for applications was presented as a requirement from early on. To address this issue, the research community had to expand from physics into chemistry and engineering, among other areas. In this process of expanding the disciplinary range, *Nature* took an active role via news and editorial notes targeted at different disciplines, and thus spreading the expectations of graphene (Interview, science journalist, June 2013).

Since its inception, the graphene field has been overflowing with speculations of the possible applications of graphene. Related to this, academic as well as private actors engaged in patent applications, leading to an overcrowded patent landscape [72]. Between 2008 and 2010 we see a number of 'design concept' videos on flexible electronics such as foldable screens, released by large firms as Nokia and Samsung and taken up in the media, indicating not only the emerging link to industry, but also early connections to the media and public.

At this stage (around 2009–2010), two important changes in the structure of the field had occurred. The first was the expansion of the graphene scientific community from a (solid-state and theoretical) physics community into other academic fields, which ranged from chemists to biologists and material scientists, making graphene a truly interdisciplinary field. This expansion was facilitated by the 'scotch tape' method as well as expectations circulated in high-level journals. Second, framing these expectations more concretely into highly relevant application domains such as electronics and linking with the expectation of the end of Moore's law attracted the early engagement of two large electronics companies, particularly research done at IBM and Intel [73] and the creation of a few graphene companies, that perceived opportunities in selling the material or graphene enabled applications. The presence of industrial actors then further reinforced ideas of the applicability and profitability of graphene, even if the landscape was still dominated by scientific actors and a science-oriented discourse.

The governance of expectations in the graphene field at this stage comes close to market-style coordination within the scientific community and increasingly also some industry actors, with a hierarchical element of some particularly relevant actors, such as renowned scientists and research labs of large industries, who have a prominent position in the definition of the promises of the field. These expectations are mediated by scientific conferences and journals, but there are no explicit attempts at coordination yet.

9.3.2 Graphene Moves into the Policy and Media Space

At the end of the decade, graphene received increasing attention from policy, thus entering the 'policy space'. In 2010, a group of European scientists started to organise around the application process for the FET (Future Emerging Technologies) Flagship, a new funding scheme that at that time was launched by the European Commission [74]. This scheme was aimed at supporting high risk, excellent science with the potential for solving societal challenges. In October of the same year, Geim and Novoselov received the Nobel Prize for their groundbreaking experiments with graphene. Although this was not the first time a Nobel

Prize was granted shortly after a discovery, the announcement was received with some surprise. While the importance of the scientific discovery was not questioned as such, the promises of graphene were seen as a possible further element taken into account by the Nobel committee [75]. This is indicated by the Nobel Prize account in Twitter that pointed out the possible applications of graphene when announcing the prize in October 2010 (Fig. 9.2).

Figure 9.2 Tweet from the feed of the Nobel Prize organisation account, highlighting potential applications of graphene. This was following the announcement of the prize in October 2010.

The Nobel Prize worked as a sort of catalyst for the graphene hype and its expansion into policy and media, and likely also for the successful application for the flagship programme (Fig. 9.3). It was a media boost for graphene. Andre Geim, as the spokesperson for the field, was invited to speak in relation to disruptive innovation (and graphene) to high-level policy circles, including the European Parliament and the World Economic Forum. In these settings, Geim positioned graphene as the next disruptive innovation [76, 77]. The prize had immediate effects in the field, since it served as a push for the UK government to provide substantial funding to Geim and his team, creating the National Graphene Institute [78, 79]. Around this time, in the United States the National Science Foundation increased the funding for research in graphene [80]. Hence, we may say, that the Nobel Prize, and the heightened expectations related to it, supported the further institutionalisation of the field.

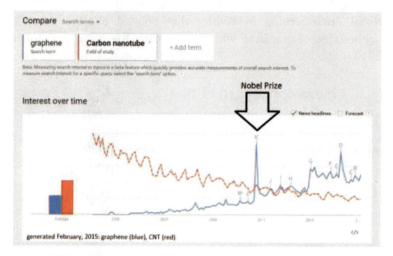

Figure 9.3 The effect of the Nobel Prize on the attention to the graphene field. The arrow indicates the moment the prize was awarded.

Supposedly, the Nobel Prize supported the application for the flagship, but the application process also required a major coordination effort within the European graphene community, in which the coordination of expectations via a roadmap process played an important role. During the process of application, from 2010 until January 2013, when the project was eventually granted, scientists and companies across Europe engaged in a series of national and supranational activities, in order to coordinate the expectations about the future of graphene. This collective effort, a form of governance of expectations, led to the development of a graphene roadmap, one of the deliverables required by the application scheme [81, 82]. This coordination process started with a series of meetings at national and European level [83, 84].

While these activities strengthened the network of graphene researchers, it also further shaped, coordinated and institutionalised expectations about the possibilities of the material. The roadmapping process was based on a Web-based, open consultation platform, through which hundreds of suggestions were received. From these, a committee selected those that would make it to the final version of the roadmap. Not only was this a selection of ideas, but also it determined which groups were in and out, respectively had better chances to acquire funding for their research (Interview 3, graphene flagship member, May 2013, and Interview 17, graphene scientist, February 2013).

As a result, the governance of expectations, but also the graphene field itself changed. The strong internal coordination of the European scientific community through the flagship amounted to a more hierarchical mode of governance with more pronounced distinctions in influence. Also, the type of actors involved in the field started to change considerably, including more and more policy and market actors (we will return to the latter in the following section). A distinction was created between those inside and outside the flagship, and those with more or less influence to define collective expectations than others. Thus it seems fair to say that the specific anticipatory practice of the roadmapping process contributed significantly to the configuration of the graphene field. We suggest that this process led to an increasing institutionalisation and structuration of the field. The flagship became a, if not the legitimate institution to produce graphene expectations in Europe, and other actors would define themselves in relation to how much they align with the flagship ideas (Interview, graphene scientist, February 2013). However, legitimising the central position of the flagship was an iterative process, and in many scientific conferences specific sessions were organised around the flagship, where expectations about the funding scheme, changes in governance and the position of European scientists were discussed [85].

This roadmap process can be interpreted as an intentional attempt at governance of and by graphene expectations, since the flagship coordinators aimed at making their roadmap a coordinating device across the whole field, and therefore—emulating a well-known roadmapping process—made it openly accessible (Interview 3 and Interview 12, graphene scientists). The roadmapping process followed a particular (anticipatory) governance model, that of the ITRS, the roadmap developed within the semiconductors industry to coordinate development efforts mentioned earlier. This transfer of an anticipatory governance tool could only succeed to some degree, given that the graphene field lacks the further context of the semiconductor industry, which enables the strict orientation to the ITRS roadmap, and also an overarching expectation, which is as firmly grounded as Moore's law. Thus, the governance effects are likely to turn out differently. Still, we may speculate that drawing on the ITRS as a well-known and widely acknowledged roadmapping approach closely linked to central expectations in the field created legitimation for the process and the methodology chosen to create the flagship roadmap. The fact that it resembled the ITRS, particularly in its openness, and that it was a collective effort helped to articulate the graphene community by giving it an identity through the roadmap (Interview 12, graphene researcher, June 2013, and Interview 12, graphene company CEO, March 2013).

At the same time, graphene developed at the global level. There were strong investments in the United States, Korea and China, and the Flagship coordinators positioned the project in the global context, both in terms of competition with emergent players and by embedding it in cooperation networks, as for instance reflected by the international strategic advisory board of the project.[5]

9.3.3 Graphene Moving into the Market Space and the Hype Becoming Reflexive

At the end of the decade, the number of private companies and startups involved in graphene was growing, indicating the take-up and move of graphene expectations in the 'Market Space'. Starting from 2008, but more clearly from 2010 on, an increasing number of spin offs were established. These companies sold graphene of different qualities and properties, or developed specific graphene-related applications.[6] However, the market demand for graphene, at least until 2013, was mainly coming from research. The material was sold to laboratories for research purposes,

---

[5]The flagship Strategic Advisory Council includes Dr. Luigi Colombo from Texas Instruments (US) and Prof. Byung Hee Hong from Seoul National University (Korea). This board was originally presented in a world map, to emphasise its international character.

[6]For an up-to-date overview of existing graphene companies, see Graphene Tracker, http://www.graphenetracker.com/companies/. The website graphenetracker. com, which follows market developments in graphene, counts about 70 graphene companies to the date.

since there was no concrete application beyond the prototype stage.[7] Still, the increasing number of companies fuelled and supported the claims about the promises of graphene, now also increasingly taken up in the media, and promoted the entrance of financial actors into the field [86].

At this stage, the graphene field looked considerably different from its origins in 2004. The European scientific community was highly coordinated via the graphene flagship, and framed its research more and more not only in terms of scientific but also societal challenges. In addition, a number of spinoffs, industrial firms and consultancy organisations entered the field. The latter represent a special type of expectation actors—intermediary organisations—that interpret and assess ongoing expectations and hype dynamics, but in so doing contribute to expectation building as well. In particular, in the graphene field Cientifica, IDTechEx, Lux Research and BCC were among the most quoted consultancies.[8]

Here we would like to highlight the ambiguous role of consultants in governing expectations, following a sort of double-sided strategy of expectation management. More and more, investors, companies and investment media wrote and talked about the promises of graphene, thus catching the public and investor's attention and actually leading to a huge interest in investing in graphene [87, 88]. At the same time, consultancy organisations attempted to assess this growing enthusiasm and go 'against the hype' by showing what they called 'realistic' expectations, promises that were produced and assessed by specific anticipatory practices such as market forecasts, market assessment reports, and expert assessments [89]. They spread these assessments through conferences, blogs, sample reports and newsletters. Most of these consultancies build their business on selling these reports, providing expert advice to companies and investors, and organising conferences. Thus, we see them

---

[7]One exception to this could be Vorbeck Materials and Graphene Labs. They focused on the low-end graphene applications (conductive inks, energy storage, 3D printing ink) and have some products in the market.

[8]IDTechEx analysts were often invited to speak in conferences (in addition to the ones they themselves organised) as experts in the graphene field. They were received positively by other actors in the field (company CEOs and scientists) who acknowledged the expertise of these organisations in the graphene field, considering their assessments accurate and legitimate.

striking a balance between on the one hand keeping expectations about graphene high, while at the same time providing critical statements about what seems to be overpromising. In doing so, they help to articulate the idea that a 'hype cycle' is happening, implicitly suggesting that disappointment, but eventually also more realistic expectations and applications may follow.

Consultancies also organised conferences, which were not limited geographically. For example, IDTechEx has organised the 'Graphene LIVE!' conferences in Berlin and Santa Barbara, California.[9] In so doing, they helped to develop the graphene market—and hype—as a global phenomenon. In contrast to the science and policy space, which were more confined geographically (particularly because the flagship is an EU project), for the graphene market geographical distinctions were secondary, and consultancy organisations had an important role in connecting expectations from different geographical spaces, and communities.

These two trends led to what we could characterise, in terms of governance modes, a return to a market mode of governance, yet with a broader constituency and attempts at careful orchestration. There were various actors strategically voicing, assessing and contesting expectations and actually selling these expectations, e.g. in the form of market reports. Around 2013 and when consultancies had positioned themselves with an active role around graphene, the 'hype-cycle' became a shared belief, and actors in the graphene field considered the need to modulate expectation dynamics (Fig. 9.4).

The Flagship and other associations, such as the Graphene Stakeholders Association (GSA) in the US, worked towards coordinating graphene actors in order to respond and avoid disappointment after the hype.[10] In particular, the GSA organised private and open meetings (so called 'summits') with market, business and policy actors; but also started to directly lobby the US government, standards organisations and nano-industry associations to push for graphene opportunities. An important aspect was the development of a strategic partnership with the National Physical Laboratory in the UK, to produce reliable graphene standards [90]. The GSA saw the development of standards as a

---

[9]The first edition of Graphene LIVE! was in 2012, and it was co-located with the Printed Electronics Show organised also by IDTechEx.

[10]A similar development has been described for a former hype around fuel cells. See [11].

milestone in the articulation of a graphene market (Interview GSA, August 2013).

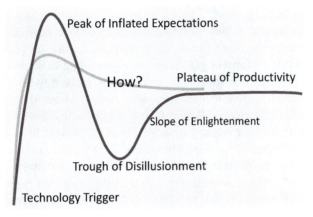

Figure 9.4 How can we avoid disappointment following the peak of the hype? This is a slide of a presentation given by Tomás Palacios from the MIT at the Graphene conference 2013, in Bilbao, Spain (photo by one of the authors).

9.3.4 The Emergence of Concerns Besides the Promises and the Call for Standardisation and New Definitions of Graphene

The initiative by the GSA and NPL was not the only one aimed at creating standards for graphene. Starting from 2012, the need for a typology for graphene materials was a topic of discussions in commercial conferences as well as in the flagship and related scientific community. Additionally, with the process of application of the flagship, the topic of environmental, health and safety (EHS) risks of graphene was introduced and given a place in the research program. This discussion was spurred by the flagship process and the European Commission's approach to EHS, but was also related to the increasing media and public attention to the material and the expectations about possible 'negative' reactions from the public.[11] Based on former experiences with adverse public reactions to genetically modified organisms and a debate

---

[11]Rip [25] has described this nano-phobia-phobia, or the fear of adverse public reaction to new technologies. This was one of the frames under which EHS discussions took place—the other one being Responsible Innovation.

on possible health risks of carbon nanotubes, the FET Flagship scheme asked its projects to include comprehensive risk research, and a work package on environmental, health and safety risks was included in the project.[12] This discussion of possible risks reached also the media, which reported about potential environmental, and health effects of graphene related materials, in particular, when in 2013 a form of graphene oxide was found to be potentially toxic [91–93]. The graphene community reacted with caution and scepticism, but it also spurred a debate about the need to differentiate between different forms of graphene. This reinforced the initiatives that worked towards the development of standards for graphene [94, 95] (Interview, standards organization, January 2014). What is important to highlight is that starting from the end of 2013 it was clearly stated in the graphene community that there was a need to distinguish between different forms of graphene in order to advance the market potential of the material, as well as to assess clearly its environmental and health risks. That is, expectations, including concerns, furthered the emergence of a rather common governance measure—standardisation—which may have broader repercussions for the understanding of what is understood as graphene.

While there is still 'hype'—in the sense that media attention is still strong and expectations about graphene high—the field has become more stabilised. This means for the governance of expectations that expectations are now coordinated in an extended network with clear leadership, provided by the Flagship in Europe and other organisations in the US and Asia that have positioned themselves as leaders in the race for graphene, and to which other actors in the field would refer to when voicing expectations. Graphene is still a rather young field; while we observe clearly some disappointment of early expectations, such as the immediate expectations regarding its role for electronics, we would not characterise this as a general disappointment with the field. Still, the idea of a hype- disappointment cycle has already mobilised significant governance efforts trying to go against a possible disappointment, thus indicating another facet of the reflexive relations between expectations and governance, when not only expectations about a technology, but also

---

[12]Available at: http://graphene-flagship.eu/?work=work-package-2-health-and-enviroment.

expectations about expectation dynamics trigger governance efforts.

9.4 Conclusion

We suggested a particular approach to conceptualising and studying forms of anticipatory governance of new and emerging science and technology, which we applied to the case of graphene, a newly emerging field within the wider realm of nanotechnologies. This approach allowed us to reveal a number of insights which are relevant both for the governance of new and emerging technologies as well as for the dynamics of expectations in science and technology.

First, we highlighted the role of expectations as an intrinsic and pervasive element in the governance patterns of a techno-scientific field. This was exemplified by the roadmapping process related to the flagship application which became more than a mere step in justifying funding, but created further repercussions in the governance of the field. Thus, we were able to trace how formal, intentional ways of governing expectations are interwoven with the de facto governance of a field and its related expectations. What emerged from this rather broad perspective, is a picture which does not only demonstrate particular instances of performative effects of mobilisation, legitimation, guidance, or coordination by specific expectations and anticipatory practices, but rather a patchwork of different expectations and anticipatory practices in various spaces, be they formal or informal, as a pervasive element of the governance of graphene.

Second, we suggested to not only follow how expectations contribute to the governance of a techno-scientific field, but how expectations themselves were governed by particular actor constellations, modes of coordination and partly intentional strategies to orchestrate expectation building and the dynamics of expectation. This could be observed once more in the ITRS-inspired roadmapping process, in editorial strategies of scientific journals, and in the strategies of consultants and other players for balancing and managing collective expectations.

Third, we were able to reveal some of the processes how governance patterns changed together with the expectation building in the field. Following the evolvement of expectations and anticipatory practices over time, we saw that the initial interest originated in the space of academic science carried by typical scientific practices, such as different formats of scientific publications or conferences. The specific content of these expectations, which were early on referring to an exceptional application potential, facilitated to get industry and increasingly also public funding and policy circles interested as well. In addition, the specific funding channels targeted by the community did not only affect the constellation of actors, and supported the general promise of graphene—while introducing the issue of safety concerns as well, but also changed the modes of governing expectations as well as the governance patterns in the field more generally. This happened most likely not independent from other dynamics as the granting of the Nobel Prize and subsequent steps supporting the institutionalisation of the field, which further fuelled the promises of graphene. Following this, additional actors entered and emerged in the graphene field, as consultancy firms, newly created industry associations and standardisation bodies, who introduced explicit strategies at expectation management. We also found clear indications that the modes of governance varied and changed, suggesting that further research should look more closely into the question, which modes may be more or less productive, for instance being more or less prone to overhyping. A former study indicated that actors in the field turn to networking strategies, in order to mitigate hype dynamics [11], and we see indications of this here as well, but further research is in order to corroborate this. Moreover, the actual effects of strategies of governing and managing expectations are likely to also depend on the overall constellation of actors, of spaces and the governance patterns therein.

Considering both the temporal and the spatial characteristics of this process, one sees a hype wave, which takes its origin in science, but then moves through, respectively expands to different spaces. In science, graphene was perceived as a hype by some, even before the media was actually taking up the topic. This suggests that different types of innovation actors may perceive different expectation dynamics at a given moment, and may be subject to different performative and governance effects, rather than being exposed to one common hype dynamic. This issue would require further research, since the scope of our study does not capture all spaces over the whole research period. By looking at multiple spaces we saw that expectations are carried by different practices in different spaces, and practices and dynamics in different spaces are often reinforcing each other. It would be a question for further research, if under different circumstances we might also observe contrasting dynamics, rather weakening expectations on a field as a whole.

With the suggested perspective, which takes account of the role of different spaces, actors and practices, we hope to have made a step towards an increased understanding how hypes emerge, neither as a quasi-natural law of current science and technology, nor as a straightforward result of intentional hyping strategies, but rather as the de facto result of diverse strategies and interlinking processes. It also sheds new light on the issue of expectation management, from a focus on coping strategies of individual actors and organisations to the modulating role of actor and institutional arrangements in a specific societal setting.

To conclude, our perspective delivered a number of new questions and insights, while opening up further directions of research. For the study of future-orientation in science and technology we hope to have shown the usefulness of an approach that captures various spaces and their interrelated dynamics for understanding hype dynamics and their performative effects. For those concerned more generally with the governance of new and emerging technologies, our study and approach suggest to take into account as relevant elements of anticipatory governance much more than the obvious formal governance measures as roadmaps, but just as well editorial practices, consultancy reports or networking strategies; and it indicates that their actual role and effects in the governance process are likely to depend on the broader patterns and dynamics in the field, rather than being mainly accountable to the tool or practice as such.

References

1. Joly, P. B. (2010) On the economics of techno-scientific promises. In: Akrich, M., Y. Barthe, F. Muniesa, and P. Mustar (eds.) *Débordements: Mélanges Offerts à Michel Callon* (Presses des Mines, Paris), pp. 203–220.

2. Rip, A. (2012) The Context of innovation journeys, *Creativity and Innovation Management*, **21**, 158–170.

3. Schaper-Rinkel, P. (2013) The role of future-oriented technology analysis in the governance of emerging technologies: The example of nanotechnology, *Technological Forecasting and Social Change*, **80**, 444–452.

4. Boyd, E., B. Nykvist, S. Borgstrom, and I. A. Stacewicz (2015) Anticipatory governance for social-ecological resilience, *Ambio*, **44**, 149–161.

5. Fisher, E., M. Boenink, S. van Der Burg, and N. Woodbury (2012) Responsible healthcare innovation: Anticipatory governance of nanodiagnostics for theranostics medicine, *Expert Review of Molecular Diagnostics*, **12**, 857–870.

6. Ozdermir, V., S. Faraj, and B. M. Knoppers (2011) Steering vaccinomics innovations with anticipatory governance and participatory foresight, *OMICS A Journal of Integrative Biology*, **15**, 637–646.

7. Barben, D., E. Fisher, C. Selin, and D. Guston (2008) Anticipatory governance of nanotechnology: Foresight, engagement, and integration. In: Hackett, E., O. Amsterdamska, M. Lynch, and J. WAajcman (eds.) *The Handbook of Science and Technology Studies, Third Edition* (MIT Press, Cambridge), pp. 929.

8. Anderson, B. (2007) Hope for nanotechnology: anticipatory knowledge and the governance of affect, *Area*, **39**, 156–165.

9. Berti, P., and L. Levidow (2014) Fuelling expectations: A policy-promise lock-in of UK biofuel policy, *Energy Policy*, **66**, 135–143.

10. Groves, C., and R. Tutton (2013) Walking the tightrope: Expectations and standards in personal genomics, *BioSocieties*, **8**, 181–204.

11. Konrad, K., J. Markard, A. Ruef, and B. Truffer (2012) Strategic responses to fuel cell hype and disappointment, *Technological Forecasting and Social Change*, **79**, 1084–1098.

12. Borup, M., N Brown, K. Konrad, and H. van Lente (2006) The sociology of expectations in science and technology, *Technology Analysis and Strategic Management*, **18**, 285–298.

13. Konrad, K. (2010) Governance of and by Expectations, paper presented at the *2010 EASST conference*, Trento, 2–4 September.

14. Te Kulve, H., K. Konrad, C. Alvial Palavicino, and B. Walhout (2013) Context matters: Promises and concerns regarding nanotechnologies for water and food applications, *NanoEthics*, **7**, 17–27.

15. Ruef, A., and J. Markard (2010) What happens after a hype? How changing expectations affected innovation activities in the case of stationary fuel cells, *Technological Analysis and Strategic Management,* **22**, 317–338.
16. van Lente, H., C. Spitters, and A. Peine (2013) Comparing technological hype cycles: Towards a theory, *Technological Forecasting and Social Change,* **80**(8), 1615–1628.
17. Brown, N. (2003) Hope against hype-accountability in biopasts, presents and futures, *Science Studies,* **16**, 3–21.
18. Fenn, J., and M. Raskino (2008) *Mastering the Hype Cycle: How to Choose the Right Innovation at the Right Time* (Harvard Business Press, Boston).
19. Ramiller, N. C. (2006) Hype! Toward a theory of exaggeration in information technology innovaton, *Academy of Management Proceedings,* A1–A6.
20. Swanson, E. B., and N. C. Ramiller (1997) The organizing vision in information systems innovation, *Organization Science,* **8**, 458–474.
21. Nerlich, B. (2013) Moderation impossible? On hype, honesty and trust in the context of modern academic life, *The Sociological Review,* **61**, 43–57.
22. Caulfied, T., and C. Condit (2012) Science and the sources of hype, *Public Health Genomics,* **15**, 209–217.
23. Rinaldi, A. (2012) To hype, or not to(o) hype, *EMBO Reports,* **13**, 303–307.
24. Will, C. M. (2010) The management of enthusiasm: Motives and expectations in cardiovascular medicine, *Health,* **14**, 547–563.
25. Rip. A. (2006) Folk theories of nanotechnologies, *Science as Culture,* **15**, 349–365.
26. Bonneuil, C., P. B. Joly, and C. Marris (2008) Disentrenching experiment: The construction of GM—crop field trials as a social problem, *Science, Technology & Human Values,* **33**, 201–229.
27. Rip, A., and P. B. Joly (2012) *Emerging Spaces and Governance* (European Forum for Studies of Policies for Research and Innovation, Paris).
28. Alvial Palavicino, C. (2016) *Mindful Anticipation: A Practice Approach to the Study of Emergent Technologies* (University of Twente, Enschede).
29. Benz, A., S. Lutz, U. Schiamank, and G. Simonis (eds.) (2007) *Handbuch Governance: Theoretische Grundlagen und empirische Anwendungsfelder* (VS Verlag für Sozialwissenschaften).

30. Treib, O., H. Bahr, and G. Falkner (2007) Modes of governance: Towards a conceptual clarification, *Journal of European Policy,* **14,** 1–20.

31. Schaeffer, G. J. (1998) *Fuell Cells for the Future-A Contribution to Technology Forecasting from a Technology Dynamics Perspective* PhD Dissertation, University of Twente, Enschede.

32. Pollock, N., and R. Williams (2010) The business of expectations: How promissory organizations shape technology and innovation, *Social Studies of Science,* **40,** 525–548.

33. Van Merkerk, R., and H. van Lente (2008) Asymmetric positioning and emerging paths: The case of point-of-care, *Futures,* **40,** 643–652.

34. Konrad, K. (2006) The social dynamics of expectations: The interaction of collective and actor-specific expectations on electronic commerce and interactive television, *Technology Analysis & Strategic Management,* **18,** 429–444.

35. van Lente, H., and A. Rip (1998) Expectations in technological developments: An example of prospective structures to be filled in by agency. In: Disco, C., and B. van der Meulen (eds.) *Getting New Technologies Together: Studies in Making Sociotechnical Order* (Walter de Gruyter, Berlin), pp. 203–231.

36. Schubert, C., J. Sydow, and A. Winderler (2013) The means of managing momentum: Bridging technological paths and organisational fields, *Research Policy,* **42,** 1389–1405.

37. Le Masson, P., B. Weil, A. Hatchuel, and P. Cogez (2012) Why are they not locked in waiting games? Unlocking rules and the ecology of concepts in the semiconductor industry, *Technology Analysis and Strategic Management,* **24,** 617–630.

38. Hajer, M. (1995) *The Politics of Environmental Discourse* (Oxford University Press, Oxford).

39. Rip, A. (2006) A co-evolutionary approach to reflexive governance- and its ironies. In: Voß, J. P., D. Bauknecht, and R. Kemp (eds.) *Reflexive Governance for Sustainable Development* (Edward Elgar, Cheltenham), pp. 82–100.

40. Rip, A. (2010) De facto governance of nanotechnologies. In: Goodwin, M., B. J. Koops, and R. Leenes (eds.) *Dimensions of Technology Regulation* (Wolf Legal Publishers, Nijmegen).

41. Brown, N., and M. Michael (2003) A sociology of expectations: Retrospecting prospects and prospecting retrospects, *Technology Analysis & Strategic Management,* **15,** 4–18.

42. Geels, F., W. Smit, N. Brown, B. Rappert, and A. Webster (2000) Lessons from failed technology futures: Potholes in the road to the future. In: *Contested Futures-A Sociology of Prospective Techno-Science* (Ashgate, Sydney), pp. 29–156.

43. Koschatzky, K. (2005) Foresight as a governance concept at the interface between global challenges and regional innovation potentials, *European Planning Studies,* **13**, 619–639.

44. Novoselov, K. S., Jiang, D., Schedin, F., Booth, T. J., Khotkevich, V. V., Morozov, S. V., and Geim, A. K. (2005) Two-dimensional atomic crystals, *Proceedings of the National Academies of Sciences, Engineering and Medicine,* **102**(30), 10451–10453.

45. van Lente, H. (2012) Navigating foresight in a sea of expectations: Lessons from the sociology of expectations, *Technology Analysis & Strategic Management,* **24**, 769–782.

46. Borup, M., and K. Konrad (2004) *Expectations in Nanotechnology and in Energy–Foresight in the Sea of Expectations* (Background Paper-Research Workshop on Expectations in Science and Technology, Risø, Denmark).

47. Kitzinger, J., and C. Williams (2005) Forecasting science futures: Legitimising hope and calming fears in the embryo stem cell debate, *Social Science and Medicine,* **61**, 731–740.

48. Felt, U., and H. Nowotny (1992) Striking gold in the 1990s: The discovery of high-temperature superconductivity and its impact on the science system, *Science, Technology & Human Values,* **17**, 506–531.

49. van Lente, H., And A. Rip (1998) The rise of membrane technology: From rhetorics to social reality, *Social Studies of Science,* **28**, 221–254.

50. Hedgecoe, A., and P. Martin (2003) The drugs don't work, *Social Studies of Science,* **33**, 327–364.

51. Van Merkerk, R. O., and D. K. R. Robinson (2006) Characterising the emergence of a technological field: Expectations, agendas and networks in Lab on a chip technologies, *Technology Analysis & Strategic Management,* **18**(3–4), 411–428.

52. Geels, F. W., and R. Raven (2006) Non-linearity and expectations in niche-development trajectories: Ups and downs in Dutch biogas development (1973–2003), *Technology Analysis & Strategic Management,* **18**(3–4), 375–392.

53. Bender, G. (2005) Technologieentwicklung als institutionalisierungsprozess. *Zeitschrift für Soziologie,* **34**, 170–187.

54. Anderson, B. (2010) Preemption, precaution, preparedness: Anticipatory action and future geographies, *Progress in Human Geography,* **34**, 777–798.

55. Morrison, M., and L. Cornips (2012) Exploring the role of dedicated online biotechnology news providers in the innovation economy, *Science Technology and Human Values,* **37**, 262–285.

56. Wustenhagen, R., R. Wubker, M. J. Burer, and D. Goddard (2009) Financing fuel cell market development: Exploring the role of expectation dynamics in venture capital investment. In: Pogutz, S., A. Russon, and P. Migliavacca (eds.) *Innovation, Markets, and Sustainable Energy: The Challenge of Hydrogen and Fuel Cells* (Edward Elgar, Cheltenham), pp. 118–137.

57. Barre, R., and M. Kennan (2008) Revisiting foresight rationales: What lessons from the social sciences and humanities? In: *Future-Oriented Technology Analysis* (Springer, Berlin), pp. 41–52.

58. Da Costa, O., P. Warnke, C. Cagnin, and F. Scapolo (2008) The impact of foresight on policy-making: Insights from the FORLEARN mutual learning process, *Technology Analysis and Strategic Management,* **20**, 369–387.

59. Georghiou, L., and M. Keenan (2006) Evaluation of national foresight activities: Assessing rationale, process and impact, *Technological Forecasting and Social Change,* **73**, 761–777.

60. Salo, A., and K. Cohls (2003) Preface—technology foresight—past and future. *Journal of Forecasting,* **22**, 79–82.

61. McDowall, W. (2012) Technology roadmaps for transition management: The case of hydrogen energy, *Technological Forecasting and Social Change,* **79**, 530–542.

62. Jeffrey, H., J. Sedgwick, and C. Robinson (2013) Technology roadmaps: An evaluation of their success in the renewable energy sector, *Technological Forecasting and Social Change,* **80**, 1015–1027.

63. Truffer, B., J. P. Voß, and K. Konrad (2008) Mapping expectations for system transformations: Lessons from sustainability foresight in German utility sectors, *Technological Forecasting and Social Change,* **75**, 1360–1372.

64. Haegeman, K., E. Marinelli, F. Scapolo, A. Ricci, and A. Sokolov (2013). Quantitative and qualitative approaches in future-oriented technology analysis (FTA): From combination to integration? *Technological Forecasting and Social Change,* **80**(3), 386–397.

65. Novoselov, K. S., V. I. Fal'ko, L. Colombo, P. R. Gellert, M. G. Schwab, and K. Kim (2012) A roadmap for graphene, *Nature,* **490**, 192–200.

66. Peplow, M. (2013) Graphene: The quest for supercarbon, *Nature*, **503**, 327–329.

67. Novoselov, K. S., A. K. Geim, S. V. Morozov, D. Jiang, Y. Zhang, et al. (2004) Electric field effect in atomically thin carbon films, *Science*, **306**(5696), 666–669.

68. Geim, A. K., and K. S. Novoselov (2007) The rise of graphene, *Nature Materials*, **6**, 183–191.

69. Shapira, P., Y. Youtie, and S. Arora (2012) Early patterns of commercial activity in graphene, *Journal of Nanoparticle Research*, **14**, 1–15.

70. Winnink, J. (2012) Searching for structural shifts in science: Graphene R&D before and after Novoselov et al. (2004) *Proceedings of the 17th international conference on science and technology indicators*, Montreal, 5–8 September.

71. Derbyshire, K. (2008) Hype overshadows graphene's real promise, *Blogging the Chip Space*. Available at: http://blog.thinfilmmfg.com/2008/04/hype-overshadows-graphene-real-promise_22.html.

72. Intellectual Property Office (2013) *Graphene: The Worldwide Patent Landscape in 2013* (IPO, Newport).

73. Van Den Brink, J. (2007) Graphene: From strength to strength, *Nature Nanotechnology*, **2**, 199–201.

74. Kinaret, J., A. Ferrari, V. Fal'ko, and J. Kivioja (2011) Graphene-driven revolutions in ICT and beyond, *Procedia Computer Science*, **7**, 30–33.

75. Heber, J. (2010) Great, the physics Nobel Prize for graphene! Now don't overhype it, *All that Matters: Exciting Developments in the Physical Sciences* Available at: http://allthatmatters.heber.org/2010/10/05/great-the-physics-nobel-prize-for-graphene-now-dont-overhype-it/.

76. Gein, A. K. (2013) Be afraid, very afraid of the tech crisis, *Financial Times*, 5 February. Available at: http://www.ft.com/cms/s/0/ad8e9df0-6faa-11e2-956b-00144feab49a.html#axzz461IgqFkh.

77. Innovation Union (2011) *Prof. André Geim, Nobel Laureate in Physics 2010—Innovation Convention 2011-Brussels*. Available at: https://www.youtube.com/watch?v=xw6iAXxmvuM.

78. Bhattacharya, A. (2012) UK Chancellor Osborne throws his weight—and a little money—behind science, *Nature News Blog*. Available at: http://blogs.nature.com/news/2012/11/uk-chancellor-osborne-throws-his-weight-and-a-little-money-behind-science.html.

79. Brumfiel, G. (2011) UK politicians embrace graphene, *Nature Newsblog*. Available at: http://blogs.nature.com/news/2011/10/graphene_gets_political_1.html.

80. Taghioskoui, M. (2009) Trends in graphene research, *Materials Today*, **12**, 34–37.

81. Boch, W. (2011) FET Flagship: Evaluation process, *FET Flagship Pilots Mid-Term Conference*, Warsaw, 24–25 November.

82. Kinaret, J. (2012) *GRAPHENE-CA WP6 Project Management of the CA Project. Deliverable 6.3 Publishable Flagship Proposal Report* (European Commission, Brussels).

83. Pascular, J. (2011) The graphene flagship, *Third NanoNetworking Summit*, UPC, 21–22 June.

84. Peplow, M. (2013) Graphene: The quest for supercarbon, *Nature*, **503**(7476), 327–329.

85. Kinaret, J. (2012) *Graphene Flagship: Working Together to Combine Scientific Excellence and Technological Impacts*. Available at: http://cordis.europa.eu/fp7/ict/programme/fet/flagship/doc/conf-july2012-06-graphene_en.pdf.

86. Eley, J. (2013) The growing graphene investment bubble, *Financial Times*. Available at: http://www.ft.com/intl/cms/s/0/9f7d2974-5d94-11e3-95bd-00144feabdc0.html#axzz2myR1rfWX.

87. Tighe, C. (2013) Graphene producer's IPO oversubscribed, *The Financial Times*, 18 November.

88. Nanoclast (2013) Two foolish ways to invest your money in graphene, *IEEE Spectrum*. Available at: http://spectrum.ieee.org/semiconductors/nanotechnology/two-foolish-ways-to-invest-your-money-in-graphene.

89. Ghaffarzadeh, K. (2013) *Graphene: What Next After the Hype?* Available at: http://www.idtechex.com/research/articles/graphene-what-next-after-the-hype-00005454.asp.

90. NPL (2013) *NPL Joins the Graphene Stakeholders Association*. Available at: http://www.npl.co.uk/news/npl-joins-the-graphene-stakeholders-association.

91. Seabra, A. B., A. J. Paula, R. De Lima, O. L. Alves, and N. Duran (2014) Nanotoxicity of graphene and graphene oxide, *Chemical Research in Toxicology*, **27**, 159–168.

92. Hamill, J. (2014) *Miracle Material Graphene Could Be Pollutant of the Future*. Available at: http://www.forbes.com/sites/jasperhamill/2014/04/30/miracle-material-graphene-could-be-pollutant-of-the-future/.

93. Johnson, D. (2014) Should we worry about graphene oxide in our water? *IEEE Spectrum*. Available at: http://spectrum.ieee.org/nanoclast/at-work/test-and-measurement/should-we-worry-about-graphene-oxide-in-our-water.
94. Wick, P., A. E. Louw-Gaume, M. Kucki, H. F. Krug, K. Kostarelos, et al. (2014) Classification framework for graphene-based materials, *Angewandte Chemie International Edition*, **53**(30), 7714–7718.
95. Bianco, A., H. M. Cheng, T. Enoki, Y. Gogotsi, R. H. Hurt, et al. (2013) All in the graphene family—a recommended nomenclature for two-dimensional carbon materials, *Carbon*, **65**, 1–6.

Chapter 10

Transactional Arrangements in the Governance of Emerging Technologies: The Case of Nanotechnology

Evisa Kica[a] and Ramses A. Wessel[b]

[a]*Governance, Risk and Compliance,*
Deloitte, The Netherlands
[b]*Centre for European Studies,*
Faculty of Behavioural Management & Social Sciences,
University of Twente, NL-7500 AE Enschede, The Netherlands

r.a.wessel@utwente.nl

10.1 Introduction

Nanotechnology, the science of controlling the structure of matter at the nanoscale, is expected to provide the platform and tools for innovative products and applications for consumers while adding value to solutions designed to address a myriad of human and environmental challenges. This has triggered agents within government and industry to invest heavily in nanotechnology research and development programs [1, 2]. The results of this investment are steadily coming to fruition, as evidenced by the increasing number of products incorporating nanomaterials making their way into commerce [3]. In 2001 the United States

---

Embedding New Technologies into Society: A Regulatory, Ethical and Societal Perspective
Edited by Diana M. Bowman, Elen Stokes, and Arie Rip
Copyright © 2017 Pan Stanford Publishing Pte. Ltd.
ISBN 978-981-4745-74-1 (Hardcover), 978-1-315-37959-3 (eBook)
www.panstanford.com

(US) National Science Foundation (NSF) predicted that by 2015, the value of products and services that will incorporate nanotechnology will go up to $US1 trillion [4]. There are also reports suggesting that the potential market value for products incorporating nanotechnology (specifically in the semiconductors and electronics sectors) could go up to $US2.6 trillion by 2014, and US$2.8 trillion by 2015 (driven by the expected commercialisation success in the healthcare and electronics sectors) [5, 6]. A detailed summary of the published market figures regarding nanotechnology is provided by Malanowski and Zweck [7:1811].

It is important to note that estimations on the potential future of the nano market have generated many debates amongst scholars. A number of them criticise these estimations on the grounds that the size of the nano market is 'over-hyped' by news media and key actors [2, 8]. Others indicate that current estimations on the potential of the nano market are ambiguous due to uncertainties related to the size of the 'nanotechnology value chain' and the '(sub)areas of nanotechnology that the market evaluation includes' [9:54]. These debates have led to a wide range of perspectives and a lack of consensus amongst scholars about the potential of the future nano market.

Concomitant to these debates have been concerns over the unintended consequences, the environmental health and safety risks that nanotechnology may pose to workers handling nanomaterials, to consumers of nanobased products, and to the public and the environment at large. Maynard and his colleagues [10], have already indicated that some engineered nanoparticles (ENPs) such as carbon nanotubes and other bio persistent-insoluble nanoparticles such as titanium dioxide may under certain conditions present toxicological hazard to humans and the environment. One of the main issues is that the unique characteristics of nanomaterials followed by rapid advancement and commercialisation of nanoscience, have challenged the application of risk and toxicological assessment methodologies, and regulatory oversight strategies outlined in current environmental, health and safety regulations [11, 12].

Scientific reviews, such as those carried out by the *United Kingdom's Royal Commission on Environmental Protection* in 2008 [13] and the *Center for International Environmental Law* in 2012 [14], emphasise that there are continues scientific and knowledge gaps on the hazardous components, the specific properties and the behaviour of nanomaterials on the environment or in living organisms. As such, formulating even small components of hard regulatory frameworks for nanotechnology remains difficult. Some jurisdictions (e.g. France, Australia and California) have already begun to tweak their existing command and control frameworks in relation to new industrial nanomaterials [15–17]. In addition, the European Parliament and Council have adopted a more wholesale approach with the introduction of nano-specific provisions for cosmetics as part of the recast of the Cosmetic Regulation [18]. However, the vast majority of countries have opted to retain the regulatory status quo. This is not surprising given the evolving state of the scientific art and the uncertainties that surround so many facets of the technology.

Scientific reports authored by Davies and Azouley have added to the broader policy and regulatory debate [12, 14]. These authors argue that the application of risk and toxicological assessment methodologies and regulatory oversight strategies outlined in current environmental, health, and safety regulations are inappropriate and too inflexible to cope with the rapid advancements and the potential risks of nanoscience. At the other hand of the spectrum, other reports such as those issued by the organisation for economic co-operation and development (OECD) [19], emphasise that existing approaches for the testing and assessment of traditional chemicals are in general adequate to deal with nanotechnology and only in some cases they may have to be adapted to the specificities of nanomaterials. Accordingly, the debate on how to embrace nanotechnology developments continues among policy makers, while the public and private sectors have voiced fears of the potential for under- and over-regulation.[1]

Whereas consensus amongst regulators and policy makers on the most appropriate regulatory response remains elusive, a number of stakeholders coming from the industry, non-governmental bodies and other public/private sectors, have joint

---

[1]Both US and European Union key bodies including, for example, the US Executive Office of the President and the European Commission claim that the existing regulations covering chemicals and materials, as well as environmental and health issues are adequate to deal with nanotechnologies.

forces to address and respond to the regulatory challenges of nanotechnology. These actors have focused on the development and implementation of voluntary governance arrangements and innovative measurement techniques. These arrangements are voluntary, non-binding and utilise the expertise of a wide range of governmental, industrial and civil society actors. The involvement of multiple actors, knowledgeable experts and epistemic communities in one regulatory setting are considered the key elements that shape the governing authority of these arrangements [20–23]. Furthermore, many scholars argue that these arrangements provide for voluntary rules or guidelines that are continuously revised to provide the most up-to-date information on technology developments and cope with situations of regulatory uncertainty [23–28]. As such they are expected to be able to respond quickly to the speed, complexity and uncertainty of nanotechnology's development.

The landscape of these arrangements is very broad. For instance, at the national level we can observe actors such as Department for Environment, Food and Rural Affairs (Defra) in the United Kingdom (UK); Environmental Protection Agency (EPA) in the US; Friends of the Earth in Australia, whose main objective has been to develop 'voluntary reporting schemes' or 'stewardship programs' to gather scientific data on the characteristics and toxicity of engineered nanomaterials from relevant organisations and assist regulators with developing appropriate risk management frameworks for nanoscale materials [29]. Voluntary initiatives have also been initiated by private actors, such as *the Responsible NanoCode* in the UK; BASF in Germany as well as DuPont-Environmental Defense in the US. The main objective of these developments has been (amongst others) to develop 'in-house' innovative regulatory mechanisms that govern the manufacture of nanoproducts; manage occupational, health and safety risks associated with the development of nanotechnology across all lifecycle phases; and ensure the responsible development, production, use and disposal of nanoscale materials [25, 29–32].

There are several voluntary initiatives taken at the European level as well. For instance, the European Commission's *Code of Conduct for Responsible Nanosciences and Nanotechnologies Research* emphasises clearly the tendency of government to broaden their collaboration with industry and civil society, and provide a 'tangible contribution to the good governance of nanotechnology' [33:478], the European Committee for Standardization (CEN) in 2005 set up a technical committee on nanotechnologies (TC352)—to develop consensus standards related to broader issues of nanotechnology, such as terminology and nomenclature, metrology and instrumentation, specifications for reference materials, test methodologies, science-based health, safety and environmental practices.

However, the low number of submissions from relevant organisations and the industry [29, 34], the failure of these initiatives to promote trust-building amongst key stakeholders, as well as the inability to disseminate effectively their activities [35], are considered major barriers for these arrangements to achieve their objectives. Furthermore, the global significance of the nanotechnology (scientific, regulatory and socio-environmental) issues and research, the evolvement of the new generations of nanomaterials and the rapid pace of commercialisation, pose additional challenges for these voluntary arrangements to deal with this emerging and complex technology [36, 37].

Since the mid-2000 a wide range of transnational governance arrangements (TGAs) have emerged in the field of nanotechnology. By the term 'transnational' we refer to 'non-territorial policy making or interactions that cross national-borders at levels other than sovereign to sovereign' [38:2; 39:4]. We use the term transnational governance arrangement to refer to a set of rules/mechanisms within an institutional setting that influence the interaction between various actors (state and non-state actors not bounded by territorial borders), to provide for voluntary rules or guidelines grounded in practical experience and expertise. For instance, the two most important TGAs in the field of nanotechnology are the OECD (despite its intergovernmental set-up) and the International Standardisation Organisation (ISO) [30]. In addition, there are other public–private and private governance arrangements in which nanotechnology is discussed. These arrangements are mostly focused on a specific sector (e.g. nanomaterial safety) and have led to a range of specific projects, workshops or dialogues. For instance, the International Council on Nanotechnology (ICON); the International Risk Governance Council (IRGC); and the International Cooperation on Cosmetic Regulations (ICCR). Intergovernmental initiatives that seek to contribute to nanotechnology related safety issues and foster the cooperation of scientists, policy-makers and industrial actors, are based on United Nations (UN) and World Health Organisation (WHO) processes. For instance, the United Nations Industrial Development Organisation's International Centre for Science and High Technology (UNIDO) and the WHO's intergovernmental forum on Chemical Safety (IFCS) [30, 37].

In many of these arrangements, states have become only one type of participating actor amongst others in the decision-making process [40]. As such, they depart from traditional forms of regulation that are based on the exclusive authority of the nation state to make collectively binding decisions. They are based on different governance actors, networking strategies, processes and structures [41]. While these arrangements have received significant attention in political science, international relation (IR) theory and (international) law [42–44], the analytical questions provided by these studies are not fully complete. Current discussions focus mostly on explaining the differences between transnational arrangements and traditional state-based forms of regulation. However, they focus less on explaining the key factors that drive the emergence of these arrangements. In these studies, it is still unclear why certain arrangements have gained a leading role at transnational level or which arrangements are likely to have the highest potential to contribute to the governance of nanotechnology.[2,3] What are their key attributes and power sources? This chapter purports to further answer these questions.

Building upon the regulatory challenges of nanotechnology, this chapter analyses the attributes and the potential of the key

---

[2]Yet, in legal science a debate was started on the reasons explaining a shift from formal legal agreements to informal arrangements, including transnational actors. It has been argued that this can partly be explained by: (a) saturation with the existing treaties and changed policy preferences of States; (b) deep societal changes that are not unique to international law but affect both international *and* national legal systems, in particular: the transition towards an increasingly diverse network society; and (c) an increasingly complex knowledge society [45].

[3]Governance refers to 'the systems of authoritative norms, rules, institutions, and practices by means of which any collectivity, from the local to the global, manages its common affairs'. In this way, as Ruggie argues, in the absence of government, transnational governance can be defined as an instance of governance [46].

transnational nanotechnology governance arrangements, which provide forums of debate at transnational level and contribute to establishing informal coordination mechanisms. In particular, our focus is on: Technical Committee on Nanotechnology (ISO/TC229); OECD Working Party on Manufactured Nanomaterials (OECD/WPMN); IFCS; IRGC and ICON.

There are several reasons that justify our decision to focus on these arrangements. To begin with, these arrangements have displayed well-defined strategies and plans to develop voluntary mechanisms that are relevant to the governance of nanotechnology. In addition, there has been no formal delegation or legal mandate for these arrangements to contribute to the field of nanotechnology or set norms which can serve as reference points. However, all of them have managed to establish internal mandates by securing resources and collaboration with influential stakeholders and experts in the field. As a result, the value and the potential of these arrangements to the governance of nanotechnology has been acknowledged and quoted in various reports [32], policy documents [1, 27, 47], and scholarly debates [24, 32].

This chapter is organised as follows. In the first section, we discuss the factors that have contributed to the emergence of TGAs and emphasise why these modes of governance are considered appropriate to respond to the nanotechnology regulatory challenges. In the second section, we introduce a typology that distinguishes governance arrangements on the basis of actors involved, as well as the functions and the regulatory stages in which the arrangements contribute. We emphasise that TGAs can be characterised not only by these attributes, but also by their degree of institutionalisation[4] as well as the normative and substantive depth of transnational outcomes. In the third section, we assess the characteristics and the potential of the five aforementioned transnational nanotechnology governance arrangements. With these cases, we demonstrate that the typology developed in this chapter is useful to study the evolution of transnational governance in the field of nanotechnology. Specifically, it allows us to understand and investigate the actions taken by various arrangements to enhance their capacity to contribute effectively to the governance of nanotechnology. The last section provides analysis and concluding remarks.

---

[4]For more information on the institutionalisation of regulatory networks, see [48].

10.2 The Transnationalisation of Nanotechnology Governance

There seems to be a general consensus amongst scholars that the internationalisation of markets, the emergence of transnational communication networks as well as new technologies have challenged the ability of national governments to define and provide public goods [49]. Hence, the creation of new forms of governance arrangements has been steadily increasing in part as a result of the limitations of the command and control regulation [41]. The proliferation of TGAs in the field of nanotechnology can be related to several political, regulatory and technological factors.

First, over the last few decades nanotechnology has emerged as a new transformative force in industrial society, covering a broad range of applications in chemicals, pharmaceuticals, electronics, energy, goods and cosmetics. Therefore, this emerging technology has attracted the attention of a wide range of actors coming from regulatory, civil society and business organisations whose activities span beyond national borders [36, 32]. Nanoscience and nanotechnology have also attracted a tremendously diverse range of skilled scientists,[5] who contribute to the creation of new products/services and advice for any innovation in nanotechnology. As a result, nanotechnology governance has become highly exposed to the direct influence and initiatives of non-state actors.

Second, commerce generally and nanotechnology specifically are increasingly global in nature [25, 36]. The experience with other technology developments on genetically modified organisms (GMOs) and regulatory failures associated with asbestos, have led to many debates on how to develop appropriate and congruent governance frameworks for nanotechnologies [50, 51]. There are many considerations that support a transnational approach to the regulation of nanotechnology. Abbot and other colleagues [36:539–541], argue that a transnational approach to nanotechnology regulation can contribute to providing better opportunities for dialogue and learning which could establish harmonised regulatory requirements for product testing, risk

---
[5]Most of these scientists have expertise in physics, chemistry, biology, information technology, toxicology, engineering and materials science.

assessment, reporting and labelling. Harmonised requirements would in turn assist producers, manufacturers and distributors to benefit at the product level, and regulators to avoid regulation that is ill informed or too stringent [36:541]. In addition, it will assist multi-national companies at the manufacturing level to deal with environmental and occupational health and safety issues. A transnational approach to these issues can lead to uniform compliance requirements, product stewardship, worker training and reporting programs [30, 36, 37, 50]. Furthermore, the global reach of nanotechnology research and trade provide additional incentives for developing regulatory frameworks at transnational level, which are expected to facilitate commerce, underpin good industrial practice and avoid regional divide [36, 37].

Third, whereas nanotechnology is surrounded by great expectations, scientific evidence indicates that with the ongoing expansion of nanotechnology, novel nanostructures causing unknown forms of hazard can be produced [30:12]. As emphasised in the previous section, regulators are facing profound challenges and uncertainties about the adequacy of the existing risk assessment and management frameworks to characterise and assess accurately the risks associated with nanotechnology. The rapid pace of commercialisation followed by the evolvement of new generations of nanomaterials pose additional challenges to the current regulatory frameworks to deal with emerging technologies [34]. Regulatory systems are expected to face several challenges, which relate mainly to their ability to (a) deal with novel materials and uncertain risks; (b) anticipate and respond rapidly to the changing technological systems; (c) develop frameworks that offer sufficient flexibility and adaptability; (d) expand the scientific capacity to include a diversity of mixed experts from public and private sectors; and (e) develop globally oriented information-gathering systems to cope with the globalisation of nanotechnology [12]. Given the fundamental nature of these challenges and the inability of the individual states to tackle these issues effectively, many scholars urge for transnational coordination and cooperation [28, 30, 37, 57].

Finally, over the last decade, nanotechnology has exploded from a relatively narrow technical field, into an arena that has to cope with constitutionally recognised interests also. The development of nanotechnology involves issues related to health, environment, occupational safety, scientific research, technological development, national security and so on [26:131]. The potential of nanotechnology to manipulate properties at the nanoscale (i.e. making materials stronger, thinner, more elastic and so forth) has made nanotechnology to impact almost every industrial sector [51]. However, the growing production and use of nanomaterials (in particular engineered nanomaterials) may increase the potential of exposure for workers, consumers and environment [52]. This has triggered representatives of various civil society/labour coalitions to become highly interested on the benefits and risks of nanomaterials, as well as on the regulatory responses addressing these issues [32, 53]. As a result, nanotechnology has experienced an evolving political landscape, with many countries, national regulators, socio-environmental actors and international organisations participating in voluntary (and often privately led) initiatives to promote the regulatory coordination of nanotechnology [20, 24, 36, 54]. These developments, we would argue, provide additional incentives for the emergence of transnational governance arrangements. In the following section, we provide a typology for understanding the characteristics and the potential of various governance arrangements at the transnational level.

10.3 Transnational Governance Arrangements Generally and Their Attributes

Transnational governance arrangements are identified mostly as voluntary, informal and flexible arrangements beyond the nation state in which private actors are systematically engaged [38, 55, 56]. These arrangements are horizontally structured, relatively institutionalised and bring together actors from various sectors to share information, best practices and harmonise rules and procedures in order to pursue certain goals in areas of limited statehood [55, 57–59]. Transnational arrangements come in different forms at transnational level. Whereas there is no single characteristic that would define transnational arrangements from the traditional modes of governance, Pauwelyn [43] indicates that new governance arrangements are characterised by

   (1) *process informality* (they build on the cross-border cooperation between public and private actors in a forum other than a traditional international organisation);

(2) *actor informality* (these arrangements build upon the cooperation of actors other than traditional diplomatic actors (e.g. regulators or agencies))[6];
(3) *output informality* (these arrangements do not result in a formal treaty or legally enforceable commitment).

These characteristics come close to the characteristics of the transnational new forms of governance that Abbot and Snidal have discussed earlier [24:521]. In their framing, new forms of governance are fundamentally distinguished from old governance models by

- *differing roles of the state in regulation* (in new governance the state is a significant player, it acts as a facilitator for supporting voluntary and cooperative programs, rather than as a top-down commander);
- *decentralisation of the regulatory authority* (in new governance regulatory responsibilities are shared among different actors coming from the state agencies and private sectors);
- *dispersed expertise* (new governance seeks to harness the expertise of a wide range of actors; it looks beyond professional regulators and seeks to incorporate also those who may have 'local' expertise on relevant issues);
- *non-mandatory rules* (new governance relies on flexible norms and voluntary rules).

In a similar vein, Börzel and Risse [22:196] argue that the more we enter the realm of new modes of governance, the more we decentralise the regulatory authority, include non-hierarchical forms of steering and share the regulatory responsibilities amongst public and private actors.[7] As a result, various forms of governance arrangements have emerged at transnational level

---

[6]In is interesting to note that in these arrangements the governance contributions are not explicitly restricted to those actors whose organisational objective lies in the provision of certain public goals (e.g. regulators, humanitarian or environmental organisations). Rather, the authority of transnational governance arrangements might also emerge from various private actors, such as business associations, industry or multinational companies [60].

[7]Building upon the constellations of state and non-state actors to induce regulation at transnational level, Börzel and Risse [22] distinguish four types of arrangements: *cooptation* (regular consultation and cooptation of private actors in international negotiation systems); *delegation* (delegation of state functions to private actors); *co-regulation* (co-regulation of public and private actors); *self-regulation* (private self-regulation in the shadow of hierarchy).

encompassing different actors, modes of steering, processes and outcomes [41:6]. Therefore, a typology of transnational governance arrangements is important to understand their key features and their potential to respond to regulatory issues [22, 61].

Scholars have proposed various typologies painting the key features of transnational governance arrangements. To begin with Andonova and colleagues [61], propose a typology mapping the realm of transnational governance according to the 'types of actors' and 'functions'. With regards to the 'types of actors' they argue that transnational arrangements involve a variety of state and non-state actors that contribute different capacities and sources of authority. They distinguish between:

- *private arrangements* (established and managed by non-state actors only);
- *public arrangements* (established by public actors acting independently from the state); and
- *hybrid arrangements* (established by public and private actors jointly).

However, the 'types of actors' are considered as a necessary but not a sufficient condition for distinguishing amongst transnational arrangements. The authors argue that these arrangements should be clustered also in terms of the 'functions' that they can or do perform. In their framing, 'functions' determine the resources and the power used within a particular arrangement to steer members to achieve certain goals [61, 62]. In principle, the 'functions' of transnational governance arrangements are divided into five categories:

(1) *information sharing* (arrangements that influence political and civil discourse through learning forums or collaborative events);
(2) *capacity building* (arrangements that provide resources or institutional support through fundraising campaigns, sponsorship);
(3) *coordination* (arrangements that coordinate state and non-state activities in a particular sector);
(4) *rule setting* (arrangements that contribute to adopting international norms, regulations or standards that respond to respective regulatory problems); and

(5) *implementation* (arrangements that provide monitoring and service provision to enable action or implementation of national or international policy goals) [62].

A different approach is taken by Abbot and Snidal [24], who propose the concept of a governance triangle to depict the involvement of various actors (i.e. states, firms and NGOs) in a respective governance arrangement. Similar to the framework employed by Andonova et al. [61], the typology of Abbott and Snidal focuses on rule-setting. These authors take a wider perspective and divide the rule setting (in the authors' words—the regulatory process of standard setting) into five distinct phases:

- *agenda-setting* (ability of the arrangement to place an issue on the regulatory agenda);
- *negotiations* (ability of the arrangement to draft and promulgate standards);
- *implementation* (ability of the arrangement to contribute to the implementation of the standards);
- *monitoring* (ability of the arrangement to monitor compliance); and
- *enforcement* (ability of the arrangement to ensure effective compliance).

Their basic premise is that in order for the transnational governance arrangements to succeed in the regulatory process they need a suite of competences, such as: independence from the targets of regulation, representativeness, expertise of several kinds and concrete operational capacity (including resources). However, since in most cases single-actor schemes do not have all the necessary competencies, they contend that collaboration with different types of actors is essential for these schemes to assemble the needed competencies and act effectively in the regulatory process. According to their line of argumentation, the potential of transnational arrangements can be understood by looking at the design choice of the governance arrangements—in particular at the relative input that states, NGOs and firms exercise in a respective arrangement and the actions taken by the governance arrangement to fulfil any competency deficit. Focusing on the regulatory standard-setting schemes of pre- and-post-1985, the authors observe a shift from old to newly emerging multi-actor schemes, characterised by high level of decentralisation and dispersed expertise [24:52–57]. Whereas these characteristics make these arrangements better suited to address regulatory gaps at transnational level, the authors suggest that some form of 'facilitative state orchestration' is important to reduce the bargaining problems between firms and NGOs to achieve socially desirable outcomes [20:573; 24].

In addition to 'actor type' and 'functions', Abbott and his colleagues [25], Liese and Beishem [63], and Martens [62] suggest a typology for mapping the realm of transnational governance arrangements based on the 'level of institutionalisation' and the 'design choice'. Martens notes that governance arrangements can be classified in low, medium and high levels of institutionalisation [62]. Whereas *high levels of institutionalisation* refer to permanent multi-stakeholder institutions that have formal membership, firmly established governing bodies, institutionalised rules of decision making, a secretariat and budget authority; *medium levels* have a clearly defined membership but not a separate legal status or formalised decision-making structures; and *low levels* are ad hoc initiatives with narrowly defined objectives, no formalised membership or governing body. According to Homkes [55] and Martens [62] institutionalisation and appropriate structural forms may be costly and time consuming to establish and maintain, but they are the key factors driving the norm-setting and decision-making powers of the governance arrangements. Scholars of transnational governance have also given increasing credence to the regulatory design—referring in particular to the stages of the regulatory process that the arrangement addresses, the relative precision of the rules (they frame this as *normative scope*)—as well as the obligatory status of the transnational outcomes (they frame this as *substantive depth*) [25, 63].

In this way, the typology of transnational governance arrangements has become a complex and multi-dimensional phenomenon, which cannot be analysed through one prism only [40]. To assess the potential of these arrangements one should understand how various attributes characterising transnational arrangements interact with each other and contribute to the efficiency of the arrangement [25]. In Table 10.1, we emphasise the key attributes of transnational governance arrangements, which cluster them into various groups and shift the balance towards greater and lesser effective arrangements. In the following section, we apply these attributes to understand the landscape and the potential of transnational governance arrangements in the field of nanotechnology.

Table 10.1   The key attributes of transnational governance arrangements

Actors involved	Functions	Regulatory process	Normative scope	Substantive depth	Degree of institutionalisation
Public actors only (single actor scheme)	Information sharing	Agenda-setting	Narrow	Significant constraints	Low level
Private actors only (single actor scheme)	Capacity building	Negotiations	Broad	Excessive flexibility	Medium level
Public and private actors (multi-actor scheme)	Coordination Rule setting Implementation	Implementation Monitoring Enforcement			High level

10.4   The Governance of Nanotechnology: A Typology of Transnational Governance Arrangements

Since the mid-2000, various transnational governance arrangements have emerged to discuss nanotechnology. In the following, we focus on five key arrangements and discuss their activities in the field of nanotechnology.

10.4.1   ISO Technical Committee on Nanotechnology (ISO/TC229)

In January 2005, the ISO Technical Management Board (TMB) established a new technical committee focused specifically in developing nanotechnology standards (TC229). A technical committee that 'would provide industry, research and regulators with a coherent set of robust and well founded standards in the area of nanotechnologies [...] whilst at the same time providing regulators, and society in general, with suitable and appropriate instruments for the evaluation of risk and the protection of health and the environment' [64].

In the first plenary meeting of the TC229 the scope of the Committee was articulated as well as the internal structure and the business plan. Kica and Bowman [54, 65], provide a detailed discussion on the internal structure of TC229. The main work in the TC229 is done by its Working Groups (WGs) [66]. The Committee allocates specific tasks to the WGs, which tasks are carried out by experts, who are individually appointed by a participating ISO member body, a liaison organisation, or both, to a particular WG when new projects are approved. TC229 consists of four WGs working on:

(1) *Terminology and Nomenclature* (WG1—develops uniform terminology and nomenclature for nanotechnologies to facilitate communication and promote common understanding);
(2) *Measurement and Characterisation* (WG2—develops measurement and characterisation standards for use by industry in nanotechnology-based products);
(3) *Health, Safety and Environment* (WG3—develops science-based standards that aim to promote occupational safety, consumer protection and environmental protection); and
(4) *Measurement and Characterisation* (WG4—develops standards that specify relevant characteristics of engineered nanoscale materials for use in specific applications) [67].

Besides the central Secretariat leading the work of the TC229, each of the WGs has its secretaries and conveners who arrange the meetings and communicate important information to the participants. The inclusion of various WG with different aims and objectives, emphasises that TC229 has shifted the focus from working only on technical issues related to defining the size and concept of nanomaterials, to addressing broader aspects of the technology such as risk management, health, environment and safety issues [65].[8] Following this evolution in the development of standards, in 2009 the former chair of the TC229 stated that ISO standards now serve three key objectives: (a) supporting commercialisation and market development; (b) providing a basis for procurement through technical, quality and environmental management; and (c) supporting appropriate legislation/regulation and voluntary governance structures [68]. Therefore, TC229 and its standards seem to have multiple functions. TC229 provides a forum for debate for various actors. Its 'plenary week' meetings organised every 10th month of the year, as well as WG meetings provide the best opportunities for experts to meet with other delegates, exchange knowledge and information on standardisation issues, and set appropriate and uniform standards.

Regarding the representation of actors in TC229, it is important to note that nanotechnology standards are developed by groups of experts under the overarching TC umbrella. ISO applies the principle of national delegation and its administrative work takes place through a Secretariat located in one of the National Standardisation Bodies (NSBs). Delegates participate in the ISO/TC meetings in negotiations and consultations that are intended to lead to the development of an international consensus. As indicated in the ISO/IEC Directives, all national bodies have the same rights to participate in the work of the committees and subcommittees [66]. TC229 has 34 participatory and 13 observatory members.

To ensure legitimacy through stakeholder representation, ISO has established procedures for including a wide range of stakeholders in the process—not only the industrial ones [28]. Within ISO the participating actors are divided into *industry and trade associations*; *consumers and consumer associations*; *governments and regulators*; and *societal and other interests*. In that sense, ISO standardisation process is considered as a multi-stakeholder process open to a variety of actors and experts.

---

[8]In 2011 ISO/TC229 took a leading role to developing a guidance document related to labelling of nanomaterials, which complements the current regulatory initiatives on the labelling of food and cosmetic products containing manufactured nano-objects. An increasing focus on health, safety, and environmental issues appear to have provided TC229 with the impetus to publish ISO/TR12885 on *Nanotechnologies-Health and Safety Practices in Occupational Settings Relevant to Nanotechnologies*.

ISO/TC229 has a number of collaborations and relationships with other organisations and standardisation bodies as well [69]. TC229 is opened to a broader range of stakeholders who are not connected with ISO through national bodies. These stakeholders are known as *liaison* members and include manufacturer associations, commercial and professional associations, industrial consortia, user groups, as well as groups concerned with the rights of consumers workers and environment (e.g. the European Consumer Voice in Standardisation (ANEC), the European Environmental Citizens Organisation for Standardisation (ECOS) and the European Trade Union Institute (ETUI)). Furthermore, as part of its outreach strategy ISO/TC229 has established two Task Groups working on Sustainability (TGS),[9] as well as on Consumer and Societal Dimensions of Nanotechnologies (TGCSDN) [67].

Regarding the outcomes, as of start 2013, TC229 has published three standards, while the majority of deliverables have been normative and informative documents developed in the form of technical specifications (TSs) and technical reports (TRs).[10] As articulated in the ISO/TC229 business plan, the Committee has given priority to developing horizontal standards that 'provide foundational support across all sectors that use nanotechnologies or nanomaterials' [67]. These deliverables have no strict legal value nor provide for excessive constraints. However, they constitute important statements, provide concrete and practical information and address a broader range of products and activities.

10.4.2 OECD Working Party on Manufactured Nanomaterials (OECD/WPMN)

OECD was established in 2006 to promote 'international co-operation in human health and environmental safety related aspects of manufactured nanomaterials (MNs), in order to assist in the development of rigorous safety evaluation of nanomaterials' [70].

---

[9]TGS have the mandate to advise the TC229 on how to include sustainability within its strategic priorities.

[10]Such documents are usually approved while the subject matter is still under development or when there is no immediate agreement to publish an International Standard.

The WPMN work programme was adopted by the Chemicals Committee in November 2006 and focuses on three key working areas:

(1) *Work Area 1*—which aims to develop working definitions for MNs for regulatory purposes within the context of environmental, health and safety (EHS) issues;
(2) *Work Area 2*—which aims to encourage cooperation and coordination on risk assessment frameworks; and
(3) *Work Area 3*—which aims to foster co-operation and share information on current and planned initiatives in risk assessment, risk management and regulatory frameworks [71].

To fulfil these overarching aims, WPMN has developed eight projects. These projects focus on

- the development of an OECD Database on EHS research for approval (*Project 1*);
- the EHS research strategies on MNs (*Project 2*);
- the safety testing of a representative set of MNs and test guidelines (*Project 3*);
- MNs and test guidelines (*Project 4*);
- co-operation on voluntary schemes and regulatory programmes (*Project 5);*
- co-operation on risk assessments (*Project 6*);
- the role of alternative methods in nanotoxicology (*Project 7*);
- exposure measurement with an initial focus on occupational settings (*Project 8*); and
- cooperation on the environmentally sustainable use of MNs (*Project 9*).

It is interesting to note that each project is carried out by specific steering groups (SGs) [72]. These groups are composed of experts nominated by the delegation heads participating in the work of the OECD/WPMN.

WPMN is a subsidiary body established under the Chemicals Committee. This Committee functions under the OECD Environment, Health, and Safety Division and consists of governmental officials from the OECD countries responsible for chemicals management. As such, WPMN encourages the participation of observers and invited experts that participate in the work of the Chemicals Committee. There are 34 OECD member countries that participate in the work of the WPMN. Member countries drive the agenda and the output of the WPMN, while financing a major part of its work and voting on proposals and policy recommendations. These countries are represented at the WPMN meetings by the delegation heads[11], each of whom is drawn from their national agencies responsible for chemicals regulation and the safety of human health and the environment. Nominated delegates are selected by consensus on the basis of merit, and their roles and duties are set up by the Committee and the WPMN.

Since its establishment in 2006, there have been ten meetings of the OECD/WPMN, which have been supplemented with several workshops, expert meetings and conferences [54]. In addition to these actors, the OECD has taken several steps to establish close relationships with non-member countries like Russia, China, Thailand, South Africa, India, the EU Commission (EC), UN bodies, ISO, WHO and other stakeholder groups such as those represented through the Trade Union Advisory Committee (TUAC) and the Business and Industry Advisory Committee (BIAC). The wide range of actors emphasises clearly the drive within the OECD to opt for a multi-stakeholder representation and secure support for its policy recommendations through a broader range of experts. This also allows us to assess the WPMN as a transnational arrangement.

With regards to the outcomes, it is important to note that WPMN does not have regulatory power, but it serves as a centre for international collaboration and policy dialogue, building 'communities of practice that promote information sharing and harmonisation' [25:291; 37]. The key achievements to date are the *Sponsorship Programme*[12], the *OECD Database on Manufactured Nanomaterials to Inform and Analyse EHS Research Activities*, and the *Preliminary Guidance on Sample Preparation and Dosimetry for the Safety of Nanomaterials* [70, 72, 73].

---

[11]These delegates serve as the main contacting point to the Working Party, and provide information on the experts that are nominated by member countries to participate in the work of the SGs [54].

[12]The *Sponsorship Programme*, as one of the key outcomes of the WPMN, gathered a number of countries and the BIAC, who volunteered to sponsor and cosponsor the testing of one or more MNs and provide test data, reference or testing materials to the lead sponsors.

10.4.3 International Risk Governance Council

IRGC is an independent foundation that was initially founded by the Swiss government, to help the understanding and management of emerging global risks [74]. Since the beginning of 2005 the Council has also been working actively on nanotechnology issues. The key objectives of IRGC in relation to nanotechnology are:

> 'to develop and make available specific advice for improving risk governance; to provide a neutral and constructive platform on the most appropriate approaches to handling the risks and opportunities of nanotechnology and to enable all actors to reach a global consensus' [74:6].

The key bodies within the IRGC are the Board Members, Advisory Committee and the Scientific & Technical Council (S&TC). Members of the Board are drawn from governments, industry, science and non-governmental organisations.[13] The Advisory Committee is the key body, which comprises of individual members (17 members) appointed by the Board to act as advisors and make proposals to the S&TC on the possible issues that need to be addressed by the IRGC. These members come from USA, Germany, France, Belgium, Korea, Switzerland, China and Canada. The S&TC is the leading scientific authority of the foundation. It comprises experts form a range of scientific and organisational background, who review the scientific quality of the IRGC work and its deliverables. The participation of these actors at the IRGC is voluntary, but there is less available information on how they are selected and how the decision-making process is structured in this arrangement.

The IRGC's nanotechnology programme is a key forum for dialogue and is supported mainly by the Swiss Reinsurance Company, EPA and the US Department of State [75]. To tackle issues of nanotechnology the IRGC, and the S&TC in particular, proposed the establishment of the working group on nanotechnology to provide an independent and cross-disciplinary approach to nanotechnology risks and hazards. The group has focused on two projects: *on the risk governance of nanotechnology* (in 2005) and *on nanotechnology applications in food and cosmetics*

---
[13]The members of the board come from USA, Portugal, Switzerland and China.

(2007). These projects were led by expert bodies consisting of recognised subject experts in the field of nanotechnology and risk governance, who prepared and reviewed the project reports [75]. For instance, the first project was led by Dr. Mihail Roco of the National Science Foundation (NSF) and a team of scientific experts coming from universities, research centres, governmental bodies, laboratories.

Over a period of 2 years, the IRGC undertook two expert workshops (May 2005 and January 2006) [75, 76]. During the second workshop, the IRGC working group also organised four surveys on the implications of nanotechnology with stakeholders coming from research organisations, standardisation organisations, nanotechnology start-ups, NGOs. The aim of the surveys was to identify the organisation interest in nanotechnology research, the governance gaps as well as measures needed to address potential risks. These activities resulted in the publication of the *White Paper on Nanotechnology Risk Governance* in 2006 and the *Policy Brief: Recommendations for a global, coordinated approach to the governance of potential risks* in 2007 [30, 75].

The *White Paper* and the *Policy Brief* suggest a regulatory framework, which anticipates two frames for four generations of nanotechnology. Frame 1 includes the first generation of nanostructures (the steady function nanostructures), which have stable behaviour and do not constitute excessive risks. Frame 2 involves the second generation (active function nanostructures), the third generation (systems of nanosystems) and the fourth generation of nanostructures (heterogeneous molecular nanosystems). In the second frame are involved nanostructures which change their design and it is more difficult to predict their behaviour [75]. It is important to note that these deliberations have been amongst the first publications to provide detailed recommendations for the risk governance of nanotechnology [75]. They recommend national and international decision makers who are involved in the nanotechnology risk issues 'to improve knowledge base, strengthen risk management structures and processes, promote stakeholder communication and collaboration, and ensure social benefits and acceptance' [75:15]. As such, the *White Paper* and the *Policy Brief* have become widely cited reference points in various reports and documents [30, 32, 77].

10.4.4 International Council on Nanotechnology

ICON was established in 2003. Based in Rice University, ICON started initially within the program of the federally funded Center for Biological and Environmental Nanotechnology (CBEN). Nowadays, its activities extend beyond CBEN to include other national and international centres. ICON has been actively involved on tackling issues related to the field of nanotechnology [76]. Its mission is to 'assess, reduce and communicate information regarding the potential environmental and health risks of nanotechnology, while maximising its societal values' [78:3].

The key bodies of ICON are the Director and the Executive Director, who are responsible for managing the internal coordination of the Council and ensuring an effective external presence. The Council is largely funded by industry[14] and it has established an Advisory Board that is composed of prominent nanomaterial safety experts coming from industry, government agencies, academic institutions and non-governmental groups. Participation in ICON is voluntary and non-compensated, and there are around 27 members participating in the Advisory Board coming from France, Japan, the Netherlands, Switzerland, Taiwan, the United Kingdom and the United States. The Executive Committee, consisting of the Director and Executive Director, has the ultimate authority over ICON's finances, the membership of the Advisory Board and of the setting of new committees [78].

ICON has been working on several projects related to nanotechnology such as the *International Assessment of Research Needs for Nanotechnology Environment, Health and Safety*; *Current Practices for Occupational Handling of Nanomaterials and the Good NanoGuide.* The main objectives of the first two projects have been to (a) facilitate the documentation of current best practices for identifying and managing risks that come during the production, handling, use and disposal of nanomaterials and (b) prioritise research needs related to the classification nanomaterials [78]. As such, they have resulted in several workshops and conferences. ICON's third project—*the GoodNanoGuide*—is an internet based collaboration platform designed to help experts in the field of nanotechnology to exchange ideas on how best to handle nanomaterials safely [79].

---

[14]The key sponsors of ICON's work are: DuPont; Intel; Lockheed Martin; L'Oreal; Mitsubishi Corporation; Procter & Gamble; Swiss Reinsurance Company.

The key objective of the *GoodNanoGuide* is to establish an open forum that complements other nanotechnology information projects by providing up-to-date information on good practices for handling of nanomaterials in an occupational setting. The *GoodNanoGuide* is freely accessible for everyone, but only experts who are members of the *GoodNanoGuide* are able to post information [79]. The forum has attracted a wide range of stakeholders to collaborate and contribute at both intellectual and financial levels. However, according to its Director the main weakness of the *GoodNanoGuide* is its lack of sustainability in a down economy, as well as its reliance on industry funds only [25]. The platform was set in 2008 and it is still in a beta version.

10.4.5 Intergovernmental Forum on Chemical Safety

IFCS was established in 1994 in the International Conference of Chemicals Safety. The main objective in establishing IFCS was to create an 'over-arching framework through which national governments, intergovernmental organisations and NGOs could work together and build consensus to promote chemical safety and address the environmentally sound management of chemicals' [80:2].

The idea to establish IFCS was created in 1991, during the preparations for the United Nations Conference on Environment and Development (UNED). The Forum is under the administration of WHO, which also provides the secretariat for IFCS. Participation in the IFCS is open to *governmental participants* (which include all member state of the UN and its specialised agencies); *intergovernmental participants* (including participants representing subregional, regional, political and economic groups of countries involved in chemical safety); and *non-governmental participants* (including NGOs concerned with science, health and workers interest). Participation is voluntary and supported by the members. The work of IFCS is organised in sessions at intervals of 2–3 years. To achieve its objectives, IFCS has established the Forum Standing Committee (FCS) to provide advice and assistance during the preparations of Forum meetings, monitor progress on the work of the IFCS and assist with regional efforts. FCS is composed of 25 participants, who serve as representatives of the views of participant countries in respective IFCS regions, NGOs or intergovernmental organisations.

Since its creation, IFCS has held six meetings/sessions. In its sixth session in 2008, IFCS considered for the first time the opportunities and challenges of nanotechnology and MNs. The final outcome of this meeting was the *Dakar Statement on Manufactured Nanomaterials* calling for more international cooperation in information sharing and risk assessment [30]. The meeting had around 200 delegates, representing 70 governments, 12 intergovernmental organisations and 39 NGOs. Amongst other issues, two main items were discussed in this session. The primary issue was whether to distinguish between nanotechnologies and MNs and to integrate them into the IFCS VI agenda.

Whereas most NGOs and developing countries argued for including both the MNs and nanotechnology, European countries supported the inclusion of only MNs in the IFCS agenda. As a result, delegates agreed to include a preambular paragraph in the Dakar statement acknowledging the need to address the safety aspect of nanotechnologies 'while limiting the focus of the statement on safety aspects of nanomaterials only' [81:5]. Amongst other recommendations, the Dakar Statement called the governments and the industry to apply the 'precautionary principle throughput the lifecycle of manufactured nanomaterials' [81:12]. The Statement recommended the evaluation of 'the feasibility of developing global codes of conduct in a timely manner' and the provision of information 'through product labelling, websites, databases [...] and cooperative actions between governments and stakeholders' [81:6]. These recommendations provided an important contribution for advancing the sound management of chemicals globally and were sent to the International Conference on Chemicals Management (ICCM) for consideration and further actions [81, 82]. Another key agenda item during the sixth meeting of the IFCS was the future of this Forum. In light of the agreement on the Strategic Approach to International Chemical Management (SAICM) in 2006, the delegates of IFCS agreed to invite the ICCM (during its second session—ICCM2) to integrate the Forum as an advisory body into the ICCM [82]. This invitation was crucial for IFCS, since the decision of the ICCM2 to reject the request of the IFCS put into question the existence and the potential of this forum to contribute to the field of nanotechnology. This is further elaborated in the next section.

10.5 Conclusion

This chapter aimed to assess the potential of different transnational governance arrangements in the field of nanotechnology and to explain the key factors that drive the emergence of these arrangements. It highlights the growing importance and relevance of five TGAs, such as ISO/TC229, OECD/WPMN, IRGC, ICON and IFCS. Building upon current debates on the modes of governance and transnationalisation, the chapter developed a framework to determine the main attributes of TGAs. This framework proved to be very useful for understanding different types of governance arrangements that have emerged in the field of nanotechnology as well as their potential to contribute to this emerging field. A comparative look at these arrangements suggests the following: First, in the field of nanotechnology TGAs have all taken various initiatives to assemble the needed competencies while combining the expertise and experiences of multiple actors. Yet, the relative input that states, NGOs and firms have in these arrangements differs considerably. We elaborate this further below. Furthermore, the arrangements differ considerably in terms of their institutional structure, organisational goals and substantive scope, all of which impact their potential to contribute effectively to the governance of nanotechnology.

Second, all of the arrangements reviewed in this chapter have engaged in agenda setting and related preliminary steps. For instance, ICON and IRGC have focused mainly to internationalise the nanotechnology safety and regulatory debate. They have served as leading fora for gathering information on the risks of nanoscale materials to inform future regulation, and supporting coordination amongst decision makers on handling these issues [30]. IFCS was a pioneer in identifying nanotechnology as an important part of the international chemical safety agenda. It aimed at sharing information and promoting coordination on nanotechnology and MNs to increase awareness on the potential benefits, challenges and risks posed by nanotechnology. In addition to health and safety issues, the leaders of the ISO/TC229 have addressed other issues that are essential to nanotechnology regulation, such as nomenclature, specifications and measurement. The OECD/WPMN has also served as the main forum for gathering and exchanging information on the risk assessment of MNs. Therefore, these arrangements differ in their functions, but also on the normative scope of the issues that they address.

This leads us to our third point. A comparative look at these arrangements suggests that some arrangements seem to be narrower, focusing almost entirely on certain products (i.e. IFCS on safety aspect of MNs), settings (ICON—and *GoodNanoGuide* in particular—on workplace) or activities (IRGC on risk governance) [25]. OECD/WPMN concentrates on human health and environmental safety implications (including risk assessment and safety testing) of MNs [74]. ISO/TC229 addresses a broader range of products, setting and activities. With its standards, TC229 provides terminology and nomenclature, measurement techniques, calibration procedures, reference materials, test methods to detect and identify nanoparticles, occupational health protocols relevant to nanotechnologies as well as risk assessment tools-which aim to support regulation, research, commercialisation, and trade of the materials and products at the nanoscale.

A number of these arrangements (such as ISO/TC229 and OECD/WPMN) have also adopted norms that call relevant actors to act in accordance to certain standards. In this way, these arrangements have started to move towards the negotiation stage. ISO/TC229, for instance, has been able to negotiate several standards, as well as technical specifications and recommendations (e.g. ISO/TS27687 and ISO/TR 12885). From a governance point of view, these deliberations may provide the 'best available options to industries requested to demonstrate product compliance with regulation' [27:17]. The *Sponsorship Programme* has also served as an incentive for countries to collaborate, share best practices, and follow a consistent approach with regards to the testing of specific endpoints of representative MNs. The substantive scope of these deliberations differs, with the TC229 standards and technical specifications providing practical information, and being more concrete and complete [25].

Fourth, a comparative look on these arrangements emphasises that ISO and OECD seem to have the highest potential to contribute to the governance of nanotechnology. In our view, there are two key factors that contribute to this. On the one hand, it is the high level of institutional structure that characterises these governance arrangements. On the other hand, it is the collaboration and the (political) support that these arrangements have ensured with key actors in Europe [27, 83]. Regarding the first point, our case studies emphasise that nanotechnology transnational governance arrangements differ considerably in terms of their structure, membership and organisational goals. TC229 and WPMN are the most organised working groups with secretariats, clear rules of membership, governance structure and decision-making procedures [28, 65]. Furthermore, they have organised regular meetings for their members to share knowledge and information and developed concrete roadmaps that guide future actions and strategies. Such a well-defined structure has helped these arrangements to contribute substantially to shaping nanotechnology regulatory agenda at transnational level, promote collaboration and harmonisation, and establish concrete regulatory governance mechanisms (e.g. standards, guidelines or other regulatory options) for nanotechnology [25, 27, 28, 65].

ISO is amongst the most recognised international organisations, which has strongest linkages with key experts and dominant industrial actors coming from more than 40 countries around the world. However, to ensure representation of other stakeholders TC229 has established *liaisons* with other actors representing government, trade unions, consumer associations, NGOs and the EU. In addition to this, the establishment of the Task Groups (i.e., TGS and TGCSDN) appears to have been one approach to opening up the membership of TC229, and thus making its actions accountable to a broader range of actors. Since 2005, ISO/TC229 has been able to broaden its activities, membership and the diversity of actors involved in the process [28, 51, 54]. ISO/TC229 plenary meetings involve a wide range of practitioners, industrial hygienists, pharmacologists, toxicologists and ecotoxicologists, chemists and physicists who exchange knowledge and contribute substantially to establishing international standards [54].

In a similar vein, OECD/WPMN and its SGs are highly structured. WPMN has strong linkages with national regulatory agencies, which is not surprising given the intergovernmental nature of the OECD. However, the inclusion of high-level experts nominated by member countries has helped SGs to proceed faster in developing well-defined strategies for tracking nanotechnology policy developments. OECD has also developed *liaisons* with other industrial actors, trade unions, NGOs as well as European Commission. Such activities are crucial for these arrangements to ensure stakeholder representation, but also the legitimacy and acceptability for their actions. While serving as a centre for policy dialogue between high-level governmental officials and non-governmental experts, there is the potential for the outputs of the OECD/WPMN to lay the groundwork for collective agreements and contribute to overcoming the uncertainties and regulatory puzzles related to nanomaterials and risk assessment practices [65, 84]. This is not without precedent; for example, it is widely acknowledged that the OECD Chemicals Committee played a leading role in promoting harmonised chemical control policies through the system of the Mutual Acceptance Data (MAD) [54, 71].

Regarding the support that these arrangements have ensured with key actors in Europe, perhaps of greatest importance is the support of the EU members and the EU Commission. In 2007, the European Commission Communication on the *Nanosciences and Nanotechnologies: An Action Plan for Europe 2005–2009*, stated that OECD/WPMN and ISO/TC229 are 'principal forums for the coordination of activities at the international level' and that 'the Commission, the European Bodies and Member States are expected to continue contributing to these international efforts' [85:10]. The Council's conclusions on *Nanoscience and Nanotechnologies* also stated that, 'the Commission needs to take into account in its policy making all activities within the OECD (e.g. definitions, nomenclature, risk management)' [26:428]. Regarding, the role of international standards in the field of nanotechnology, in 2010 the EU Commission addressed a mandate to the European Standardisation Bodies (ESOs) (i.e., CEN, CENELEC and ETSI) to develop European standards related to the characterisation and toxicity testing of the nanomaterials, as well as to the occupational handling and exposure [27].[15] An important element of the Mandate is that the EC requests the

---

[15]In the mandate, the EC stated that nanotechnology standardisation is crucial and is viewed as 'a means to accompany the introduction on the market of nanotechnologies and nanomaterials, and a means to facilitate the implementation of regulation'.

ESOs to develop and adopt European standards in support of the European policies and legislations, while taking into account and giving priority to the existing ISO standards [27]. Furthermore, the Mandate asks the ESOs to work in close collaboration with ISO and OECD. These statements indicate clearly that the EU not only is aware of the work undertaken by ISO and OECD, but it also suggests that these arrangements and their deliverables are relevant and can contribute to the nanotechnology regulatory debate in the EU.

Other governance arrangements analysed in this chapter, have also been able to ensure collaboration with influential stakeholders. IFCS, for instance, managed to provide equal representation to state actors, NGOs and intergovernmental actors. Regarding its structure, IFCS has the most informal structure. IFCS operates under the intergovernmental regime of the WHO, but it considers itself as a 'non-institutional arrangement', a forum that builds on the loose grouping of interested parties and experts, who come together to integrate national and international efforts to promote chemical safety [86:886]. However, even though being one of the key actors to consider the issue of nanotechnology within the international chemicals agenda, the rejection of the ICMM2 to include IFCS as an advisory body put into question the ability of this forum to contribute effectively to nanotechnology governance [82]. Furthermore, in the final resolution on the emerging issues, the ICMM2 recognised the potential health and environmental issues related to nanotechnologies and MNs, but no reference was made to the Dakar Statement. In light of these events, in the last session of the Forum (Forum IV) the FCS agreed to suspend its work for the foreseeable future [82].

ICON and IRGC have a moderate level of institutionalisation with both having established several workshops and a network of growing stakeholders. Both of these arrangements focus on nanotechnology risk governance, but none of these arrangements aspires to go beyond mere information exchange and international coordination. The IRGC in the initial phases of its work on nanotechnology developed an ad hoc working group on nanotechnology to provide an independent and cross-disciplinary approach to nanotechnology risks and hazards. However, this group does not have the same structure with clear rules for membership, formalised decision-making structures as well as strategies for future work like ISO/TC229 for instance. Its Advisory Committee is representative of a less number of European countries, but it has the support of EPA. In 2006, IRGC organised a Conference to promote stakeholder dialogue and feedback on the IRGC *White Paper* [76, 87]. What we can observe here is the participation of actors from regulatory agencies (e.g., DEFRA) as well as industrial actors (such as DuPont).

The inclusion of these actors combined with the support that EPA has for IRGC can contribute to increasing the relevance of the Council's recommendations in the field of nanotechnology. ICON, on the other hand has also a moderate level of institutionalisation. Compared to ISO and OECD, its working groups are less structured with few members and less formalised decision-making strategies. The Council started initially as an affiliate programme of the CBEN centre at Rice University, but it has been able to build a network of growing stakeholders. Council has also been working with EPA to review the best practices for nanomaterial safety [76]. *GoodNanoGuide* is one of the main outcomes of ICON which still continues to be in a beta version [76]. Whereas the relevance of IRGC and ICON has been mentioned in some documents [32, 88, 89], none of them has been involved formally by the EU institutions. Furthermore, compared to ISO/TC229 and OECD/WPMN, these arrangements have not established any formal collaboration with the EU institutions or Commission.

In conclusion, our analysis on nanotechnology transnational governance arrangements suggests that IRGC and ICON have the potential to become important actors on the transnational governance of nanotechnology risk regulation. However, the institutional structure, the actors, the normative scope, the political support and the strategies incorporated by ISO/TC229 and OECD/WPMN places these arrangements in a better position to take a lead on the transnational debates of nanotechnology governance. The huge potential of these arrangements, which operate beyond the state level, brings forward many concerns. In these arrangements, the rule making authority rests on the hands of those who operate beyond the state level and are 'neither elected nor managed by elected officials' [90]. They build on non-hierarchical steering principles and are characterised by interaction amongst various public and private actors. As such, they have become the 'hard case' for legitimacy [21], raising therefore many questions over the clear lines of accountability, stakeholder representation, roots of decision making and reasons for social acceptability. Such questions are beyond the scope of this chapter, but pose an urgent need for further research.[16]

References

1. European Commission (2006) *Press Release—Council Approves EU Research Programmes for 2007-2013*, 18 December (European Commission, Brussels).
2. Hullman, A. (2006) *The Economic Development of Nanotechnology-An Indicators Based Analysis* (European Commission, Brussels).
3. Woodrow Wilson International Centre for Scholars (2012) *An Inventory of Nanotechnology-Based Consumer Products Introduced on the Market*. Available at: http://www.nanotechproject.org/inventories/consumer/.
4. Roco, C. M., C. A. Mirkin and C. M. Hersam (2010) *Nanotechnology Research Directions for Societal Needs in 2020: Retrospective and Outlook* (National Science Foundation/World Technology Evaluation Center Report, Washington, D.C.).
5. Australian Office of Nanotechnology (2008) *National Nanotechnology Strategy Annual Report 2007-08* (Australian Government, Canberra).
6. Smith, P. (2008) Can't afford to waste a nano in this industry, *Australian Financial Review*, 25 August, p. 32.
7. Malanowski, N. and A. Zweck. (2007) Bridging the gap between foresight and market research: Integrating methods to assess the economic potential of nanotechnology, *Technological Forecasting & Social Change*, **74**, 1805–1822.
8. Ebeling, M. (2008) Mediating uncertainty: Communicating the financial risks of Nanotechnologies, *Science Communication*, **29**(3), 335–361.

---

[16]A first start was made by pointing to the 'thick stakeholder consensus' in these types of arrangements, rather than the 'thin state consent' in more traditional intergovernmental forms of cooperation. In fact, it was argued that both new and traditional arrangements can offer legitimate forms of cooperation and that the conventional dividing line between formal and informal legal/regulatory arrangements—with only the former being effective, needing control or deserving legitimacy—no longer holds.

9. Seear, K., A. Petersen and D. M. Bowman (2009) *The Social and Economic Impacts of Nanotechnologies: A Literature Review* (Monash University, Melbourne).

10. Maynard, D. A., D. B. Warheit and M. Philbert (2011) The new toxicology of sophisticated materials: Nanotoxicology and beyond, *Toxicological Sciences*, **120**, 1–64.

11. Brown, S. (2007) Nanotechnology environmental health and safety standards. *The Magazine of the International Organization for Standardization*, **4**(4), 14–15.

12. Davies, J. C. (2006) *Managing the Effects of Nanotechnology* (Woodrow Wilson, Washington, D.C.).

13. Royal Commission On Environmental Pollution (2008) *Twenty-Seventh Report: Novel Materials in the Environment: The Case of Nanotechnology* (UK Government, London).

14. Azoulay, D. (2012) *Just Out of REACH: How REACH is Failing to Regulate Nanomaterials and How It Can be Fixed.* Available at: http://www.ciel.org/Publications/Nano_Reach_Study_Feb2012.pdf.

15. Australian Government (2010) Adjustments to NICNAS New Chemicals Processes for Industrial Nanomaterials, *Chemical Gazette No.C.10.* (October 5). Available at: http://www.nicnas.gov.au/Publications/Chemical_Gazette/pdf/2010oct_whole.pdf.

16. French Ministry (2012) *Décret n° 2012-232 du 17 février 2012 Relatif à la Déclaration Annuelle des Substances à L'état Nanoparticulaire Pris en Application de L'article L. 523-4 du code de L'environnement.* Available at: http://www.legifrance.gouv.fr/affichTexte.do?cidTexte=JORFTEXT000025377246&dateTexte=&categorieLien=id (available in French only).

17. Ryan, B. et al. (2011) Towards safe and sustainable nanomaterials: Chemical information call-in to manufacturers of nanomaterials by california as a case study, *European Journal of Law and Technology*, **2**(3), 1–11.

18. Bowman, D. M., G. van Calster and S. Friedrichs (2009) Correspondence, nanomaterials and the regulation of cosmetics, *Nature Nanotechnology*, **5**, 92.

19. Organisation for Economic Co-Operation and Development (2013) *Recommendation of the Council on the Safety Testing and Assessment of Manufactured Nanomaterials* (OECD, Paris).

20. Abbott, K. W. and D. Snidal (2009) Strengthening international regulation through transnational new governance: Overcoming the orchestration deficit, *Vanderbilt Journal of Transnational Law*, **42**, 501–578.

21. Black, J. (2008) Constructing and contesting legitimacy and accountability in polycentric regulatory regimes, *Regulation and Governance*, **2**, 137–164.

22. Börzel, T. and T. Risse (2005) Public-private partnerships: Effective and legitimate tools of international governance? In: Grande, E., and W. L. Pauly (eds.) *Complex Sovereignty: On the Reconstitution of Political Authority in the 21st Century* (University of Toronto Press, Toronto), pp. 195–217.

23. Quack, S. (2010) Law, expertise and legitimacy in transnational economic governance: An introduction, *Socio Economic Review*, **8**(3), 5–7.

24. Abbott, W. K. and D. Snidal (2009) The governance triangle: Regulatory standards institutions and the shadow of the state. In: Mattli, W., and N. Woods (eds.) *The Politics of Global Regulation* (Princeton University Press, Princeton), pp. 44–88.

25. Abbott, K., G. E. Marchant and E. Corley (2012) Soft law oversight mechanisms for nanotechnology, *Jurimetrics*, **52**, 279–312.

26. Dorbeck-Jung, B. R. and M. van Amerom (2008) The hardness of soft law in the United Kingdom: State and non-state regulatory conceptions related to nanotechnological development. In: Schooten, V. H., and J. Verschuren (eds.) *International Governance and Law. State Regulation and Non-State Law* (Edgar Elgar, Cheltenham), pp. 129–147.

27. European Commission (2008) *Commission Mandate M/409 Addressed to CEN, CENELEC and ETSI, Report From CEN/TC 352 Nanotechnologies 17* (European Commission, Brussels).

28. Forsberg, E. M. (2010) *The Role of ISO in the Governance of Nanotechnology* (Work Research Institute of Norway, Oslo).

29. Bowman, D. M. and G. A. Hodge (2009) Counting on codes: An examination of transnational codes as a regulatory governance mechanism for nanotechnologies, *Regulation & Governance*, **3**, 145–164.

30. Breggin, L., R. Falkner, N. Jaspers, J. Pendergrass and R. Porter (2009) *Securing the Promise of Nanotechnologies: Towards Transatlantic Regulatory Cooperation* report (Chatham House, London).

31. Falkner, R. and N. Jaspers (2012) Regulating nanotechnologies: Risk, uncertainty and the global governance gap, *Global Environmental Politics*, **12**(1), 30–55.

32. Mantovani, E. et al. (2010) *Developments in Nanotechnologies Regulation and Standards*. Available at: http://www. steptoe.com/

assets/htmldocuments/ObservatoryNano_Nanotechnologies_Regulation%20and%20Standards_June%202010.pdf.

33. Bowman, D. M. and G. A. Hodge (2008) Governing' nanotechnology without government?, *Science and Public Policy*, **35**(7), 475–487.

34. United States Environmental Agency (2007) *Nanotechnology White Paper* (EPA, Washington, D.C.).

35. Dorbeck-Jung, B. and C. Shelley-Egan (2013) Meta-regulation and nanotechnologies: The challenge of responsibiltisation within the European Commission's Code of Conduct for Responsible Nanosciences and Nanotechnologies Research, Science and Engineering Ethics, *NanoEthics*, **7**(1), 55–68.

36. Abbott, K. W., J. D. Sylvester and G. E. Marchant (2010) Transnational regulation of nanotechnology: Reality or romanticism? In: Hodge, G. A., D. M. Bowman, and A. D. Maynard (eds.) *International Handbook on Regulating Nanotechnologies* (Edward Elgar, Cheltenham), pp. 525–545.

37. Falkner, R. and N. Jaspers (2012) Regulating nanotechnologies: Risk, uncertainty and the global governance gap, *Global Environmental Politics*, **12**(1), 30–55.

38. Hallström T. K. and M. Boström (2010) *Transnational Multistakeholder Standardization* (Edward Elgar, Cheltenham).

39. Hale, T. and D. Held (2011) *Handbook of Transnational Governance: New Institutions and Innovations* (Polity Press, New York).

40. Djelic, M. L. and S. K. Andersson (2006) Introduction: A world of governance: the rise of transnational regulation. In: Djelic, M. L., and S. K. Andersson (eds.) *Transnational Governance: Institutional Dynamics of Regulation* (Cambridge University Press, Cambridge), pp. 1–31.

41. Handl, G. (2012) Introduction. In: Handl, G., J. Zekoll, and P. Zumbansen (eds.) *Beyond Territoriality: Transnational Legal Authority in an Age of Globalization* (Martinus Nijhoff Publishers, Nijhoff), pp. 3–13.

42. Koppell, S. (2010) *World Rule, Accountability, Legitimacy, and the Design of Global Governance* (University of Chicago Press, Chicago).

43. Pauwelyn, J. (2012) Informal international lawmaking: Framing the concept and research questions. In: Pauwelyn, J., R. A. Wessel, and J. Wouters (eds.) *Informal International Lawmaking* (Oxford University Press, Oxford), pp. 13–34.

44. Slaughter, M. A. (2004) *A New World Order*. (Princeton University Press, Princeton).

45. Pauwelyn, J., R. A. Wessel and J. Wouters (2014) When structures become shackles: Stagnation and dynamics in international lawmaking, *European Journal of International Law*, **25**(3), 733-763.

46. Ruggie, G. H. (2014) Global governance and new governance theory: Lessons from business and human rights, *Global Governance: A Review of Multilateralism and International Organizations*, **20**(1), 5-17.

47. European Commission (2008) *Towards an Increased Contribution from Standardisation to Innovation in Europe* (European Commission, Brussels).

48. Berman, A. and R. A. Wessel (2012) The international legal status of informal international law-making bodies: Consequences for accountability. In: Pauwelyn, J., R. A. Wessel, and J. Wouters (eds.), *Informal International Lawmaking* (Oxford University Press, Oxford), pp. 35-62.

49. Knill, C. and D. Lehmkuhl. (2002) Private actors and the state: Internationalization and changing patterns of governance, *Governance*, **5**(1), 41-64.

50. Bonny, S. (2003) Why are most Europeans opposed to GMOs? Factors explaining rejection in France and Europe, *Electronic Journal of Biotechnology*, **6**(1), 50-58.

51. Forsberg, M. E. (2011) Standardization in the field of nanotechnology: Some issues of legitimacy, *Science and Engineering Ethics*, **18**, 1-21.

52. National Research Council (2012) *A Research Strategy for Environmental, Health and Safety Aspects of Engineered Nanomaterials* (National Academies Press, Washington, D.C.).

53. ETC Group (2007) *An Open Letter to the International Nanotechnology Community at Large* (ETC Group, Ottawa).

54. Kica, E. and D. M. Bowman (2012) Regulation by means of standardization: Key legitimacy issues of health and safety nanotechnology standards, *Jurimetrics*, **53**, 11-56.

55. Homkes, R. (2011) *Analysing the Role of Public-Private Partnerships in Global Governance: Institutional Dynamics, Variation and Effects* (PhD Thesis, London School of Economics, London).

56. Risse, T. (2006) Global governance and communicative action. In: Held, D., and M. Koenig-Archibugi (eds.) *Global Governance and Public Accountability* (Wiley Publishing, New York), pp. 167.

57. Cadman, T. (2012) *Quality and Legitimacy of Global Governance* (Palgrave Macmillan, New York).
58. Koenig-Archibugi, M. (2006) Introduction: Institutional diversity in global governance. In: Koenig-Archibugi, M., and M. Zürn (eds.) *New Modes of Governance in the Global System: Exploring Publicness, Delegation and Inclusiveness* (Palgrave Macmillan, Basingstoke), pp. 1–30.
59. Ruggie, G. H. (2014) Global governance and new governance theory: Lessons from business and human rights, *Global Governance: A Review of Multilateralism and International Organizations*, **20**(1), 5–17.
60. Knill, C. and D. Lehmkuhl (2002) Private actors and the state: Internationalization and changing patterns of governance, *Governance*, **5**, 42.
61. Andonova, L., M. Betsill and H. Bulkeley (2009) Transnational climate governance, *Global Environmental Politics*, **9**(2), 52–73.
62. Martens, J. (2007) *Multistakeholder Partnerships-Future Models of Multilateralism?* (Friedrich-Ebert-Stiftung, Berlin).
63. Liese, A. and M. Beisheim (2011) Transnational public-private partnerships and the provision of collective goods in developing countries. In: Risse, T. (ed.) *Governance Without a State–Policies and Politics in Areas of Limited Statehood* (Columbia University Press, New York), pp. 115–144.
64. International Organization for Standardization Central Secretariat (2005) *Proposal For A New Field Of Technical Activity* (ISO, Geneva).
65. Kica, E. and D. M. Bowman (2013) Transnational governance arrangements: legitimate alternatives to regulating nanotechnologies?, *NanoEthics*, **7**(1), 69–82.
66. International Organization for Standardization/International Electrotechnical Commission (2011) *Part 1: Procedures For The Technical Work 11* (ISO/IEC, Geneva).
67. International Organization for Standardization (2012) *ISO TC/229 Business Plan* 8 (ISO, Geneva).
68. Hatto, P. (2008) *Innovative Approaches to Nanotechnology Environmental Governance: Standardization-In Support of Nanogovernance*. Available at: http://www.nanogovernance.com/images/Hatto.pdf.
69. David, S. E. (2007) Update on ISO nanotechnology standards activities, *Clean Rooms*, **12**, 12–13.

70. Organisation for Economic Co-Operation and Development (2013) *Recommendation of the Council on the Safety Testing and Assessment of Manufactured Nanomaterials* (OECD, Paris).
71. Visser, R. (2007) A sustainable development for nanotechnologies: An OECD Perspective. In: Hodge, G. A., D. M. Bowman, and K. Ludlow (eds.), *New Global Frontiers In Regulation: The Age Of Nanotechnology* (Edward Elgar, Cheltenham), pp. 320–333.
72. Organisation for Economic Co-Operation and Development (2011) *Safety of Manufactured Nanomaterials*, Environmental Health and Safety News (OECD, Paris).
73. Organisation for Economic Co-Operation and Development (2011) *Nanosafety at the OECD: The First Five Years 2006–2010* (OECD, Paris).
74. Renn, O. and M. Roco (2006) *White Paper on Nanotechnology Risk Governance* (International Risk Governance Council, Geneva).
75. International Risk Governance Council (2007) *Policy Brief: Nanotechnology Risk Governance, Recommendations for a Global, Coordinated Approach to the Governance of Potential Risks* (IRGC, Geneva).
76. International Risk Governance Council (2006) *Survey on Nanotechnology Governance: The Role of NGOs* (IRGC, Geneva).
77. Pelley, J. and M. Saner (2009) *International Approaches to the Regulatory Governance of Nanotechnology* (Carleton University, Ottawa).
78. International Council on Nanotechnology (2009) *Governance Structure and Operational Plan* (ICON, Houston).
79. Kulinowski, M. K. and J. P. Matthew (2009) The GoodNanoGuide: A novel approach for developing good practices for handling engineered nanomaterials in an occupational setting, *Nanotechnology Law & Business*, **6**(37), 37–44.
80. Intergovernmental Forum on Chemical Safety (1997) President's Progress Report 1994–1997, *Forum II Second Session of the Intergovernmental Forum on Chemical Safety*, 1–14. Available at: http://www.who.int/ifcs/documents/forums/forum2/pres_e.pdf.
81. Intergovernmental Forum on Chemical Safety (2008) *Sixth Session of the Intergovernmental Forum on Chemical Safety: Final Report*. Available at: http://www.who.int/ifcs/documents/forums/forum6/f6_finalreport_en.pdf.
82. Earth Negotiations Bulletin (2012) *International Institute for Sustainable Development*. Available at: http://www.iisd.ca/chemical/iccm3/compilatione.pdf.

83. European Parliament (2006) Nanoscience and nanotechnology. European Parliament resolution on nanosciences and nanotechnologies: An action plan for Europe 2005–2009, *Official Journal of the European Union* (European Parliament, Brussels).

84. Bowman, D. M. and G. Gilligan (2007) How will the regulation of nanotechnology develop? clues from other sectors. In: Hodge, G. A., D. M. Bowman, and K. Ludlow (eds.) *New Global Frontiers In Regulation: The Age Of Nanotechnology* (Edward Elgar, Cheltenham), pp. 353–385.

85. European Commission (2007) *Nanosciences and Nanotechnologies: An Action Plan for Europe 2005–2009* (European Commission, Brussels).

86. Mercier, J. M. (1995) Chemical safety: A global challenge. *Environmental Health Perspectives,* **103**(10), pp. 886–887.

87. International Risk Governance Council (2006) *The Risk Governance of Nanotechnology: Recommendations for Managing a Global Issue 6–7 July 2006* (IRGC, Geneva).

88. European Commission (2007) *Towards a Code of Conduct for Responsible Nanosciences and Nanotechnologies Research* (European Commission, Brussels).

89. European Commission (2007) *Opinion 21 on the Ethical Aspects of Nanomedicines* (European Group on Ethics in Science and New Technologies to the European Commission, Brussels).

90. Thatcher, M. and A. Stone Sweet (2002) Theory and practice of delegation to non-majoritarian institutions, *West European Politics*, **25**(1), 1–22.

Chapter 11

Co-Regulation of Nanomaterials: On Collaborative Business Association Activities Directed at Contributing to Occupational Health and Safety

Aline Reichow

Federal Institute for Occupational Safety and Health,
Friedrich-Henkel-Weg 1-25, D-44149 Dortmund, Germany

11.1 Introduction

The regulation of employee health and safety in work with nanomaterials is a precarious matter. In most jurisdictions worldwide, employers are legally obliged to protect the health of their employees through having risk assessment and management in place. But since scientifically coherent risk data for nanomaterials is scarce, it remains unclear whether existing health and safety frameworks are evidently effective at protecting employee health. To resolve the uncertainty surrounding nanomaterials occupational health and safety (OHS) calls have been made for regulators to cooperate with nano-companies in order to uncover and avoid risk [1, 2]. Likewise, industry at large has been asked

---

Embedding New Technologies into Society: A Regulatory, Ethical and Societal Perspective
Edited by Diana M. Bowman, Elen Stokes, and Arie Rip
Copyright © 2017 Pan Stanford Publishing Pte. Ltd.
ISBN 978-981-4745-74-1 (Hardcover), 978-1-315-37959-3 (eBook)
www.panstanford.com

to engage in global networking for the purpose of sharing scientific data and knowledge which would allow for improvements to traditional risk assessment frameworks in the context of nanomaterials [3].

Collaboration between business associations, firms and government is said to offer mutual benefits to the involved parties [4, 5]. In systems of co-regulation, government agencies retreat from direct intervention and rely on third parties, like companies or business associations, as 'surrogate regulators' who decide how to regulate specific activities in the best way [6]. While the government directs regulatory responsibilities to third parties, who then develop and administer their own arrangements,[1] it still provides legislative backing by enforcing the arrangements; as both the interests of state and private actors are met, one can speak of co-produced regulation [8]. Co-regulation is expressed most commonly with regard to activities designed to develop and implement non-legally binding soft instruments[2] aimed at achieving public policy goals [4, 11–17].

Business associations in the chemical sector are actively engaged in the discussion on nanomaterials OHS. Both in the EU and US leading associations have made this topic one of their top priorities. At the same time, a clear understanding of the nature of business association involvement in the regulation of nanomaterials OHS is lacking. One of the most active associations in the EU context is the German Chemical Industry Association (VCI). Since 2003 the VCI has been engaged in the topic of workplace health and safety regarding nanomaterials [18].

For the purposes of this chapter the VCI, as nationally and internationally renowned business association, serves as an example to explore the supposed role of business associations to be vital partners to government in the co-regulation of nanomaterials OHS. Based on the analysis of publicly available documents, key collaborative activities of the VCI directed at nanomaterials OHS will be mapped and analysed for the time period of 2003 until 2013. In this regard, the following question will be discussed: With which approach can collaborative

---

[1] Instead of referring to 'regulatory arrangements' some authors refer to the term 'self-regulation' [7].

[2] Soft regulation instruments comprise, e.g. voluntary reporting schemes, voluntary risk management systems and benchmarks, codes of conducts, guidelines and auxiliaries [9] and they might be established by public as well as private organisations [10].

business association activities directed at contributing to nanomaterials OHS regulation be described? And relatedly, by means of which activities does the VCI actually contribute to nanomaterials OHS regulation and which position towards nanomaterials OHS is taken accordingly? Lastly, how can the involvement of the VCI in the regulation of nanomaterials be evaluated? As to the latter question a conceptual discussion on the effectiveness of VCI activities in the regulation of nanomaterials OHS will be provided, which serves as basis for an informed deliberation on (future) challenges and opportunities of nanomaterials OHS co-regulation.

Key regulatory bodies such as the European Commission (EC) consider existing regulations that cover chemicals and OHS to be adequate for nanomaterials [19]. Their line of reasoning is that nanomaterials may not be regulated specifically, but they are regulated generally as 'chemical substances'.[3] With regard to OHS practices, general statements rather than precise definitions are used to describe principles and duties of care with the aim of integrating health and safety policy into general business policy. In the EU (and most other jurisdictions worldwide) employers are legally obliged to care for safe workplaces through having risk assessment and management in place.[4]

---

[3]A chemical substance is defined as 'chemical element and its compounds in the natural state or obtained by any manufacturing process, including any additive necessary to preserve its stability and any impurity deriving from the process used, but excluding any solvent which may be separated without affecting the stability of the substance or changing its compositions' (see Regulation (EC) No 1907/2006 concerning the Registration, Evaluation, Authorisation and Restriction of Chemicals (REACH) [2006] OJ L396/11, Article 3).

[4]For instance, in the US Nanomaterials are regulated under the Toxic Substances Control Act (TSCA); Section 2 states that manufacturers and processors of chemical substances are required to develop 'adequate data' with respect to the effect of chemical substances and mixtures on health and the environment prior to manufacturing or introducing them to commerce. The Environmental Protection Agency (EPA) can require testing after finding that (1) a chemical may present an unreasonable risk of injury to human health or the environment, and/or the chemical is produced in substantial quantities that could result in significant or substantial human or environmental exposure, (2) the available data to evaluate the chemical are inadequate, and (3) testing is needed to develop the needed data (EPA 2011). In Australia the Model Work Health and Safety Regulations Act, Part 3(1) requires employers to identify and assess workplace risks to health and safety of employees and to implement adequate control measures to manage risks. The Act was adopted in 2012 and by the end of December 2016 all workplace chemicals must be classified and updated according to the new provisions [20].

Chemicals handled in workplaces are regulated by various European Union (EU) Directives,[5] which are generally applicable to chemical substances regardless of their particle size or particular mode of production, and hence apply equally to conventional materials as well as nanomaterials. One of the essential pieces of legislation with regard to health and safety at work is the Framework Directive 89/391/EEC on the safety and health of workers at work.[6]

The Framework Directive does not mention nanomaterials, but because of its broad remit, it is deemed to cover the potential risks of nanomaterials and lays down the specific tasks of the employer obligation to care for safe workplaces. To that end the employer shall (a) be in possession of an assessment of the risks to safety and health at work, (b) decide on the protective measures and equipment to be taken, (c) keep a list of occupational accidents, and (d) draw up for the responsible authorities on occupational accidents.[7] Additionally the Registration, Evaluation, Authorisation and Restriction of Chemicals (REACH)[8] regulates all chemical substances meeting certain thresholds (e.g. volume of production) stipulated in the legislation, and is therefore considered to cover nanomaterial chemical substances too. While it should be noted that REACH will not be fully implemented until 2018, it progressively replaces various

---

[5]Directives relevant to risks that may result from possible exposure to nanomaterials are: Directive 2004/37/EC of the European Parliament and of the Council of 29 April 2004 on the protection of workers from the risks related to exposure to carcinogens or mutagens at work; Council Directive 98/24/EC of 7 April 1998 on the protection of the health and safety of workers from the risks related to chemical agents at work; Council Directive 89/655/EEC of 30 November 1989 concerning the minimum safety and health requirements for the use of work equipment by workers at work; Council Directive 89/656/EEC of 30 November 1989 on the minimum health and safety requirements for the use of personal protective equipment at the workplace and, Directive 1999/92/EC of the European Parliament and of the Council of 16 December 1999 on minimum requirements for improving the safety and health protection of workers potentially at risk from explosive atmospheres.

[6]Council Directive 89/391/EEC on the introduction of measures to encourage improvements in the safety and health of workers at work [1989] OJ L 183/1.

[7]Ibid, Article 9(1).

[8]Regulation (EC) No 1907/2006 concerning the Registration, Evaluation, Authorisation and Restriction of Chemicals (REACH) [2006] OJ L396/11.

Directives and Regulations applicable to chemicals. REACH requires manufacturers and importers of chemical substances, to submit a technical dossier with information regarding the substance properties in question and proposed tests in order to establish the chemicals' safety.[9] Similar to the Framework Directive, under REACH the risks associated with chemical substances are to be characterised and exposure scenarios must be estimated for purposes of risk assessment.[10]

However, conducting risk assessment for nanomaterials is currently a challenge. Overall, for nanomaterials the scientific basis of risk assessment suffers from considerable limitations [3]. According to traditional OHS methods, risk assessment follows the process of hazard identification and characterisation, exposure assessment and risk characterisation [21, 22]. As to the first step, a challenge is that there exists no model to predict hazard solely based on the physicochemical characteristics of nanomaterials, which could be used for risk assessment [23]. Undertaking exposure assessments for nanomaterials also suffers various difficulties. While traditional risk assessment relies on the mass of chemical substances as the exposure dose, with nanomaterials various other exposure parameters (e.g. the surface area and its reactivity) have been identified as potential key elements; traditionally these parameters are not considered in exposure measurement [24].

An additional problem is the lack of standard techniques to measure exposure in relation to toxicity [25]. These issues with risk assessment culminate in the final step of risk characterisation. The characterisation of nanomaterials risks is marked by concerns about unidentified or poorly quantified adverse effects [24]. Despite these issues, the traditional risk assessment framework has been identified as adequate for nanomaterials under the premise that 'some amendments' will

---

[9]Ibid, Article 10.
[10]Council Directive 98/24/EC on the protection of the health and safety of workers from the risks related to chemical agents at work [1998] OJ L 131, Article 4 (this Directive is within the meaning of Article 16(1) of Directive 89/391/EEC on the introduction of measures to encourage improvements in the safety and health of workers at Work [1989] OJ L 183, 29.6.1989, p. 1); Regulation (EC) No 1907/2006 concerning the Registration, Evaluation, Authorisation and Restriction of Chemicals (REACH) [2006] OJ L396/11, Annex 1 General Provisions for Assessing Substances and Preparing Chemical Safety Reports.

be necessary [19, 26]. In light of this, risk assessment to be applicable to nanomaterials will need to be reconsidered. Such a task requires the generation of fundamental scientific data that is currently lacking.

Business associations are said to possess scientific data and knowledge crucial for the understanding and regulation of potential risks of emerging technologies such as nanomaterials [27]. However, as will be argued in the next section, even though business associations might potentially be supportive in the generation and collection of scientific data, such activities need to be understood in relation to other activities that follow a logic of association self-sustainment since, in the economic sector, business associations and companies are profit-oriented organisations [4]. In the next section, this approach to business association activities will be discussed.

11.2 Approach to Business Association Activities

Before discussing any specific association activities, it is worth outlining the different categories of business associations. The most comprehensive discussion on forms of business associations[11] is provided by Bennett [31]. He articulates six forms of business associations distinguishable by their types of members (see Fig. 11.1 for an overview):

(1) *Associations of companies* with mainly large companies (>200 employees) as members,
(2) *Associations of owner-managers* with chiefly small companies (2–10 employees) as members,
(3) *Associations of self-employed professionals* with predominantly individuals as members,
(4) *Professional associations of individuals* with employees from all classes of companies as members,

---

[11]One should note that in the literature at large terminology is applied incoherently. Different authors speak of 'trade associations', 'industry associations' or 'professional associations' by referring to the same phenomenon. In this chapter Bennett's [28–30] definition of 'business associations' is followed broadly. He is one of few scholars who provide a coherent distinction across various forms of business associations.

(5) *Mixed associations* (but rather small companies, owner managers, self-employed, individuals), and

(6) *Federations* which are associations of associations.

Collective goods	Selective goods
Business reputation & representation	Specific member status
Lobbying	Accredited standards (technical/trading)
Soft regulation	Marketing support
Information exchange (scientific/technical/political)	Business advice

Figure 11.1 Overview of typical activities of business associations.

It is useful to distinguish associations according to their types of members as this provides a first understanding of specificities of association activities and dynamics reflected in their demand orientation [28], as well as their potential for influence in public policy [32]. The type of business association most active in the context of nanomaterials OHS in Germany is *Associations of companies*. This type of business association is most often discussed in the regulatory literature at large, albeit in different terms. Before going into the particularities of activities by the German VCI, first a general discussion of association 'logic of existence' in the context of Germany is required so as to develop an approach by which collaborative association activities can be described.

Generally speaking, business association activities centre on the provision of collective and individual or selective services to their members [33, 34]. goods benefit all businesses (relatively) equally [30]. Associations emerge and are maintained when a certain collective interest of involved businesses is perceived, that is when all, or some, companies in a sector are better off acting collectively rather than individually [35]. A well-known and widely discussed example is the protection of business reputation or representation; especially in potentially more hazardous industries like the chemical industry, companies are 'hostages to each other' [36] since the major accident occurring company usually damages the reputation of the rest thereby prompting new government legislation applicable to the sector as a whole [37, 38].

Further, through lobbying associations aim at supplementing or complementing government regulation [39] or they intend to minimise the impact of unfavourable regulation through bargaining or manipulating information and publicity [40]. Associations may also forestall government legislation by developing and implementing sector-wide non-legally binding soft regulation (e.g. codes of conducts, guidelines, technical standards) in the light of threats of future government regulation [41]. On the other hand, soft regulation can play an important risk management role by providing education and information to member businesses or by interacting with government over the best ways to achieve required (risk) standards [42–45]. Associations communicate such standards to businesses collectively and in that way coordinate a standard of behaviour [46].

In comparison to collective services, selective goods benefit only individual businesses [30]. Examples are specific business advice or targeted marketing support, which serve as incentives for association membership. Technical standards may also constitute selective goods if they are provided by associations solely to member companies. Providing individual goods allows associations to obtain income from membership fees and other sources so as to support their collective activities [31].

There is a need for associations to balance collective and selective activities: on the one hand, they need to fulfil the needs and demands of their members in order to 'survive' (in terms of income by member fees). On the other hand, associations need to improve their relations with government (in order to stabilise and increase their influence in regulatory processes) [30, 30].

11.2.1 Business Associations in Germany

In Germany, business associations are known to follow a tradition of economic self-organisation, which supports the collective organisation of business interests significantly [47]. German associations have been, and constantly are, an important feature of economic life that is independent from changes in political regimes through forming a highly centralised and hierarchical system of organisation [48–50]. German business associations are said to offer both collective as well as selective goods.

Prime selective goods are research and development services to their members as well as the provision of individual (business) advice and consultancy services [46]. The systematic development of soft regulation (e.g. best practices) is a particularly noteworthy collective good of German associations. Another meaningful selective good offered by German associations are technical standards. Looking back in history, both the production and the implementation of technical norms have been important association activities all throughout the 19th century [51]. By offering technical standards for the protection of workers some associations gained nearly a monopoly position in their industry, which secured generally high levels of membership [49]. Many German companies even render their membership as de facto compulsory [30]. As a result, associations possess a high degree of authority to create rules of behaviour common to a particular industry.

The ability of business associations to attract and keep members determines their material resources. German associations obtain continuous income from both the fees of member firms and their provision of selective goods. Additionally many associations operate separate businesses through which income can be generated independent from membership [51]. Their relatively healthy financial situation enables German associations to establish and maintain a high level of professionalisation. Typically they employ qualified staff with graduate and postgraduate education [47, 52]. Well-resourced associations are said to have advantages over those that are not: they can lower administrative costs, improve compliance with regulation, apply knowledge more efficiently and they can facilitate immediate response of regulations to technical and market developments in the light of new knowledge [29, 53].

Overall, the organisation and structure of German business associations tends to be highly formalised. Associations control a high degree of resources as well as membership and they can offer a mixture of collective and selective goods to members while they can be useful collaborators to policy-making agencies. The next section discusses whether this general picture of German business associations applies to the most active association in the context of nanomaterials OHS. While business associations are expected to collect and share scientific data that would support the characterisation of nanomaterials risks through active collaboration with public policy makers, it has been emphasised in this section that business associations wear many hats. Even though associations might be willing and able to support nanomaterials OHS, at the same time, these organisations follow and are organised around complex collective and individual interests. Considering this general approach to business association activities, by which kind of activities does the German VCI actually contribute to nanomaterials OHS regulation?

11.2.2 The Chemical Industry Association (VCI)

Compared to other European business associations populated predominantly by 'companies', the Chemical Industry Association (Verband der Chemischen Industrie, VCI) is the most active in supporting health and safety at workplaces where nanomaterials are handled. The VCI was founded in 1877 and has more than 1600 member companies accounting for more than 90% of all German chemical firms [54]. Membership in the VCI is possible via two routes: A company may become either an ordinary or an associated member and must be a manufacturer in the chemical or a directly related industry with its headquarters being situated in Germany (also German branch offices of foreign countries are permissible) [55].

Since 2003, the VCI has participated in discussions related to nanomaterials OHS and has initiated a large number of activities accordingly. As will be shown in the course of this section, the VCI delivers a broad range of collective goods related to nanomaterials OHS ranging from information exchange to reputation, representation and lobbying. The most significant VCI collective goods are soft regulation and information exchange.

In the period from 2003 through to 2013, the VCI has initiated in total 19 activities directed at supporting nanomaterials OHS. Based on the publicly available documents related to each these activities the VCI position towards nanomaterials OHS regulation will be elicited over time (see the complete overview of VCI activities and positions towards nanomaterials OHS in the appendix). To that end, the VCI activities are grouped into three main periods with regard to the predominant type of activities in terms of collective and/or selective association goods:

(1) 2003–July 2007
(2) August 2007–September 2011
(3) October 2011–January 2013

Period 1. In the time from 2003 to July 2007, the VCI initiated five activities directed at supporting nanomaterials OHS all of which represent collective association goods. The activities in this period can be understood as 'setting the scene' for an informed discussion on nanomaterials OHS involving business associations, companies, policy makers and researchers. In 2003, by forming a joint working group with the Society for Chemical Engineering and Biotechnology (DECHEMA), namely the group Responsible Production and Use of Nanomaterials, the VCI advocated the need to obtain scientific data and best practices on safety aspects of nanomaterials [56].[12] This activity constitutes a collective association good directed at representing responsible behaviour by the VCI towards nanomaterials OHS.

In September 2005, the VCI organised a workshop on uses of nanomaterials in the workplace. The association pointed to the need for safety research and initiated a broad stakeholder discussion on future steps to be taken in this respect. As such, the activity was primarily aimed at information exchange among stakeholders. Next, in February 2006, the VCI conducted a survey on nanomaterials OHS practices among association member companies so as to collect and exchange information on OHS methods applied in the chemical industry. Based on the results of this survey the VCI emphasised the need to develop soft regulation that would guide companies in the safe handling and use of nanomaterials.

In April 2007, another workshop (as an extension to the one held in 2005) was organised to exchange information on nanomaterials OHS. Consequently the VCI stressed the need to further develop soft regulation for nanomaterials OHS, which should pay special attention to exposure measuring techniques,

---

[12]The working group consists of academic and industrial experts and is regularly joined by representatives from German public authorities [56]. The acronym DECHEMA stands for Society for Chemical Engineering and Biotechnology, which is a German interdisciplinary organisation with the aim to stimulate exchange and transfer of scientific knowledge and industrial applications in the area of engineering, biotechnology and process engineering.

protection methods and issues of communication across involved actors. Lastly, in July 2007 a Roadmap for Safety Research on Nanomaterials was developed, based on which the VCI questioned the appropriateness of existing exposure measurement techniques and toxicological testing strategies for potential risks of nanomaterials. The association lobbied for the development of internationally harmonised standards on analytical measuring methods.

To sum up, during the period from 2003 to July 2007 the VCI commenced five activities directed at nanomaterials OHS. All activities constituted collective association goods with a focus on information exchange (three activities). To a limited extent activities were characterised as representation/reputation (one activity) and lobbying (one activity). In the following period, from August 2007 to September 2011 a different focus of VCI activities emerges.

Period 2. Between August 2007 and September 2011, the VCI initiated nine activities directed at supporting nanomaterials OHS, and again all activities constituted collective goods. On a general level, this period is characterised by association activities that began to formulate more and more concrete advice on how to handle nanomaterials at workplaces safely. Also legislators started to engage in this debate. Against this background, based on the results of the 2006 survey, the VCI developed a guideline for handling and use of nanomaterials at workplaces [57].

In August 2007, this guidance was drawn up together with the Federal Institute for Occupational Safety and Health (BAuA), in line with the principles of the global Responsible Care Program[13] of the chemical industry [56]. Again the VCI stressed the need to develop soft regulation so as to provide orientation for companies in matters of nanomaterials OHS measures, based on the latest state of science. The association voiced concerns

---

[13]The Responsible Care Program (RCP) was launched in 1988 by the Chemical Manufacturers Association in the US and has widely been discussed in the regulatory literature [38, 58]. The program contains more than 100 specific management practices to be followed voluntarily by the member companies of most chemical business associations worldwide. There has been ample discussion as to the need for third party validation and enforcement of company adherence to the RCP principles, but principally, if firms fail to meet the requirements they will be expelled [59]. By March 2012, more than 150 of the world's largest chemical companies ascribed to the program [60].

that exposure to nanoparticles might have effects different from those of larger particles and that the chemical industry has a decisive role in commencing research accordingly. In October 2007 the VCI strengthened this position further with the publication of a strategy paper on the standardisation of nanomaterials. Here the responsibility and reputation of the chemical industry to provide expertise on matters of chemical risk assessment was emphasised.

This call was put into practice with the development of five soft regulatory instruments (guidance documents) throughout 2008. Without going into the details of each one instrument, it suffices here to note that the VCI paid most attention to the appropriateness of existing risk assessment frameworks for nanomaterials in contexts such as REACH or the Responsible Care Program. Essentially, the VCI postulated that existing risk assessment methods may require adaptation for nanomaterials and the chemical industry would be suited and willing to take a leading position therein.

Against this background, the association acknowledged that there was much uncertainty on potential hazardous effects of nanomaterials, which should be resolved through exchange of knowledge among key stakeholders [61]. To this end, the VCI organised a workshop in September 2008 in which actors from academia, politics and consumer organisations participated; the VCI stressed the need for ongoing exchange of scientific data for the future of nanomaterials safety. Building on the soft regulation instruments from 2008, the VCI developed another guidance document on nanomaterials exposure measurement and assessment in collaboration with many research institutes. Here the association put forward the opinion that a harmonised and pragmatic approach to nanomaterials exposure measurement was required to be applicable by small and medium-sized enterprises (SMEs) and large chemical operations alike. While exposure-monitoring techniques exist, these would require improved applicability in routine workplace operations.

In sum, between 2007 and 2011, the VCI initiated nine activities contributing to nanomaterials OHS. All of these activities constitute collective association goods with a clear focus on soft regulation (seven activities), while other activities are characterised as information exchange (one activity) and representation/reputation (one activity). This focus does not come as a surprise against the background of political action. On 17 June 2008, the European Commission published its first regulatory review of nanomaterials in a Communication on regulatory Aspects of Nanomaterials. In its Communication, the Commission stresses that Framework Directive 89/391/EEC on safety and health at work, which places obligations on employers to take measures for the protection of safety and health of their employees, fully applies to nanomaterials. This means that employers must carry out risk assessments for nanomaterials handled at workplaces. In this light the VCI appeared most active in supporting companies in their day-to-day work with nanomaterials and by helping to comply with their legal duty to conduct risk assessment for nanomaterials through providing extensive soft regulation instruments. Here it is noteworthy to mention that the VCI carries out own studies related to nanomaterials OHS by collaborating with scientists and participating in cross-sectoral, publicly funded EU-research projects (including, for example, NanoCare, Nanoderm, Nanosafe, NanoNature) [62].

Period 3. Between October 2011 and February 2013, VCI activities show a focus on direct participation in political discussions. A hallmark event of this period was the European Commission Recommendation for a definition of a nanomaterial from 18 October 2011.[14] Even though the EC definition is not in itself legally binding, it may be enforced by recourse to legally binding policy instruments such as Directives. Immediately after its publication the content of the EC nanomaterial definition[15] was widely criticised by industry and the VCI published a reaction

---

[14]Commission Recommendation No 2011/696 on the definition of nanomaterials [2011] OJ L275/38.

[15]According to the EC Recommendation, a nanomaterial is '[a] natural, incidental or manufactured material containing particles, in an unbound state or as an aggregate or as an agglomerate and where, for 50% or more of the particles in the number size distribution, one or more external dimensions is in the size range 1 nm–100 nm. In specific cases and where warranted by concerns for the environment, health, safety or competitiveness the number size distribution threshold of 50% may be replaced by a threshold between 1% and 50%. By derogation from the above, fullerenes, graphene flakes and single wall carbon nanotubes with one or more external dimensions below 1 nm should be considered as nanomaterials'.

in which the position is taken that the proposed definition was too broad; in result the definition would apply to many chemicals that are on the market for decades already thereby posing an unnecessary burden to companies. Further, the VCI advocated that the EC definition needed to be linked with exposure measuring methods and, in its 2011 version, would hinder worldwide harmonisation of nanomaterials regulation in that the EC is internationally the only political body that has decided for a definition of nanomaterials to be incorporated into legislation [63].

In November 2011, the association reacted to a report on nanomaterials risks by the German Advisory Council of the Environment. In addition to calls for safety research on nanomaterials (which would substantiate any claims that nanomaterials harm human health), the VCI lobbied for a harmonised definition of nanomaterials, for the rejection of nano-specific legislation and for more the development of appropriate risk assessment and management for nanomaterials in the context of REACH. The latter standpoint was taken up again in May 2012, with the publication of the revised 2007 soft regulation (guidance) in updated form [64]. Here the VCI stressed that nanomaterials have new characteristics, which require more research and specific safety measures and the conduction of risk assessment. The association emphasised the complexity of existing exposure measurement techniques that oftentimes cannot be conducted by SMEs but merely experienced research institutes. In October 2012, the association published a reaction to the EC Communication on the 2nd Regulatory Review of Nanomaterials[16] in which they lobbied for the adequacy of traditional risk assessment methods for nanomaterials [54, 61]. While existing regulations might have to be adapted for nanomaterials, current legislation (i.e. REACH) is in principal thought to be adequate. Lastly, in February 2013 another lobbying activity was commenced with the VCI position paper on the EC REACH report of the same month [65]. The association agreed with the essential viewpoint of the EC as to the appropriateness of REACH for nanomaterials, but at the same time it demanded

---

[16]Communication from the Commission to the European Parliament, the Council and the European Economic and Social Committee *Second Regulatory Review on Nanomaterials* COM (2012) 572 final.

provision of additional, nano-specific guidance for companies in the conduction of risk assessment for nanomaterials.

Thus, between 2011 and 2013, the VCI initiated five activities with a clear focus on lobbying (four activities). While the importance of soft regulation for their member companies in conducting nanomaterials risk assessment still had been emphasised by the VCI (one activity), at the same time the importance of becoming involved in the political discussion around nanomaterials OHS regulation had also been stressed. Again, all activities represented collective association goods (see Fig. 11.2 for an overview).

Time	Activity	Type
2003–2007	Information exchange (3×) Representation/reputation (1×) Lobbying (1×)	Collective goods
2007–2011	Soft regulation (7×) Information exchange (1×) Lobbying (1×)	Collective goods
2011–2013	Lobbying (5×) Soft regulation (1×)	Collective goods

Figure 11.2 Characteristics of 19 VCI activities between 2003 and 2013.

Taking into account all VCI activities throughout 2003 and 2013, interestingly the VCI had actively supported OHS and risk assessment at workplaces where nanomaterials are handled exclusively by means of collective association goods. The provision of selective goods related to nanomaterials OHS has not been observed. However, delivering selective goods that benefit only individual members are said to be vital incentives for association membership by companies which, in turn, allows associations to obtain income through membership fees [30, 31]. One explanation could be that the VCI in Germany has a near monopoly position, which might underlie the high number of members even in the absence of any provision of selective goods: almost 90% of all chemical companies in Germany are VCI members [54] with the topic of nanomaterials OHS merely being one amongst a great variety of other topics in which the association is actively engaged and for which they might provide selective goods.

This could give the impression that the VCI has sufficient other sources of income and therefore is not dependent on income constituted by selective nanomaterials OHS-related activities. However, drawing any conclusions in this respect should not be based solely on the analysis of publicly available documents but other data sources shall be considered. Furthermore, whether or not offering selective goods in support of nanomaterials OHS could be explained by member companies actually not requiring them can only be speculated about here.

What can be derived from the information available here is the VCI position towards nanomaterials OHS. Accordingly, five key-observations can be made:

(1) In the early years (2003–2007) the VCI saw a need for more research and collaboration on nanomaterials and potential risks in the context of OHS; in this regard, activities were initially directed at collecting and generating scientific data. While the focal interest was to support member companies, the association clearly saw the need to collaborate with policy makers.

(2) In the time from 2007 to 2011, as a result of the persistent uncertainty on the applicability of traditional risk assessment frameworks for nanomaterials, the VCI developed various soft regulatory instruments.

(3) In recent years (2011–2013) this focus on company supportive activities developed into a more concrete VCI position that some knowledge on nanomaterials would be available already, suggesting that there are no recognised health effects specific to nanomaterials (nanomaterials are like any other chemical).

(4) Based on this emerging knowledge on nanomaterials OHS, the VCI postulated that general, non-nano specific regulation would be adequate. The association sees the need to be involved in the regulatory process so as to represent the interests of the German chemical industry.

All in all, the VCI appears actively involved in the regulation of nanomaterials OHS by collecting and generating scientific data on nanomaterials and by collaborating with policy-makers such as suggested in the literature. In this light, questions as to the value of the involvement become relevant. Hence, how can collaborative business association activities directed at nanomaterials OHS regulation be evaluated in terms of their effectiveness? Rather than providing empirical answers to this question in the following section the evaluation of effective regulation is discussed on a conceptual level.

11.3 Discussion: Effective Nanomaterials OHS Regulation

Typically, regulatory effectiveness is defined as the degree to which policy goals have been achieved [66]. From this perspective the focus is on the *outcome* of regulation. Policy goals are said to be achieved when their underlying rules are followed. In that way, effectiveness relates to regulatee compliance with rules. Rule compliance depends on the willingness (or motivation) and the capacity of regulated parties to follow rules [67]. For each of these two elements of effectiveness various conditions have been developed [1, 68–72].

In the context of nanomaterial OHS, the desired regulatory outcome is the protection of employee health. However, since the risk endpoints of nanomaterials are largely unknown, one cannot know whether any rules in order to mitigate risks are evidently effective from a scientific point of view.[17] As there is no one scientific standard at the moment, the issue of rule compliance is not feasible; first and foremost it shall be investigated whether collaboration among key-stakeholder who hold fundamental scientific data on nanomaterials, can contribute to the development of evidence-based rules. Therefore, researching effectiveness from a traditional regulatory perspective is not useful. Instead of focusing on the outcome of regulation, it is perhaps more useful to evaluate *processes of collaboration*

---

[17]Such a situation poses a typical 'technology control dilemma': there is too little knowledge on new or emerging technologies available to be able to develop evidently effective regulatory controls; when problems with such technologies emerge regulatory control becomes difficult because the technology is fixed or settled in society already and cannot be changed without significant disruptions [73]. The control dilemma, as suggested by the Royal Commission on Environmental Pollution (RCEP), suggests that such new technologies require a governance framework that is adaptive, having the capability to develop and grow when dangers to human health become apparent over time [74:8].

among actors aimed at supporting OHS regulation related to work with nanomaterials.

In a process-oriented perspective on effectiveness the aspect of collaboration is attributed a central role: when key actors (i.e. public policy maker and business associations) work together or cooperate in specific activities to support the provision of safe workplaces, and when this collaboration is rendered 'successful', activities contribute to effective regulation of nanomaterials OHS. Successful collaboration is here defined broadly as: both actors, in the *process* of their teamwork, acquire new knowledge, which means they *learn* how to improve traditional risk assessment frameworks for nanomaterials. As to the issue of learning Abbot [13] emphasises that in co-regulation of nanotechnologies non-state actors hold key resources (such as scientific expertise and data) that state lacks and likewise, state holds resources (including legal authority and legitimacy) that non-state actors lack; exchange of such resources furthers mutual learning among the two parties. In this 'new architecture', the regulatory focus shifts from rules to frameworks for the creation of rules in which capacities for problem-solving are exchanged, deliberated and bargained between state and non-state actors [75:307].[18]

For the purpose of investigating processes of collaboration that induce learning among public-private actors to solve policy challenges (such as the one of effective nanomaterials OHS regulation), the network governance literature provides useful points of entry. This strand of literature allows for considering individual actors interests as being connected with those of other actors during processes of collaboration [76–78]. Processes of collaboration are conceived as having two general dimensions. On the one hand, actor behaviour takes shape in complex and unpredictable 'games' of collaboration among actors in relation to broader networks characterised by uncertainty that results from incomplete knowledge. On the other hand, uncertainty derives from the presence of multiple actors each having own perceptions, goals and strategies in view to problem definitions and solutions [79]. Both of the two dimensions of uncertainty

---

[18]Sabel and Zeitlin [75] actually focus on the issue of accountability in regulation; however, their general approach of a 'new destabilisation regime' appears also useful in the context of effective nanomaterials OHS regulation.

are relevant in the context of effective nanomaterials OHS regulation: On the one hand, there is a lack of scientific data on nanomaterials that causes collaboration in networks involving business associations and public policy makers. On the other hand, since business associations 'wear many hats', games of decision-making with regulators can be expected.

To decrease uncertainty and provide solutions to policy challenges various actor key resources such as knowledge, skills, money and decision-making authority are employed. In collaborative processes in networks these resources are exchanged and negotiated depending on general and individual actor strategies. In this way, collaboration among actors is organised around relations of resource dependency [80].[19] Collaboration brings about exchange of information, perceptions and goals and, if successful, it may create certainties 'through mutual perceptions and binding decisions and agreements, that enable common action and joint solutions' [79].

Taking a process-oriented approach to studying collaborative business association activities contributing to effective nanomaterials OHS regulation is promising because, on the one hand, it considers the substantive (knowledge) uncertainty that is characteristic for nanomaterial risks and OHS. On the other hand, this approach offers a broader scope of analysis by accompanying, yet at the same time not taking for granted the notion of co-regulation among private and public parties who supposedly collaborate meaningfully on the basis of shared interests to solve complex policy challenges. The private actor 'business association' has been conceived as possessing promising potential for the governance of nanomaterials through acting as a vehicle for the dissemination of industry best practices while also providing the basis for a normative framework [81]. But, to date, in-depth empirical studies that shed light on the specific activities and related processes of collaboration among business associations and policy makers in order to deal with the complex challenge of effective nanomaterials OHS regulation are still missing in the regulatory debate.

---

[19]In Scharpf's [80] typology on dependency among actors, he distinguishes two dimensions of dependency, namely the importance of the resource and the substitutability of the resource. Actors are dependent on other actors if the relation to that actor is based on low substitutability and a high level of importance.

Based on the literature discussion and the publicly available business association documents, it can be speculated that networks involving business associations and public policy-makers have the potential to serve as forum for cooperation and mutual learning based on the exchange and deliberation on scientific knowledge on nanomaterials. Areas of mutual learning could comprise the decision on which scientific parameters are relevant for nanomaterials risk assessment, how nanomaterial hazard can be predicted, or with which techniques nanomaterial exposure can be measured in routine workplace operation. In the past many voluntary information disclosure regimes/voluntary reporting schemes for nanotechnologies that have been initiated by states [82–84] were considered a 'failure' in that response rates by nano-companies were low as they were not willing to share and exchange scientific information on nanomaterials [85]. It may be that companies are more willing to submit data to programs or arrangement initiated by business associations. Based on the analysis of documents by the German Chemical Industry Association there are indicators that this might indeed be the case. Business associations act first and foremost in the interest of their members (to keep up member fees and thereby secure association existence) and only secondly they might act as regulators in co-regulatory arrangements. Because of this logic of existence business associations can create a 'safe space' for companies in which exchange of scientific knowledge on nanomaterials is furthered because actors trust that others refrain from opportunistic behaviour, i.e. do not use resources of other actors for their own advantage. Processes of collaboration, based on trust relations among industry and state actors, may enhance the potential for developing innovative solutions to the problem of risk assessment for nanomaterials thereby contributing to effective nanomaterials OHS regulation.

11.4 Conclusions: An Outlook to the Future of Nanomaterials OHS Regulation

From this analysis, three conclusions as to the future of nanomaterials OHS regulation and the involvement of business associations can be drawn. First, as suggested in the literature, business associations actually are actively involved in discussions on nanomaterials OHS. In line with the general picture of German business associations from the literature, the VCI has developed a range of soft regulatory instruments by interacting with government over the best ways to achieve required obligations, i.e. the employer duty to conduct risk assessment for nanomaterials. Based on such activities the association has increasingly lobbied for the adequacy of existing general legislation, thereby rejecting proposals for nano-specific legislation. Against the background of wider developments in nano-specific legislation, taking into account the recent introduction of mandatory nano-specific labelling provisions for cosmetic products containing nanoparticles enacted by the Regulation (EC) 1223/2009[20] and the proposal for nano-specific rules on food for infants in the Regulation (EU) 609/2013,[21] the chemical industry is likely to see a need in the future to push forward its position that no nano-specific OHS legislation is required. Linked to the near monopoly position of the VCI in the German chemical industry, as well as their scientific expertise, the association has a high degree of authority to develop rules of behaviour to care for nanomaterials OHS. It is likely that German chemical companies will take into account and follow VCI advice on nanomaterials OHS in the future.

But even though the VCI has been, and probably will be, involved in discussions on nanomaterials OHS, it appears difficult to distinguish clearly whether the particular association activities are directed at contributing to *regulation* or whether they have a broader *governance* function or a self-serving interest. For instance, the business association activity of collecting and sharing scientific risk data related to specific nanomaterials may be directed at supporting policy makers in the regulation of OHS; but likewise the activity may serve a self-interest of business associations (such as keeping their members up-to-

---

[20]Regulation (EC) No 1223/2009 on cosmetic products [2009] (recast) OJ L 342/59.
[21]Regulation (EU) No 609/2013 of on food intended for infants and young children, food for special medical purposes, and total diet replacement for weight control and repealing Council Directive 92/52/EEC, Commission Directives 96/8/EC, 1999/21/EC, 2006/125/EC and 2006/141/EC, Directive 2009/39/EC of the European Parliament and of the Council and Commission Regulations (EC) No 41/2009 and (EC) No 953/2009 [2013] OJ L 181/35.

date on matters of potential risks of nanotechnologies). Publicly available documents of business associations often do not provide clear information as to the concrete purpose of specific activities and it remains unclear which aim is followed. Empirical studies need to shed light on this issue so as to be able to evaluate association activities and intentions adequately and to be able to ascribe responsibilities to all actors involved in the co-regulation of nanomaterials OHS.

Second, business associations oscillating between providing collective goods (reputation and representation, lobbying, soft regulation, information exchange) and selective goods (business advice, marketing, and accreditation) so as to satisfy their member firms require closer attention. It is crucial to understand which role business associations play for their member firms in the nanomaterials sector. To date little is known about the perception of firms on the topic of nanomaterials OHS and which capacities and needs such firms have in order to fulfil their legal obligation to provide safe workplaces. Such insights would provide crucial information to public policy-makers regarding the applicability of general OHS legislation.

Third, this analysis points out a need to reconsider the traditional conceptualisation of regulatory effectiveness in the light of new technologies characterised by unknown risks. As has been demonstrated in the case of nanomaterials OHS with its high level of (for now) unknown 'risks', traditional understandings of regulatory effectiveness are not feasible. Instead of starting from the presumption that regulatory effectiveness depends exclusively on *a priori* set rules regulatees should comply with, processes of learning on how to provide safe workplaces should be evaluated. In this line, one should be aware that learning typically goes along with the growth of trust relations among collaborators, in this case policy makers and business associations representing companies. Trust relations can further the efficient application of knowledge, enhance compliance levels and can provide a stable foundation for the sustainable future regulation of nanotechnologies. Researching these processes in empirical studies will provide new entrance points to the effective regulation of new technologies.

Appendix: Overview VCI activities and related positions to nanomaterials OHS 2003–2013

Time	Activity	Position
2003	Founding joint working group (with DECHEMA) Responsible Production and Use of Nanomaterials [18].	There is a need for scientific findings and best practices of safety aspects of nanomaterials.
September 2005	Workshop Nanomaterials at the workplace [86].	There is a need for safety research and discussion at national and international levels; results of stakeholder discussions help formulating the next steps.
February 2006	Survey on occupational health and safety in the handling and use of nanomaterials among VCI member companies (in collaboration with BAuA) [87].	Need to obtain an overview of nanomaterials OHS methods applied in the chemical industry; soft regulation as to the handling and use of nanomaterials should be developed.
April 2007	Workshop Nanomaterials at the workplace [87].	Soft regulation which is already in place needs to be further developed; topics requiring discussion are measuring and protection methods and communication.
July 2007	Roadmap for Safety Research on Nanomaterials [18].	It is questionable whether existing exposure measurement techniques and toxicological testing strategies are appropriate to assess potential hazards and risks of nanomaterials; in cooperation with academia the chemical industry should continue to conduct research on the safety and risks of nanomaterials; internationally harmonised standard analytical measuring methods need to be developed.

Time	Activity	Position
August 2007	Guidance for Handling and Use of Nanomaterials at the Workplace (in collaboration with BAuA) [56].	There is a need to develop soft regulation to guide companies and provide orientation on nanomaterials OHS measures regarding the latest state of science; exposure to nanoparticles might have effects different to the ones of larger particles; the chemical industry has a decisive role in participating and initiating research projects.
October 2007	Strategy Paper of the German Chemical Industry on the Standardisation of Nanomaterials (in the context of ISO/TC 229) [88].	The German chemical industry can and should provide expertise to ISO/TC229 (exposure measurement; evaluation of risk management measures).
February 2008	Guidance 'Anforderungen der REACH-Verordnung an Stoffe, welche auch als Nanomaterialien herge-stellt oder eingeführt werden' [89].	Producer and importer of nanomaterials should provide safety data sheets for nanomaterials.
February 2008	Guidance for a Tiered Gathering of Hazard Information for the Risk Assessment of Nanomaterials [90].	The existing test methods for gathering toxicological data may require adaptation as to the physiochemical properties; research activities to further develop risk assessment approaches for nanomaterials should be continued.
March 2008	Guidance for the Passing on of Information along the Supply Chain in the Handling of Nanomaterials via Safety Data Sheets [91].	Guidance as to the development and use of safety data sheets in the handling of nanomaterials is required to ensure that correct & complete information on chemical substances are provided (crucial for risk assessment).

(*Continued*)

Appendix: (*Continued*)

Time	Activity	Position
March 2008	Guidance Responsible Use and Production of Nanomaterials [92].	Ensuring workplace safety is at the heart of the of the German chemical industry culture; there is a gap of uncertainty on possible hazardous effects of nanomaterials; safety research needs to be continued; exchange of experiences is necessary; need for dialogue with all stakeholders, based on openness, transparency & trust.
April 2008	Guidance 'Umsetzung von Responsible Care® für eine verantwortliche Herstellung und Ver-wendung von Nanomaterialien' [93].	The German chemical industry has decades of experiences in the assessment of chemical substances which ensures that also nanomaterials at workplaces undergo proper risk assessment.
September 2008	Workshop 'Verantwortlicher Umgang mit Nanomaterialien' (in collaboration with the industry association IG Bergbau, Chemie, Energie (IG BCE)) [94].	Continuing open and transparent exchange of knowledge & arguments are significant for the future of nanomaterials safety research.
August 2011	Guidance 'Tiered approach to an Exposure Measurement and Assessment of Nanoscale Aerosols Released from Engineered Nanomaterials in Workplace Operations' [95].	A harmonised and pragmatic approach to nanomaterial exposure measurement is required usable by SMEs and large chemical operations with global operation; Exposure measurement of nanoscale aerosols released from ENMs in the workplace is possible and exposure assessment methodologies exist.

Time	Activity	Position
		However, methodologies are not yet standardised and more difficult to apply as in routine operations; a tiered approach to exposure assessment of nanoscale aerosols released from ENMs in workplace operations is most effective—main advantage is the most efficient use of limited, qualified resources to ensure a high level of employee protection.
October 2011	Reaction to EC definition of a nanomaterial [96].	The EC definition of a nanomaterials is too broad (would apply to many daily chemical products that are on the market for decades); the EC definition hinders worldwide harmonisation of nanomaterials regulation; the EC definition must be linked with measuring methods.
November 2011	Reaction to a report on nanomaterials risks by the German Advisory Council on the Environment [96].	There is no scientific evidence that nanomaterials harm human health; nanoscaled substances as such are not dangerous; as with all chemicals, risk assessment of nanomaterials must be conducted on a case-by-case basis; safety research must continuously be supported; a harmonised, definition of nanomaterials is required; nano-specific legislation is to be rejected; REACH provides adequate risk assessment & management for nanomaterials; the results of publicly-funded safety research should be made more transparent.

(*Continued*)

Appendix: (*Continued*)

Time	Activity	Position
May 2012	Revised Guidance 'Empfehlung für die Gefährdungsbeurteilung bei Tätigkeiten mit Nanomaterialien am Arbeitsplatz' (2007/2012) [64].	Nanomaterials have new characteristics, therefore those materials that have not been researched sufficiently must be dealt with by taking specific safety measures; companies need orientation as to the existing measures for nanomaterials OHS that reflect the latest state of science & technology; often it is not possible to clearly identify whether nanomaterials are handled at workplaces; in this case risk assessment should be based on a precautionary approach; exposure to nanoparticles at workplace is more likely in SMEs than in bigger companies; the principles for nanomaterials OHS developed by the German 'NanoKommission' should be followed; as with all chemical substances nanomaterials may affect human health (through inhalation, dermal take-up); exposure measurement of nanoparticles is currently complex, expensive, not standardised and can only be conducted by experienced research institutes.
October 2012	Reaction to EC Communication on the 2nd Regulatory Review on Nanomaterials [55, 61].	Agreement with EC that REACH applies also to nanomaterials; the nanoscale of a substance is not per se dangerous; therefore

Time	Activity	Position
		nano-specific legislation is not necessary (but specific regulations might have to be adapted); nano-specific risk assessment methods are not necessary (case by case test methods might have to be adapted).
February 2013	VCI/VCH-Position zum Gesamtbericht der Europäischen Kommission zu REACH vom 5. February 2013 [65].	Agreement with EC that REACH should approach nanomaterials like any other chemical substances of which some are toxic, some are not; traditional risk assessment is principally applicable to nanomaterials, but some specific adjustments are necessary; there should be additional guidance for risk characterisation and assessment for nanomaterials in context of the REACH appendix.

References

1. Coglianese, C., and E. Mendelson (2010) Meta-regulation and self-regulation. In: Baldwin, R., M. Cave, and M. Lodge (eds.) *The Oxford Handbook of Regulation* (Oxford University Press, Oxford), pp. 146–167.
2. Bartis, J. T., and E. Landree (2006) *Nanomaterials in the Workplace. Policy and Planning Workshop on Occupational Safety and Health* (RAND Corporation, Santa Monica).
3. Savolainen, K., U. Backman, D. Brouwer, B. Fadeel, T. Fernandes, et al. (2013) *Nanosafety in Europe 2015–2025: Towards Safe and Sustainable Nanomaterials and Nanotechnology Innovations* (Finish Institute of Occupational Health, Helsinki).

4. Gunningham, N., and J. Rees (1997) Industry self-regulation: An institutional perspective, *Law & Policy*, **19**(4), 363–414.
5. Sethi, S. P. (2005) The effectiveness of industry-based codes in serving public interest: The case of the international council on mining and metals, *Transnational Corporations*, **14**, 55–99.
6. Gunningham, N., M. Phillipson, and P. Grabosky (1999) Harnessing third parties as surrogate regulators: Achieving environmental outcomes by alternative means, *Business Strategy and the Environment*, **8**, 211–224.
7. Ayres, I., and J. Braithwaite (1992) *Responsive Regulation. Transcending the Deregulation Debate* (Oxford University Press, Oxford).
8. Bowman, D. M., and G. Gilligan (2010) The private dimension in the regulation of nanotechnologies: Developments in the industrial chemical sector, *UCLA Journal of Environmental Law & Policy*, **28**(1), 77–101.
9. Meili, C., and M. Widmer (2010) Voluntary measures in nanotechnology risk governance: The difficulty of holding the wolf by the ears. In Hodge, G. A., D. M. Bowman, and A. D. Maynard (eds.) *International Handbook on Regulating Nanotechnologies* (Edward Elgar, Cheltenham UK), pp. 446–461.
10. Zumbansen, P. (2011) Neither 'public' nor 'private, 'national' nor 'international': transnational corporate governance from a legal pluralist perspective, *Journal of Law and Society*, **38**(1), 50–75.
11. Sethi, S. P., and O. Emelianova (2006) A failed strategy of using voluntary codes of conduct by the global mining industry, *Corporate Governance: The International Journal of Effective Board Performance*, **6**, 226–238.
12. Trubek, L. G., P. Cottrell, and N. Nance (2006) Soft law, hard law and EU integration. In: De Burca, G., and J. Scott (eds.) *Law and New Governance in the EU and the US* (Hart, Portland OR), pp. 65–94.
13. Abbott, K. W., and D. Snidal (2000) Hard and soft law in international governance, *International Organization*, **54**(3), 421–456.
14. Braithwaite J. (1982) Enforced self-regulation: A new strategy for corporate crime control, *Michigan Law Review*, **80**, 1466–1507.
15. Sinclair, D. (1997) Self-regulation versus command and control? beyond false dichotomies, *Law & Policy*, **19**, 529–559.
16. Webb, K. (2004) Understanding the voluntary code phenomenon. In: Webb, K. (ed.) *Voluntary Codes: Private Governance, the Public Interest, and Innovation* (Carleton University, Ottawa), pp. 3–32.

17. Nash, J., and J. Ehrenfeld (1997) Codes of environmental management practice: assessing their potential as a tool of change, *Annual Review of Energy and the Environment*, **22**, 487–535.

18. German Society for Chemical Engineering and Biotechnology and German Chemical Industry Association (2007) *Roadmap for Safety Research on Nanomaterials.* Available at: https://www.vci.de/Downloads/122303-Roadmap%20Safety%20Research%20on%20Nanomaterials_05%20July%202007.pdf.

19. European Commission (2008) *Regulatory Aspects of Nanomaterials. Summary of Legislation in Relation to Health, Safety and Environment Aspects of Nanomaterials, Regulatory Research Needs and Related Measures.* Available at: http://ec.europa.eu/nanotechnology/pdf/comm_2008_0366_en.pdf.

20. Safe Work Australia (n.d.). *Hazardous Chemicals* (Australian Government, Canberra).

21. Organisation for Economic Co-operation and Development (2012) *Important Issues on Risk Assessment of Manufactured Nanomaterials* Environmental Directorate OECD, Paris.

22. National Research Council (1983) *Risk Assessment in the Federal Government: Managing Process* (National Academy Press, Washington DC).

23. Oberdörster, G., V. Stone, and K. Donaldson (2007) Toxicology of nanoparticles: A historical perspective, *Nanotoxicology*, **1**(1), 2–25.

24. Shatkin, J. A., L. C. Abbott, A. E. Bradley, R. A. Canady, T. Guidotti, et al. (2010) Nano risk analysis: Advancing the science for nanomaterials risk management, *Risk Analysis*, **30**(1), 1680–1687.

25. Kuhlbusch, A. J., C. Asbach, H. Fissan, D. Göhler, and M. Stintz (2011) Nanoparticle exposure at nanotechnology workplaces: A review, *Particle and Fibre Toxicology*, **8**(22), 1–18.

26. Holdren, P. J., R. C. Sunstein, and A. I. Siddiqui (2011) *Policy Principles for the US Decision-Making Concerning Regulation and Oversight of Applications of Nanotechnology and Nanomaterials* (Memorandum from the Heads of Executive Departments and Agencies, Executive Office of the President).

27. Abbot, C. (2012) Bridging the gap–non-state actors and the challenges of regulating new technology, *Journal of Law and Society*, **39**(3), 329–358.

28. Bennett, R. J. (1999) Explaining the membership of sectoral business associations, *Environment and Planning*, **31**, 877–898.

29. Bennett, R. J. (1998) Business associations and economic development: Re-exploring an interface between the state and market, *Environment and Planning A*, **30**, 1367–1387.
30. Bennett, R. J. (ed.) (1997) *Trade Association in Britain and Germany: Responding to Internalization and the EU* (Anglo-German Foundation, London).
31. Bennett, R. J. (2000) The logic of membership of sectoral business associations, *Review of Social Economy*, **58**(1), 17–42.
32. Boleat, M. (2000) *The Changing Environment for Trade Associations and Strategy for Adaptation* (Boleat Consulting, Washington, D.C.) Available at: http://www.boleat.com/materials/CETA_2000.pdf.
33. van Waarden, F. (1991) Two logics of collective action? Business associations as distinct from trade unions: The problem of associations of organizations. In: Sadowski, D., and O. Jacobi (eds.) *Employers' Associations in Europe: Policy and Organisation* (Nomos, Baden Baden) pp. 51–84.
34. Streeck, W., and P. C. Schmitter (1985) *Private Interest Government: Beyond Market and State* (Sage, London).
35. Olson, M. (1971) *The Logic of Collective Action: Public Goods and the Theory of Groups*, 2nd ed. (Harvard University Press, Cambridge).
36. Rees, J. V. (1996) *Hostages of Each Other: The Transformation of Nuclear Safety since Three Mile Island* (University of Chicago Press, Chicago).
37. Coglianese, C. (2010) Engaging business in the regulation of nanotechnology. In: Bosso, C. J. (ed.) *Governing Uncertainty: Environmental Regulation in the Age of Nanotechnology* (RFF Press, Washington, DC), pp. 12–07.
38. King, A., M. Lenox, and M. Barnett (2001) Strategic responses to the reputation commons problem. In Hoffman, A. J., and M. J. Ventresca (eds.) *Organizations, Policy and the Natural Environment: Institutional and Strategic Perspectives* (Stanford University Press, Stanford), pp. 393–406.
39. Gupta, A. K., and J. L. Lawrence (1983) Industry self-regulation: An economic, organizational, and political analysis, *The Academy of Management Review*, **8**(3), 416–425.
40. Veljanovski, C. (2010) Strategic use of regulation. In: Baldwin, R., M. Cave, and M. Lodge (eds.) *The Oxford Handbook of Regulation* (Oxford University Press, Oxford) pp. 87–103.

41. Lenox, M. J. (2006) The prospects for industry self-regulation of environmental externalities. Available at: http://faculty.darden.virginia.edu/LenoxM/pdf/isr_theory2.pdf.
42. Hutter, B. M. (2011) Negotiating social, economic and political environments: Compliance with regulation within and beyond the state. In: Parker, C., and V. L. Nielsen (eds.) *Explaining Compliance: Business Responses to Regulation* (Edward Elgar, Cheltenham), pp. 305–321.
43. Hutter, B. M., and C. J. Jones (2007) From government to governance: External influences on business risk management, *Regulation & Governance*, **1**, 27–45.
44. Gunningham, N. (2002) Regulating small and medium sized enterprises, *Journal of Environmental Law*, **14**, 3–32.
45. Henson, S. J., and M. Heasman (1998) Food safety regulations and the firm: Understanding the process of compliance, *Food Policy*, **23**, 9–24.
46. Bailey, I., and S. Rupp (2006) The evolving role of trade associations in negotiated environmental agreements: The case of United Kingdom climate change agreements, *Business Strategy and the Environment*, **15**, 40–54.
47. Lane, C., and R. Bachmann (1997) Co-operation in inter-firm relations in Britain and Germany: The role of social institutions, *The British Journal of Sociology*, **48**(2), 226–254.
48. Bundesverband der Deutschen Industrie e.V. (2012) Der BDI-Spitzenverband der deutschen Wirtschaft, *Bdi.eu*. Available at: http://www.bdi.eu/Ueber-uns.htm.
49. Henneberger, F. (1993) Transferstart: Organisationsdynamik und Strukturkonservatismus westdeutscher Unternehmerverbände, *Politische Vierteljahreszeitschift*, **34**, 640–673.
50. Rampelt, J. (1979) *Zur Organisations-und Entscheidungsstruktur in westdeutschen Unternehmerverbanden. Ein Literaturbericht* (Wissenschaftszentrum Berlin für Sozialforschung GmbH, Berlin).
51. Weber, H. (1987) *Unternehmerverbände zwischen Markt, Staat und Gewerkschaften. Zur intermediären Organisation von Wirtschaftsverbänden* (Campus, Frankfurt).
52. Abromeit, H. (1993) *Unternehmerverbände*. In: Andersen, U., and W. Woyke (eds.) *Handwörterbuch des Politischen Systems der Bundesrepublik Deutschland* (Leske and Budrich, Opladen).

53. Izushi, H. (2002) The 'voice' approach of trade associations: Support for SMEs accessing a research institute, *Environment and Planning C*, **20**, 439–454.
54. Verband der Chemischen Industrie (2013) *Mitglieder*. Available at: https://www.vci.de/Der-VCI/Mitglieder/Seiten/Startseite.aspx.
55. Verband der Chemischen Industrie (2013) *Werden Sie Mitglied*. Available at: https://www.vci.de/der-vci/mitglieder/mitglied-werden/Seiten/Startseite.aspx.
56. Verband der Chemischen Industrie and German Federal Institute for Occupational Safety and Health (2007) *Guidance for Handling and Use of Nanomaterials at the Workplace* (VCI and BAuA, Berlin).
57. Plitzko, S., E. Gierke, and H. G. Schäfer (2007) *Fragebogenaktion der BAuA und des VCI. Umgang mit Nanomaterialien*. Available at: https://www.vci.de/Downloads/121338-Plitzko_BAuA.pdf.
58. Prakash, A. (2000) Responsible care: An assessment, *Business Society*, **39**, 183–209.
59. Evangelinos, K. I., I. E. Nikolaoub, and A. Karagiannisc (2010) Implementation of responsible care in the chemical industry: Evidence from greece, *Journal of Hazardous Materials*, **77**(1–3), 822–828.
60. International Council of Chemical Associations (2012) *ICCA Responsible Care® Progress Report* (ICCA, New York).
61. Verband der Chemischen Industrie (2012) *Chemie Report* Available at: https://www.vci.de/Downloads/Publikation/chemie-report/nanotechnology_cr_special_2012_print_version.pdf.
62. Verband der Chemischen Industrie (2013) *Nanomaterialien*. Available at: https://www.vci.de/Downloads/Top-Thema/OP_Nanomaterialien.pdf.
63. Verband der Chemischen Industrie (2011) *EU macht fast alle Alltagsprodukte zu 'Nano'*. Available at: https://www.vci.de/Presse/Pressemitteilungen/Seiten/EU-macht-fast-alle-Alltagsprodukte-zu-„Nano".aspx.
64. Verband der Chemischen Industrie and German Federal Institute for Occupational Safety and Health (2012) *Empfehlung für die Gefährdungsbeurteilung bei Tätigkeiten mit Nanomaterialien am Arbeitsplatz* (VCI and BAuA, Berlin).
65. Verband der Chemischen Industrie and Verband Chemiehandel (2013). *VCI/VCH-Position zum Gesamtbericht der Europäischen Kommission zu REACH vom 5*. Available at: https://www.vci.de/Downloads/PDF/VCI-Position%20zum%20REACH-Review.pdf.

66. Opschoor, H., and K. Turner (1994) *Economic Incentives and Environmental Policies: Principles and Practice* (Kluwer Academic Publishers, Dordrecht).

67. Karlsson-Vinkhuyzen, S. I., and A. Vihma (2009) Comparing the legitimacy and effectiveness of global hard and soft law: An analytical framework, *Regulation & Governance*, **3**, 400–420.

68. Kagan, R. A., and J. Scholz (1984) The criminology of the corporation and regulatory enforcement styles. In: Hawkins, K., and J. Thomas (eds.) Enforcing Regulation (Kluwer-Nijhoff, Boston), pp. 67–69.

69. Braithwaite, V. (1995) Games of engagement: Postures within the regulatory community, *Law & Policy*, **17**(3), 225–255.

70. Baldwin, R., and J. Black (2008) Really responsive regulation, *The Modern Law Review*, **71**(1), 59–94.

71. Vogel, D. (2009) The private regulation of global corporate conduct. In Mattlie, M., and N. Woods (eds.) *The Politics of Global Regulation* (Princeton University Press, Princeton NJ), pp. 151–188.

72. Gunningham, N. (2010) Enforcement and compliance strategies. In: Baldwin, R., M. Cave, and M. Lodge (eds.) *The Oxford Handbook of Regulation* (Oxford University Press, Oxford), pp. 120–144.

73. Collingridge, D. (1980) *The Social Control of Technology* (Francis Pinter, New York).

74. Royal Commission on Environmental Pollution (2008) *Novel Materials in the Environment: The Case of Nanotechnology* (RCEP, London).

75. Sabel, C. F., and J. Zeitlin (2008) Learning from difference: The new architecture of experimentalist governance in the EU, *European Law Journal*, **14**(3), 271–327.

76. Kenis, P., and V. Schneider (1991) Policy networks and policy analysis: Scrutinizing a new analytical toolbox. In: Marin, M., and R. Mayntz (eds.) *Policy Networks* (Campus Verlag, Frankfurt), pp. 25–59.

77. O'Toole, L. J. (1997) Treating networks seriously: Practical and research-based agendas in public administration, *Public Administration Review*, **57**, 45–52.

78. Provan, K. G., and P. Kenis (2007) Modes of network governance: Structure, management, and effectiveness. *Journal of Public Administration Research and Theory*, **18**(2), 229–252.

79. Koppenjan, J., and E. H. Klijn (2004) *Managing Uncertainties in Networks. A Network Approach to Problem Solving and Decision-Making* (Routledge, London).

80. Scharpf, F. W. (1978) Interorganizational policy studies: Issues, concepts and perspectives. In: Hanf, K., and F. Scharpf (eds.) *Interorganizational Policy Making; Limits to Coordination and Central Control* (Sage, London).

81. Bowman, D. M., and G. A. Hodge (2009) Counting on codes: An examination of transnational codes as a regulatory governance mechanism for nanotechnologies, *Regulation & Governance*, **3**, 145–164.

82. Department for Environment, Food and Rural Affairs (2008) *Voluntary Reporting Scheme for Engineered Nanoscale Materials* (Defra, London).

83. Environmental Protection Agency (2009) *Nanoscale Materials Stewardship Program*. Available at: http://epa.gov/oppt/nano/stewardship.htm.

84. Environmental Protection Agency (2011) *Chemical Testing and Data Collection*. Available at: http://www.epa.gov/oppt/chemtest/pubs/data.html.

85. Stokes, E. (2013) Demand for command: Responding to technological risks and scientific uncertainties, *Medical Law Review*, **21**(1), 1–28.

86. Verband der Chemischen Industrie (2005). *Nanomaterials at the Workplace Stakeholder Dialog on Industrial Health and Safety* (VCI, Berlin).

87. Verband der Chemischen Industrie (2007) *Nanomaterials at the Workplace. Stakeholder Dialog on Industrial Health and Safety* (VCI, Berlin).

88. Verband der Chemischen Industrie (2007) *Strategy Paper of the German Chemical Industry on the Standardisation of Nanomaterials* (VCI, Berlin).

89. Verband der Chemischen Industrie (2008) *Anforderungen der REACH-Verordnung an Stoffe, welche auch als Nanomaterialien hergestellt oder eingeführt werden*. Available at: https://www.vci.de/Downloads/122418-PP%20Nanomaterialien%20und%20REACH%20_26.02.2008_.pdf.

90. Verband der Chemischen Industrie (2008) *Guidance for a Tiered Gathering of Hazard Information for the Risk Assessment of Nanomaterials*. Available at: https://www.vci.de/Downloads/122300-Hazard%20Information%20for%20RA%20of%20Nanomaterials_28%20February%202008.pdf.

91. Verband der Chemischen Industrie (2008) *Guidance for the Passing on of Information along the Supply Chain in the Handling of Nanomaterials via Safety Data Sheets.* Available at: https://www.vci.de/Downloads/122313-Leitfaden_Sicherheitsdatenblatt_03.2008.pdf.

92. Verband der Chemischen Industrie (2008) *Responsible Use and Production of Nanomaterials.* Available at: https://www.vci.de/Downloads/Responsible-Production-and-use-of-Nanomaterials.pdf.

93. Verband der Chemischen Industrie (2008) *Umsetzung von Responsible Care® für eine verantwortliche Herstellung und Verwendung von Nanomaterialien.* Available at: https://www.vci.de/Downloads/122490-Umsetzung%20von%20Responsible%20Care%20bei%20Nanomaterialien%20_16.04.2008.pdf.

94. Verband der Chemischen Industrie and Industriegewerkschaft Bergbau, Chemie, Energie (2008) *Verantwortlicher Umgang mit Nanomaterialien.* Available at: https://www.vci.de/Themen/Chemikaliensicherheit/Nanomaterialien/Seiten/Workshop-Gespraechsstoffe-Verantwortlicher-Umgang-mit-Nanomaterialien.aspx#0.

95. Air Quality and Sustainable Nanotechnology, Institute of Energy and Environmental Technology e.V., Federal Institute for Occupational Safety and Health, German Social Accident Insurance, Institution for the Raw Materials and Chemical Industry, German Chemical Industry Association, Institute for Occupational Safety and Health of the DGUV and Technical University Dresden (2011) *Tiered Approach to an Exposure Measurement and Assessment of Nanoscale Aerosols Released from Engineered Nanomaterials in Workplace Operations.* Available at: https://www.vci.de/Downloads/Tiered-Approach.pdf.

96. Verband der Chemischen Industrie (2011) *Gutachten des SRU, Vorsorgestrategien für Nanomaterialien".* Available at: from https://www.vci.de/Themen/Chemikaliensicherheit/Nanomaterialien/Seiten/Nano-Gutachten-des-SRU.aspx.

Part 3

Looking to the Future of Disruptive Technologies

Chapter 12

The 'Metamorphosis' of the Drone: The Governance Challenges of Drone Technology and Border Surveillance

Luisa Marin

Centre for European Studies,
Faculty of Behavioural Management and Social Sciences,
University of Twente, NL-7500 AE Enschede, The Netherlands

l.marin@utwente.nl

12.1 Introduction: Let's Face It! They Are Here to Stay

Unmanned aerial vehicles (UAV) or remotely piloted aerial systems (RPAS), commonly called drones, are mostly known for the various 'targeted killing programs' carried out by Israel and the United States (US) in the Middle East. Targeted killings, a practice not defined in international law, have been practiced in a variety of contexts: first in secrecy, and then admittedly since 2000 by Israel toward alleged terrorists in the Occupied Palestinian Territories; and by the US after 9/11, in the context of the global 'war on terror', initiated under the G. W. Bush Administration and later continued by President Obama [1]. The Central

---

Embedding New Technologies into Society: A Regulatory, Ethical and Societal Perspective
Edited by Diana M. Bowman, Elen Stokes, and Arie Rip
Copyright © 2017 Pan Stanford Publishing Pte. Ltd.
ISBN 978-981-4745-74-1 (Hardcover), 978-1-315-37959-3 (eBook)
www.panstanford.com

Intelligence Agency (CIA) has used drones since November 2002, in a first drone strike in Yemen, and then reportedly in over 120 drone strikes [1]. The CIA 'personality strikes' and 'signature strikes', which have also targeted American citizens,[1] have triggered a number of reactions: on the one side, for their alleged breaches of American constitutional guarantees (in particular, Fourth and Fifth Amendments rights, on the freedom from unreasonable searches and the right not to be deprived of life without due process of law); on the other side, for their alleged lack of compliance with international law, in particular international human rights law and international humanitarian law (the right to territorial sovereignty, the right to self-defence, the principles of proportionality, precaution and necessity in relation to the civilian casualties) [1, 3]. Most notably, the targeted killing program has been criticised for changing warfare in its nature, by lowering the threshold to engage in warfare [1]. This criticism has been captured in the expression 'play-station mentality' toward killing and warfare, put forward by Philip Alston, UN Special Rapporteur on extrajudicial, summary or arbitrary executions [1].

It is recurrent in history that innovation developed for the military is subsequently transferred, and further developed, into civilian assets, giving rise to what are described as 'dual-use' technologies. The examples are countless, and span from nuclear energy to information technologies (IT). The Internet, considered today as a tool of democratisation, constitutes one of the more emblematic examples of de-militarisation of an innovation [1]. Thus, the 'war industry' is a strong and powerful motor for innovation, but is and remains first a source of death and collateral (even when targeted) destruction.

Drones are just another example of this phenomenon. First developed by US Air Force (USAF) after World War II in the military context,[2] drone technology has potential applications

---

[1]Such as Anwar al-Awlaqi, a dual US-Yemeni citizen, killed in Yemen on 30 September 2011. See [2].

[2]Actually the first concept of unmanned aerial vehicle can be traced back to the unmanned balloons, filled with bombs, developed in Treviso by the Austro-Hungarian Empire and used to attack Venice on August 22, 1849. The success of their deployment, however, was highly dependent on favorable wind conditions. Drones are the product of experimental evolutions, but it can be considered that Reginald Denny developed the first prototype target drone in 1935. UAV evolved from Remote Piloted Vehicle, used originally both to train antiaircraft gunners and to fly attack missions.

in a number of non-military contexts, be it for commercial or governmental purposes. In the latter category, the examples of drone technology range from those used in civil protection (post-crisis management systems for natural disasters), security (coastal or pre-border surveillance) and environmental protection/ preservation [4:29]. So, paraphrasing Ben Emmerson (UN Special Rapporteur on Counter-Terrorism and Human Rights, investigating the deployment of drones in the military context), it can rightly be argued that the application of drone technology to civil areas is a highly topical issue today, and is 'here to stay'.

This phenomenon, which I have termed the 'metamorphosis' of the drone, extending from military to civil applications, creates new scenarios and, thus, many ethical, regulatory and legal questions. How should we look at this metamorphosis? Should regulators interpret drones as new tools performing old functions, as claimed by drone-users, or do the new tools have the potential to change the nature of those old functions? Consider, for example, the act of killing before the invention of firearms: the killer presumably had physical contact with the victim, was close to him/her, and could appreciate he/she was as human as the killer. Firearms, such as guns or rifles, are distancing technologies, enabling killing without requiring a physical proximity with the target [5]. Ought firearms (and, by extension, drones) to be considered a new tool performing an old function (in the case of drones, functions of killing, or surveillance) or should they be viewed as new tools creating potentially new and unknown issues, which should be approached with more caution? I argue that the transformation experienced in the case of drones (from military to civilian) creates potentially new situations and (types of) risks, which should be investigated, assessed and regulated as such in the appropriate fora. And even considering that new challenges posed by drone technology will develop 'along the way', the regulatory approach should be careful to set up mechanisms capable of adapting to those new challenges, as and when they emerge.

The globalisation of markets has contributed to this shift from military to civil deployments. For example, US and Israel are the two big users and producers of drones for military purposes, and their industries have led innovation and growth of new markets.[3] Building on their experiences on military drones, it is unsurprising that both US and Israel already deploy drones in the civil context, together with Japan and South Korea [8]. These developments have placed pressure on the European Union (EU) to act and develop industry, markets and regulations in the area of drone technology [6].[4]

Currently, the EU is holding discussions on the exploitation of drones for civil purposes. The debate is motivated by the alleged potential for economic growth and the need to establish a harmonised legal framework at EU level. US and Israel are already leading the industry and Europe cannot lag behind, which provides the main motivation behind the EU discourse on drones [6]. This debate is led by industry representatives,[5] and aims at moving the Commission, the Council and the Member States

---

[3] For the US: 'Reflecting a growing awareness and support for UAS, Congress has increased investment in unmanned aerial vehicles annually. The FY2001 investment in UAS was approximately $667 million. For FY2012, Department of Defense (DOD) has asked for $3.9 billion in procurement and development funding with much more planned for the outyears. DOD's inventory of unmanned aircraft increased from 167 to nearly 7,500 from 2002 to 2010' [7]. The Staff Working Document of the European Commission [6], states that: 'Today, military RPAS applications are driving technology development and market expansion, leading the way in terms of research and development, standards, certification and pilot training. RPAS are currently almost exclusively used for military applications (±95%) although their potential for civilian applications has been widely recognised. Teal Group estimates that the global RPAS procurement and R&D expenditures reached $6 billion in the year 2011, with about 40% spent on R&D'.

[4] See the Staff Working Document of the European Commission [6:8]: 'With respectively 66% and 10% of the worldwide RPAS sales, the U.S. and Israel dominate the sector. The production of European countries, all together, does not represent more than 10%. Teal Group estimates that the worldwide RPAS market will double over the next decade to represent an annual procurement and R&D market of US$ 11.3 billion in 2020 with European and Asian manufacturers falling behind. Overall, it is estimated that 35,000 RPAS will be produced worldwide in the next 10 years. The European market should experience the same growth trend but at lower scale. If Europe's ambition is maintained at current levels, the United States together with Israel will remain, in the foreseeable future, the dominant players in a growing RPAS market. This is why it is imperative for the EU to take action now.'

[5] Cf. at this purpose, the structure of the European Commission Working Document [6], based on workshops organized by companies and their representatives.

towards a study of current policy and the enactment of a harmonised and—depending on the size of the drones in question—single legal framework, in order to open up EU airspace (which is currently segregated for such purposes) and therefore, EU markets, to drone technology.

The deployment of drone technology in civil contexts raises ethical, societal, and regulatory issues. Targeted killings and the debate on armed drones remain outside the scope of this chapter. Rather, the aim here is to explore some of these questions by focusing on just one of the main non-military applications of drone technology: that is, border surveillance. The chapter will proceed as follows: Section 12.2 presents the current trend of advocating the deployment of drone technology into non-military contexts. Section 12.3 explores the US experience of deploying drones for border surveillance. Section 12.4 is devoted to the European experience, whereby the EU's external borders agency Frontex, together with other (European and international) agencies, are investing in the testing of drone technology for border surveillance. Section 12.5 then explores the regulatory and ethical questions raised by drone technology used for border surveillance, while Section 12.6 draws conclusions on the importance of enacting appropriate regulations, in order to strengthen EU's capacity to set itself as a leading regulatory player in the global arena, vis-à-vis global partners.

This chapter takes an exploratory approach. It aims to compile an inventory of some of the issues that can and are likely to arise in this context. It aims to present the current legal constraints on border surveillance and asks whether deploying drones for border surveillance might help to fulfil those legal obligations. It is nevertheless not an easy enterprise to carry out research in this area, perhaps more so for lawyers, for two reasons. First, the little information available on plans and initiatives carried out by Frontex is fragmented, is operational in nature and therefore needs to be carefully interpreted. Second, the regulatory dimension of drone technology is still largely in the process of being developed. Therefore, this chapter seeks to offer tentative comments on some of the issues in this emerging domain.

12.2 Drones and Border Surveillance

Speaking about armed drones, UN Special Rapporteur, Ben Emmerson, eloquently stated:

> Let's face it, they're here to stay. This technology, as I say, is a reality. It is cheap, both in economic terms and in the risk to the lives of the service personnel who are from the sending state. And for that reason there are real concerns that because it is so cheap, it can be used with a degree of frequency that other, more risk-based forms of engagement like fixed-wing manned aircraft or helicopters are not, and the result is there's a perception of the frequency and intensity with which this technology is used is exponentially different, and as a result, there is necessarily a correspondingly greater risk of civilian casualties [9].

Drones are already a reality for border surveillance in the US; in Europe, there are media reports indicating that, occasionally, states have used drones to patrol their borders. For example, during the emergency of the Arab Spring in 2011, Switzerland deployed small Drones ADS 95 Ranger at its borders with Italy during night-time in order to prevent unauthorised border crossing by irregular migrants and prospective asylum-seekers.

Border surveillance is only one of the possible uses of drones for government purposes. In this domain, UAVs are deemed to be an asset for several reasons. From a technical point of view, they enable the surveillance coverage of vast and remote areas that would be more difficult to reach by other aircraft or border guards on foot. Bearing in mind that a US helicopter's average time flight is a bit over 2 hours, a drone such as the Predator B can fly for more than 30 hours without refuel. Drones can be equipped with electro-optical sensors (cameras) and thermal detection sensors which can detect small objects at the distance of 60,000 feet, and detect humans moving across woods, under foliage [8]. Piloting at a distance may also help to protect the lives of pilots from risks related to weather and other natural hazards. This will benefit the agency deploying the drones, in their capacity as an employer responsible for the safety of its employees. At policy level, the proponents of drone technology in border surveillance (in Europe: industry stakeholders, Frontex, European Defence Agency and European Commission) argue that this technology is economically affordable, and potentially cheaper than surveillance carried out with traditional manned aircraft. Second, once developed and implemented, the technology would enable savings on the costs of border surveillance operations, in terms of both human resources and material assets deployed [7].

However, one cannot understand the reasons for deploying drone technology in border surveillance without some background on border surveillance and how it became increasingly securitised and militarised. How did it become acceptable to deploy drone technology for border surveillance? The investment of the EU in border surveillance is a consequence of the Schengen process, which has removed controls at frontiers between Member States of the EU, and by contrast has required the strengthening of the controls at the external ones. Currently, given that the EU is a unique international organisation with a quasi-federal structure, the borders of a Member State are also the borders of the EU. In this context, next to coordinating efforts on irregular and regular migration, the EU has enacted a policy on border controls, whose main legal document is the Schengen Borders Code.[6] Hence Member States have extensively resorted to the EU for their immigration policy. It is commonly acknowledged by scholars that the Europeanisation of national migration policy is caused by national failures in the domain. EU migration policy can also be explained through the theory of securitisation, according to which migration and migrants are framed in political discourses [10] by security actors [11], and through practices [12] as security threats, prompting Member States and the EU to react by defending the internal security from alleged external threats. So, if globalisation has turned the world into a 'global village', where goods, capitals, information circulate freely (or without significant hindrance) across the globe, and given that there are serious and ongoing problems of poverty in less economically developed parts of the world, states have become increasingly interested in regulating issues of human mobility. This policy, which aims to control migration by imposing legal limits, has attracted criticism, as indicated by the description of Europe as a 'fortress'.

---

[6]Regulation (EC) No 562/2006 of the European Parliament and of the Council of 15 March 2006 establishing a Community Code on the rules governing the movement of persons across borders (Schengen Borders Code), OJ L 105 of 13.4.2006.

As part of this process, border surveillance has gained relevance and has been highly politicised within the EU. The EU and Member States are investing in technological applications, ranging from biometrics and databases, to drones, in order to deploy the most effective technological means of dealing with the security threats allegedly coming from elsewhere. Drones are part of this process, referred to as the EU's digital fix [13], or the transformation of Europe into a high-tech fortress, also through a quasi-militarisation of border surveillance [14].

Hence, drone technology can contribute to the attainment of the objectives of EU's border controls, by reducing the number of migrants entering illegally into the EU, by preventing undocumented migration and by contributing to fight against cross-border crime. Drones could provide information to border guards present on the ground, be it at sea or on land. These ground patrols could then respond to the situation, for example in the case of migration by sea, by re-directing migrants to international seas or to the authorities of cooperating third countries, if agreements so provide.

In the EU political discourse, the Commission is particularly careful to pay attention to the humanitarian dimension of the phenomenon. The Commission has, for example, argued that one of the policy objectives of EU's border surveillance is to reduce the human death toll of migrants [15], which is one of the most serious consequences of the EU's and Member States' restrictive immigration policies. For this same reason, Frontex has been equipped with the European Border Surveillance System, better known as EUROSUR, with the aim of creating a technically interconnected information sharing environment, in order to support Frontex in its intelligence and risk analysis functions.[7]

So, will Frontex need drones in order to carry out border surveillance? Will drones be deployed to transform Europe into a humanitarian fortress, or will they increase the intelligence dimension of border surveillance, by preventing migration? The next section will look at the US experience, in order to learn how drone technology has been operating there.

---

[7]Regulation (EU) No 1052/2013 of the European Parliament and of the Council of 22 October 2013 establishing the European Border Surveillance System (Eurosur), OJ EU L 295 (2013).

12.3 The US Experience of Border Surveillance with Drones

12.3.1 The American Drone

The US has experience of deploying drones for both military and non-military purposes. It (but other states too) has been working on drone technology in a number of classified projects since the Cold War and has deployed reconnaissance drones since the Vietnam war. Attack drones are the armed variant of regular drones, which are and remain intelligence, surveillance, target acquisition, reconnaissance ('ISTAR') tools. Since 2004, the US has increasingly employed drones for border surveillance [8:3]. Within the Department of Homeland Security drones are operated by the Customs and Border Protection's Office of Air & Marine to patrol both the US-Canada border and the US-Mexico border for a corridor of 25 miles, the Caribbean Sea and the Gulf of Mexico.

According to the *Washington Post*, the US in 2011 deployed 10 drones, seven Predator B drone and three Predator B 'Guardians', the maritime variant.[8] Both Predator B and Guardian are substantially close to the armed Reaper drone. The plans are to increase the fleet to 24 between Predators and Guardians by 2016, the target being the deployment of a drone anywhere in the continent within 3 hours. The purpose of maritime surveillance is to cover illicit activity, to support joint counter narcotics operations and to tackle serious crimes, whereas the scope of land border surveillance is to counter terrorism and illegal cross-border activity, and other serious crimes. According to official information of Customs and Border Protection, their Unmanned Aerial Systems program focuses operations on the office's priority mission of anti-terrorism, by help to identify and intercept potential terrorists and illegal cross-border activity, among others, such as disaster relief.

The Office of Air & Marine uses the Guardian to conduct long-range surveillance in support of joint counter-narcotics operations, in the area of the Gulf of Mexico. A maritime radar

---

[8]The Predator drone is manufactured by General Atomics Aeronautical Systems industry. The Guardian is the process of cooperation between US Coast Guards and OAM (Office of Air Marine) [16].

is necessary to detect a number of threats. In June 2012, in the Caribbean Sea, the Office of Air & Marine has held a successful advance capabilities demonstration, paving the way for Unmanned Aerial Systems operations with the purpose of detecting illicit activity in Puerto Rico and surrounding area.

In the US, so-called Predator drones have been labelled the 'new, sexy, futuristic fix for immigration control', and gained the support of both Republicans and Democrats in Congress. But what is behind the expanding deployment and increasing investment in the drones program? Currently there is a trend, both in the US and in Europe in the direction of opening up additional airspaces to drones and to increase the fleet, as noted above. The advantages of deploying drones are (theoretically) to safely conduct missions in areas that are difficult to access or high-risk areas for manned aircraft, besides a general improvement on patrolling capabilities with fewer agents [8:1–2].

However, there are disadvantages, too. First, the accident rate experienced by the Customs and Border Protection is problematic, and definitively an issue from the perspective of deploying drones in domestic airspaces [17]. The same criticism has been levelled at the targeted killings program, deploying the Predator and the Global Hawk before their development programs were completed [8:7]. In 2006, a Predator crashed in the Arizona desert, near Nogales, when its remote pilot turned off the engine by mistake. The remote pilot was a contractor for plane manufacturer General Atomics. The drone missed a residential area and there were no casualties [18]. Weather conditions have a different impact on piloting drones compared to manned aircraft, as the remote pilot cannot benefit from his/her own sensorial perceptions, in case of a difficult landing. Equipping drones with the technologies required to overcome these problems, such as synthetic aperture radar, and moving target indicator radar, will increase significantly the costs of the operations [8].

Third, while the cost comparison between manned and unmanned aircraft is difficult and dependent on whether one calculates the cost per hour, drones are also an expensive solution. While the cost of drones can be lower than of manned aircraft, the costs of operating UAV are higher, as they require logistical support, specialised operators and maintenance training [8:4].

On this point, one should not forget that drones acquire information, but then it will be a law enforcement agent who will have to use that information and act on the basis of the information received.

Several criticisms have been made of the deployment of drone technology in border surveillance: in the US, the plan is to deploy more drones, although in 2012 the fleet was unused for the 63% of the time. This was due to a lack of budget for drone operations, and associated costs of maintenance and drone-related equipment [19].

Putting these administrative issues to one side, another aspect deserves more attention. Customs and Border Protection is increasingly flying border-patrol drones for domestic surveillance operations. Between 2010 and 2012, more than 700 missions have been carried out by the Customs and Border Protection for the Coast Guard, the Drug Enforcement Administration and immigration authorities, but also for a range of administrations not involved in border protection. While the disclosure of Customs and Border Protection has been minimal, American NGO Electronic Frontier Foundation covering internal mass-surveillance brought legal suits under the Freedom of Information Act against Customs and Border Protection and the Federal Aviation Administration for the potential breach of privacy created by this domestic surveillance flights [20]. The *Los Angeles Times* reported that Predators were used in North Dakota to help police find three crime suspects in a cow pasture [19]. Moreover, at a policy level, there is active lobbying to integrate drones into the US sky. The Federal Aviation Administration Modernization and Reform Act, has called on the Federal Aviation Administration to accelerate the process of integration of UAV into the national airspace system by 2015. Efforts to reach the target seem currently to be delayed, because of technical difficulties.

12.3.2 Legal Framework for Border Surveillance and Privacy in the US

This section discusses the US approach to the protection of personal freedoms and rights in areas where drone technology could be deployed, focusing primarily on the regulation of borders. It is useful to understand whether there might be some overlaps with the European experience, and to determine whether the European legal framework is radically different in this regard. As noted above, drones are, first, a sense-and-detect technology, and therefore they perform surveillance functions.

It is also important to be aware that 'in addition to benefits, there is also the potential for drone technology to enable invasive and pervasive surveillance without adequate privacy protections'. Drone technology is a surveillance tool that can put new pressures on the old tension between security and privacy. At the same time, this new technology might construct new possibilities and new ways to carry out surveillance, and, consequently, might affect the nature and the extent of surveillance itself, thus (co-)creating also radically new challenges. In US law, the core provision to protect rights and liberties from state surveillance is the Fourth Amendment of the Constitution, stating that 'the right of people to be secure in their persons, houses, papers, and effects, against unreasonable searches and seizures, shall not be violated'.[9] Thanks to the adjudication of the Supreme Court, the Fourth Amendment is safeguarding privacy and prevents excessive government intrusion by prohibiting 'unreasonable searches and seizures [21].

Interestingly, the Fourth Amendment regulates the conditions and possibilities for the government to conduct searches and seizures. However, it does not apply to all government acts, but only to those that constitute a search. At the same time, the Fourth Amendment does not constitute a comprehensive framework for the protection of privacy of US citizens. The US approach to issues of privacy, defined as in early 1890 as 'the right to be let alone' [22], does not rest upon a comprehensive treatment of data protection, as is the case in the EU, but consists of sector-based regulations that are narrowly applicable [23].

With this proviso, the cornerstone of Fourth Amendment protection of privacy rests upon *Katz v. United States*,[10] where the Court held that an FBI agent violated Mr. Katz's Fourth Amendment Rights by listening to his private conversations while in a telephone booth using a telephone bug. In *Katz v. United States*, the Supreme Court broadened the scope of the

---

[9]U.S. Const. Amend. IV.
[10]*Katz v. United States*, 389 U.S. 347, 351 (1967).

Fourth Amendment to privacy. More precisely, Justice Harlan's judgment in *Katz v. United States* provides one of the tests most employed in assessing whether government surveillance constitutes a search, using a subjective expectation test and society-based acceptance of that expectation. Justice Harlan noted that Fourth Amendment 'protects people, not places',[11] but in determining the protection it affords to people, it requires reference to a place. So the place becomes the decisive factor in defining whether the person has a legitimate expectation of privacy in the area to be searched and whether the society is prepared to deem that expectation to be reasonable. So, according to the scholars' interpretation [21], the traditional property-based test has been amended by a reasonable expectation test, which varies according to the place of the search.

Which are the powers and the constraints for law enforcement agencies originating from the Fourth Amendment? Looking at borders, one should be aware that they are distinctive from homes, curtilages and open fields. Borders are typically subject of overt surveillance [24]. Federal law enforcement agencies have extensive search powers at the border. Immigration officers can conduct warrantless searches of any vessel within a reasonable distance from the US and any vehicle within 25 miles of a border for the 'purpose of patrolling the border to prevent the illegal entry of aliens into the United States',[12] through the Immigration and Nationality Act [21:10]. Similar powers are accorded to custom officers.

The Supreme Court has acknowledged the federal interest in the borders, recognising that the Fourth Amendment balance of interests is qualitatively different at the borders.[13] This is because 'the Government's interest in preventing the entry of unwanted persons and effects is at its zenith in the international border.'[14] Hence, routine searches are not subject to any requirement of reasonable suspicion, probable cause or warrant. Routine searches include pat downs for weapons and contraband,[15] resort to

---

[11]Harlan, J., concurring, 351.
[12]Section 287 of the Immigration and Nationality Act, codified at 8 U.S.C. 1357.
[13]*In United States v. Flores Montano*, 541 U.S. 149, 152 (2004).
[14]Ibid.
[15]*United States v. Beras*, 183 F.3d 22, 24 (1st Cir. 1999).

drug iffing dogs,[16] luggage inspections.[17] Non-routine searches include prolonged detentions, strip searches, and body cavity searches.[18]

The legal framework for the protection of privacy provided by the Fourth Amendment indicates that there is no overall framework for privacy and data protection to which US authorities are bound while deploying drones in border surveillance. This seems to suggest that, if no distinction will be made between drone surveillance and other public surveillance and monitoring actions and tools, a federal court will upheld, under the current jurisprudence, drone surveillance conducted for purposes other than law enforcement and with no individualised suspicion [21:7]. As held by Chief Justice Rehnquist in *Mich. Dep't of State Police v. Sitz*, 'for purposes of Fourth Amendment analysis, the choice among such reasonable alternatives remains with the governmental officials who have a unique understanding of, and a responsibility for, limited public resources, including a finite number of police officers'.[19] Having in mind that in the US the deployment of drone technology in domestic skies is high in the agenda, several proposals have been presented in order to protect privacy and constrain surveillance [21].

In my assessment this brief analysis of the US legal framework underpinning the deployment of drone technology in border surveillance and privacy, is expressive of a legal mindset where society is more ready to adapt its perception of privacy at the pace of technological innovation. This example is an interesting case. However, one should look at it also by reflecting on the EU's approach, where both private and public authorities, without distinction, are bound to a data protection legal framework codifying precise obligations and rights, which affect their domestic ambit of actions, but also the international one, for example, in the case one Member State or the EU should undertake agreements with a third country [25]. Let us turn now our attention to the EU, in order to explore what is the current state of affairs of deployment of drone technology in border surveillance.

[16] *United States v. Kelly*, 302 F.3d 291, 294-95 (5th Cir. 2002).

[17] *United States v. Okafor*, 285 F.3d 842 (9th Cir. 2002).

[18] *Montoya De Hernandez*, 473 U.S. at 541; *United States v. Asbury*, 586 F.2d 973, 975 (1978); *United States v. Ogberaha*, 771 F.2d 655, 657 (2d Cir. 1985).

[19] *Mich. Dep't of State Police v. Sitz*, 496 U.S. 444, 453-54 (1990).

12.4 The Deployment of Drone Technology into Border Surveillance in the EU

12.4.1 Border Surveillance in the EU: Frontex[20]

While the responsibility for the control and surveillance of the EU's post-Schengen external borders remains with Member States, the EU in 2004 established Frontex,[21] which is tasked, among other things, with the coordination of the operational cooperation of Member States' action in border management, the organisation of return operation, and risk analysis. The Agency was reformed in 2007 and 2011, under the RABIT Regulation and the Frontex recast Regulation.[22] Since its establishment, it has grown in budget and resources.[23] Its powers and role are slowly evolving from an emergency-driven approach, typical of the first years [26], answering mainly to support requests by states, toward a more militarised and intelligence-based approach to border surveillance. This is demonstrated by the powers of

(1) initiation of joint operations at Frontex initiative, and not only at Member States' requests;

(2) the coordination and organisation of joint return operations;

(3) newly set-up European Border Guards Team, whose members have also the right to carry weapons;

---

[20]In the process of publication of this chapter, Regulation No. 1624/2016 on the European Border and Coast Guard has been approved. OJ L 251 16/09/2016, p. 1–76.

[21]Council Regulation (EC) No. 2007/2004 of 26 October 2004 establishing a European Agency for the Management of Operational Cooperation at the External Borders of the Member States of the European Union, OJ L 349/1 (hereinafter: Frontex Regulation).

[22]Regulation (EC) No 863/2007 of the European Parliament and of the Council of 11 July 2007 establishing a mechanism for the creation of Rapid Border Intervention Teams and amending Council Regulation (EC) No 2007/2004 as regards that mechanism and regulating the tasks and powers of guest officers; Regulation (EU) No 1168/2011 of the European Parliament and of the Council of 25 October 2011 amending Council Regulation (EC) No 2007/2004 establishing a European Agency for the Management of Operational Cooperation at the External Borders of the Member States of the European Union, OJ L 304/1 (hereinafter: Frontex Recast).

[23]For example, the Border Surveillance Development Programme counted for €230,000 in 2012 and for €580,000 in 2013.

(4) through EUROSUR (European Surveillance System) the agency is reinforcing its intelligence nature, thanks to the development of this network of information systems, reinforcing the information at the disposal of Frontex through the National Situational Pictures, the European Situational Picture and, last but not least, the Common Pre-Frontier Situational Picture.

We will now turn to the current research and progress in the context of the deployment of drone technology in the field of border surveillance.

Accessing information in this domain is an investigative operation, since it requires the construction of information from cooperation projects that are currently underway, and involve security industries and the European Defence Agency. This information, when not classified, is spread across sectorial and technical projects. It suggests that the situation is in continuous evolution and that this chapter can realistically only attempt to sketch some scenarios and regulatory issues arising from this context.

At present, some EU Member States including, for example Italy, are deploying drone technology in the context of border surveillance, namely the Predator. Furthermore, the information available indicates that there are a number of projects of research and development going on in the fields of maritime and land surveillance, and that Frontex, together with industry and other stakeholders, namely the European Defence Agency, have been already working in the area since some years [27].

At the same time, regulations have been enacted in order, first, to enable Frontex to buy technical assets, and second, to allow Frontex to collect and use the information acquired through drones and also satellite imagery. Article 12 of the EUROSUR Regulation provides for common application of surveillance tools. This gives rise to an important regulatory framework, requiring the Agency to 'coordinate the common application of surveillance tools in order to supply the national (...) centres and itself with surveillance information on the external borders and on the pre-frontier area on a regular, reliable and cost-efficient basis' (Article 12, EUROSUR Regulation). The surveillance information will be provided combining and analysing data which may be collected from ship reporting systems, satellite imagery and sensors mounted on any vehicle, vessel or other craft. It is indeed interesting to observe that this Regulation has paved the way for the collection, analysis and distribution of information collected, among others, also thanks to Intelligence, Surveillance and Reconnaissance devices. At another level, due to the current limitation imposed on UAV flying in the commercial airspace in Europe, and while developments are taking place also at the level of aviation regulation [4], Frontex has also expressed interest in OPA, Optionally Piloted Aircraft.

Frontex is cooperating with other agencies and bodies in the construction of a Common Information Sharing Environment, of which EUROSUR is part of, and is regularly involved or adjudicated projects on security research and drone technology through the financing of the 7th Framework Programme. Being drones first of all an Intelligence, Surveillance and Reconnaissance technology, it is important to realise that they will play an important role also in the future. In relation to border surveillance within the EU, one can only conclude that 'they are here to stay'. Hence, it makes sense to 'face them'.

While an overview of all the funded research projects on drones is not a goal of this chapter [27], I would like to present those closely directed to border surveillance. This illustrates that it is realistic to expect that in some years Frontex will be ready to deploy drone technology, or optionally piloted aircraft, in border surveillance operations.

12.4.2 Frontex and Drones

In the 2013 Frontex work program, we can read that the agency has been involved with drones since, at least, 2009. For example, the All Eyes[24] project 2012 continues the Remote Sensing and Detection technology project. It focuses on developing sensors and platforms, broadband communication and data fusion systems for the benefit of Frontex' joint operations. It includes, in particular:

- Aerial Border Surveillance Trial with manned aircraft with optionally piloted aircraft capability equipped with multi-intelligence sensors and report

---

[24]Aerial, Ground and Sea Surveillance–sensors and Platforms and advanced system solutions. Budget: EUR 450.000.

- Demo of MALE (Medium Altitude Long Endurance) Remotely Piloted Aircraft (RPA) in an operational environment and report (...) and
- Elaborated CONOPS [concept for Operations] for detecting & tracking of small boats, in relation with the EU FP7 initiatives, aiming to optimise synergies with activities that are supported by other EU funding.

Another project where Frontex is involved is CLOSEYE. The project is a 'collaborative evaluation of border surveillance technologies in maritime environment by pre-operational validation of innovative solutions' [28] and has received over €9 million in funding from the security strand of the EU's 7th Framework Programme. The project was launched in Spain in 2013 and will feature the deployment of drones, satellites and aerostats over the Southern Mediterranean. The aim is to provide the EU 'with an operational and technical framework that increases situational awareness and improves the reaction capability of authorities surveying the external borders of the EU' [28].

CLOSEYE is closely connected with EUROSUR and is meant to reinforce the Spanish SIVE system (*Sistema Integrado de Vigilancia Exterior*, Integrated System for External Surveillance), operational since 2002, using radar and surveillance cameras to detect incoming vessels. In this process, drones such as the Camcopter S-100 and the UAV Predator have been tested in 2013. The CLOSEYE project will have its definition in 2013, in 2014–15 its execution, with experiment planning, execution and supervision and it will finish with evaluation in 2015–2016. So it is reasonable to expect that drones might soon be employed for border surveillance. In addition the SAGRES project aims at validating and further testing of tools designed for the tracking of vessels on the high seas as part of the EUROSUR project, including monitoring of a specific third country port and tracking the identified vessels over high seas. The work carried out in this project will contribute to the collection of information due to constitute the Common Pre-Frontier Intelligence Picture of EUROSUR. Besides these projects, Frontex regularly exchanges contacts and organises events with drone manufacturers (such as Aeronautics, Selex, Israel Aircraft Industries, Sagem, Lockheed Martin, Aerovisión, BAE Systems, Finmeccanica) and it has held a number of workshops and events[25] where leading security and military industries have had the opportunity to meet with end-users and to synchronise with their needs.

All in all, the last decade has been characterised by a rush to devise, test and acquire new tools and instruments for the governance of EU's external borders. The EU's apparent transformation into a high-tech 'fortress' is thus proceeding at a pace [14]. The EU, though its agencies, such as European Defence Agency and Frontex, is cooperating on Intelligence, Surveillance and Reconnaissance systems, of interests and concern also for border surveillance, among others. Currently, drones for border surveillance are being studied, devised and tested in pilot projects. At the same time, the EU legislature has already paved the way for their deployment with the EUROSUR regulation, which is meant to strengthen the intelligence dimension of border surveillance, also through the deployment of drone technology in border surveillance. The next section of the chapter explores some of the regulatory issues, which arise in the context of the deployment of drone technology in border surveillance.

12.5 Policy Issues, Ethical and Regulatory Challenges Underlying the Political Choice of Deploying Drones for Border Surveillance

Currently there are several regulatory issues around drones. They could be divided into general (or horizontal) and specific (or sectorial) questions. In the former, we consider the broader regulatory perspective of how to insert drones into the civil and commercial airspace, since currently drones can only fly into so-called 'segregated' airspaces. The range of relevant issues is

---

[25]2009, workshop on UAVs and Land Border surveillance, Imatra, Finland; 2011, workshop in Warsaw, focussing on integrated sensor platforms for border surveillance at the EU external borders but also for other types of tools/platforms. Drones rank among the latter; 2011, event in Finland, on UAVs, aerostats and sensor systems allegedly potentially filling the gap in the existing surveillance systems; 2011: research and development workshop in France and Greece on the surveillance of the sea on MALE and also small UAVs with Long Endurance. It featured demonstrations by, among others, Israel aircraft Industries, aeronautic defence systems, Sagem, Lockheed Martin, Aerovisión; 2012: drone workshop in Bulgaria.

extremely broad, relating for example to aviation law, insurance and tort law, privacy law, safety law, will not be covered in this chapter. This has resulted in interesting dialogue among stakeholders, under the coordination of the EC [4, 6, 29]. The focus of this section is narrower; it examines the specific or sector-based problems of border surveillance: how the deployment of a new technology might and will affect the nature of the function performed.

12.5.1 Policy Issues Underlying the Deployment of Drones in Border Surveillance

A preliminary issue tackled here concerns the metamorphosis of the drones and the transparency and democracy of this process, including also the financing of the whole transformation. One of the arguments of the supporters (industries, state agencies and European institutions) of the deployment of drone technology in the civil governance is that it will enable savings of public expenditure.[26] Arguments relating to cost savings and resource rationalisation can be especially persuasive in times of economic crisis. Equally, however, there are criticisms of the economic efficiency of drones.

First, deploying drones in border surveillance operations means that drone technology will have to be researched, tested and adjusted before being operated. So, there are research and development costs that should be clearly quantified, even though spread across myriads of programs. Second, operating drones or Remotely Piloted Aircraft is expensive, as the American experience demonstrates.[27] In practice, it requires staff, a ground-

---

[26]The typical argument is that deploying a manned aircraft is more expensive than an unmanned one, and secondly, that the unmanned aircraft can stay airborne for longer time. For example, the data of the Italian Navy and Aeronautics, indicate that a drone Predator cost €3k/hour whereas other types of helicopters cost from €4k to €7k/hour. See also Gertler [7] eloquently stating that 'Conventional wisdom states that UAS offer two main advantages over manned aircraft: they are considered more cost-effective, and they minimize the risk to a pilot's life.' And offering data and evidence on the management costs of UAS.

[27]In Gertler [7:15], one can read: 'Congress has noted that, 'while the acquisition per unit cost may be relatively small, in the aggregate, the acquisition cost rivals the investment in other larger weapon systems.' Quoting U.S. Congress, 107th Congress, 2nd Session, House of Representatives, Bob Stump National Defense Authorization Act for Fiscal Year 2003.

pilot, instead of a pilot in the aircraft; depending on regulatory choices, a certified pilot might be required.[28] Then, another agency will use the information collected by Remotely Piloted Aircraft.

It seems thus that another operational layer is going to be added, compared to border surveillance performed without drones. Thirdly, bases are necessary in order to enable drones to take off and land. It might be necessary to accommodate this new technology into current aviation bases, or even build new bases, as a consequence of air traffic restrictions.[29] It is true that mutualisation[30] might allow some economic efficiencies, but one should then take into account that significant investments have been made first by the military. Evidence based on the US experience (and reiterated in Frontex' officials declarations and presentations) indicates that this is everything but a cheap way of doing border surveillance, certainly not at this stage. Deploying drones in border surveillance will simply entail an increased militarisation of border surveillance: is this necessary? Is this proportionate? With the choice of deploying (long endurance) drones for border surveillance, the EU and the Member States are not opting for a cheap solution, as the experience with the Italian operation *Mare Nostrum* demonstrates. Drones are a new and costly technology, in which investments of public money are made for the benefit of private and public industries, in times of deep financial and social crisis, and which mark a clear shift toward an increased militarisation of border surveillance.

Furthermore, there is little information on these costs and no public debate on the suitability of deploying this technology into this context. In part, this can be explained by the

---

[28]As an example, a Predator air vehicle costs US$ 4.5 million, while the Predator system, including four air vehicles and control equipment, costs over US$ 20 million [7].

[29]This might be necessary because drones can currently fly only into so-called segregated airspaces, due to current aviation regulations.

[30]The term mutualisation designates 'the operation of military RPAS assets by the military for non-military governmental applications. The basic advantages of mutualisation appear to be the following: (1) Offers the military additional RPAS flight training opportunities; (2) supplies added value to military flight training exercises; (3) permits to increase the return on investment for military RPAS by using them for non-military governmental missions with societal benefits (incl. improved European external border surveillance)—from UVSI Discussion Paper on RPAS Industry & Market Issues' [6].

securitisation process in which the broader debate on the governance of migration is framed. The process is technology-driven and is dealt with as a purely technical issue, even though it has implications beyond that. No question or policy choice has been debated beforehand; research, studies are first done at technical level, through research financing, pilot projects and then, regulation will ratify what has been already achieved into a totally technical and undemocratic process [27].

This raises serious questions about democracy being replaced by technocracy, and the case of drones is not the first one. EUROSUR is a good example of this trend. First, EUROSUR has been studied (BORTEC and MEDSEA feasibility projects, published in partially declassified versions), then it has been set-up and made operational in pilot projects (since 2011); in the meanwhile the Regulation providing for EUROSUR has been debated and adopted (October 2013). Within this process, also the information on costs is vitiated. At the beginning, the forecasts relating to the cost of a project is under-estimated in order to get political support. For EUROSUR for example, the first estimate was 800,000 Euros, but up to 2013 approximately 6.54 million Euros have been spent on the project [30:4]. Also the Schengen Information System is an emblematic case of this epiphany of the securitisation dominating the governance of migration.

Another argument concerns the costs of these programs. Initiatives are spread across different programs, and estimating costs, scrutinising and debating them democratically is increasingly difficult. For example, under FP7 program there are numbers of initiatives supporting research for and by security industry, which contributes to the operational part of the technical and operational aspects mentioned above, but which are conducted under different frameworks. These are mainly public and private partnerships, and imply a high involvement of the industry. Another issue, also affecting the process, concerns the involvement of stakeholders who bear financial interests in the process. Considering that it is often about security industries, participated in by states, it is important, besides the formal respect of public procurement law, to ensure the maximum transparency of the whole process.

Last but not least, one could see a problem of political opportunity. It can be argued that the surveillance of the external borders is increasingly autonomous, with its own actors (Frontex), technological toolkit (EUROSUR, drones, but also satellites) and regulations, disentangled from other policies. Confronted with humanitarian crises originated from past and present wars and civil unrest situations (Arab Spring, Afghanistan, Syria), the European Union is lacking the solidarity, relocation and burden-sharing mechanisms for refugees and asylum-seekers. Second, in spite of the reception of search and rescue provisions in recent border surveillance regulation, rescuing and allowing disembarkation to boats of irregular migrants still means taking responsibility for the application of asylum requests:[31] makes Southern European states reluctant to engage in active humanitarian policies, but also to fully abide to their obligations under the law of the sea, toward the phenomenon of irregular migration at sea (with the exception of *Mare Nostrum*).

12.5.2 Ethical and Regulatory Questions

Turning our attention to the regulatory questions for border surveillance, this issue could be unpacked from the following threefold perspective. First, the new technology will be implemented into the current regulatory framework. Second, the deployment of drone technology might affect or change, to some extent, the nature of border surveillance; so, from this perspective, the same existing regulations might happen to become obsolete, or not accurate enough or might require adjustment. Third, to another extent the new technology could create new situations, which might postulate and require new regulatory frameworks.

Detailing the regulatory framework of border surveillance in detail remains out of the scope of the chapter, which focuses on the challenges arising from the deployment of drone technology in border surveillance. However, the core provisions on border surveillance is provided by Article 12 of the Schengen Borders Code, according to which the purpose of border surveillance is 'to prevent unauthorised border crossings, to

---

[31] According to the well-known Dublin Regulation. See Regulation (EU) No 604/2013 of the European Parliament and of the Council of 26 June 2013 establishing the criteria and mechanisms for determining the Member State responsible for examining an application for international protection lodged in one of the Member States by a third-country national or a stateless person (recast), OJ L 180/31.

counter cross-border criminality and to take measures against persons who have crossed the border illegally.' Furthermore: 'Surveillance shall be carried out by stationary or mobile units which perform their duties by patrolling or stationing themselves at places known or perceived to be sensitive, the aim of such surveillance being to apprehend individuals crossing the border illegally. Surveillance may also be carried out by technical means, including electronic means.'[32] The Annex VI of the Schengen Borders Code provides detailed rules for the various types of border and the various means of transport used for crossing the Member States' external borders, be it land border, air border or sea border.

Considering border surveillance and drones in Europe, one cannot omit to refer to the social phenomenon of irregular migration through the Mediterranean Sea: in this context other regulations come into play, namely the law of the sea, codified in the UN Convention on the Law of the Sea (UNCLOS). Within this context, the safety of life at sea, as further defined by the Safety of Life at Sea (SoLaS) Convention and by the Search and Rescue (SAR) Convention, is of paramount importance and constitute the background regulation in which border surveillance takes place.

How can drones have an impact on border surveillance? First, drone technology has the potential radically to change the way this function is performed. By enabling a widespread and diffuse surveillance of remote areas, beyond the borders, drone technology will entail another step in the direction of 'moving outside' the borders of Europe, potentially affecting also the sovereignty of third states. Second, the collection of information of migrants' routes and fluxes will reinforce the intelligence dimension of border surveillance.[33] From this perspective, the deployment of long endurance drones is thus functional also to a policy aiming at radically preventing the phenomenon of migration instead of waiting for migrants to arrive on the European coasts, by acquiring prompt information in order to

---

[32]Article 12 of the Schengen Borders Code.
[33]For example, the Predator drone deployed in Mare Nostrum is probably operating close to Libyan shores, i.e., migrants' departure country [31].

enable ground vessels to approach migrants' boats and intercept and divert them back to the country of departure.[34]

This leads us to a second issue: that this approach is functional to a policy of externalisation of border controls, which is based on the cooperation and involvement of third country authorities in the achievements of the objectives of such preventive border surveillance practices.[35] It might be useful to involve third country authorities in order to have their cooperation on preventing migrants from leaving, accepting them if diverted back and so forth, in a process where border surveillance is increasingly externalised toward third countries. In other words, deploying drones in border surveillance will entail a significant policy shift toward the prevention of migration, a human phenomenon.

But if with drones the EU and Member States are reinforcing their toolkit to prevent such a human phenomenon, one should also consider that, in order to be effective, this policy requires some forms of cooperation also with third country authorities. Which regulatory and ethical issues arise from cooperating on migration controls to third countries? Is it desirable that the Member States and the EU cooperate with a third country without a clear democratic record? How can the EU and Member States monitor that in a third country the fundamental rights standards are adequate? Besides the question of political opportunity, actual legal challenges undermine this policy. Increased surveillance might entail a form of control, which can also trigger obligations and duties [3].

Third, from another perspective, the deployment of drones has the potential to contribute to dehumanising border

---

[34]This is enabled by the Regulation establishing rules for the surveillance of the external sea borders in the context of operational cooperation coordinated by the European Agency for the Management of Operational Cooperation at the External Borders of the Member States of the European Union, for example Articles 7 and 10. Southern European states have traditionally bilateral cooperation agreements with north-African counter-parts. For example, Italy and Libya have bilateral cooperation agreements on, among others, irregular migration, since 2000.

[35]Cf. Article 20 of the EUROSUR Regulation enabling the collection in EUROSUR of information gathered on the basis of cooperation with third countries. Frontex can conclude working arrangements with third country authorities on the basis of Article 14.2 of the Frontex Regulation.

surveillance, as has been suggested with reference to warfare [32]. The manned staff remain on the mainland and at distance from migrants, whereas the RPA patrols the high seas and perhaps also the waters and the lands of the third country: the unmanned gets in proximity of the migrants. With drones we have a machine, instead of a man, carrying out border surveillance. Can a drone help to distinguish situations where different legal frameworks are at stake, for example? Can a drone determine whether there are persons in need of medical assistance on board? This information is important as different situations determine different legal frameworks. The answer is certainly affirmative, as the surveillance technology they carry, create 'epistemic bridging' overcoming the remoteness created by the drone [5].

However, will the 'increased situational awareness' (in Frontex' jargon) *per se* imply better assistance to migrants in distress? This is unlikely, in the absence of a strengthening of the search and rescue capacities of the Member States and of Frontex itself, and without any new approach in the rules governing asylum requests. Also this aspect seems to indicate that border surveillance is turning into a highly technological and quasi-militarised operation, but no progress is achieved on reinforcing the humanitarian dimension of border surveillance [14]. Deploying drones for border surveillance seems functional to a policy oriented toward the prevention of migration, a policy inspired by a fragmented reading and interpretation of the legal framework governing border surveillance and migration controls [33]. Migration and border surveillance are increasingly separated from one another. In this scenario, borders become the source of the (security) threat, which is illegal migration and other criminal phenomena, the place which needs to become the object of total surveillance: Frontex' objective is the 24/7 blue/green situational awareness.

Fourth, drones are usually coupled with sensors and other Intelligence, Surveillance and Reconnaissance technologies. Thus, they have a tremendous potential in terms of surveillance, they could basically acquire the same information that otherwise would be acquired with a seizure of a boat, seizure that would be justified according to the international law of the sea, only in some cases [33]. How border surveillance by drones ought to be framed? Given the surveillance capacities of drone technology on the legal sphere and privacy of the individuals, will a warrant be necessary in order to carry out surveillance in a legitimate manner? Considering that surveillance of boats on the high seas does entail the surveillance of private boats, should the limitation of a private sphere take place with a warrant or simply by a statute that authorises beforehand an extensive invasion of the private sphere? This is a case where a new technology might entail a new regulatory situation. Considering that Member States and the EU are bound to the respect of their international and human rights obligations, for examples under the Charter of Fundamental Rights of the EU and under the European Convention of Human Rights, will current privacy and data protection regulations apply to drones patrolling high seas? As stated by the Court of Human Rights of Strasbourg, though the jurisdiction is in principle territorial,[36] Member States cannot be exempted from the respect of their Conventional obligations because they act extraterritorially.[37] The multifunctional character of drone technology suggests that, next to their potential uses, the regulator should take a close look at all the challenges they pose, also to privacy and to data protection: therefore, their deployment shall be firmly guaranteed by the respect of adequate legislations, starting from current data protection rules. Drones are not simply a technique. Their deployment changes border surveillance and makes it more pervasive and subtle. The next section will look at the regulatory constraints originating from data protection regulations in the EU.

12.5.3 Legal Framework for Border Surveillance and Privacy in the EU

The deployment of a dual-use technology, more precisely of a powerful surveillance tool such as drone technology, to the law enforcement domain triggers several questions under data protection and privacy, where strict rules apply, in contrast to the military. The following sketches the main issues.

---

[36]European Court of Human Rights, 12 December 2001, Decision, *Bankovic and Others v. Belgium and Others*, Application No. 52207/99.
[37]European Court of Human Rights, 23 February 2012, Judgment, *Hirsi Jamaa and Others v. Italy*, Application No. 27765/09.

In the EU, data protection regulation has a fundamental rights status and it is a crucial means of protection for privacy (Article 7 Charter). According to Article 8 of the Charter of Fundamental Rights of the EU, having the same legally binding status as the EU treaties, individuals have the right to protection of personal data, based on the principles of consent, access to data, rectification, and more recently including also the right to be forgotten. Long before the Charter, the EU legal system referred to the European Convention of Human Rights (ECHR), and its Article 8, ensuring that everyone has the right to have their private and family life, home and communication respected.[38]

In addition to these provisions, enshrining data protection among the fundamental rights of the EU's legal order, there is European regulation binding in all Member States: these are the Data Protection Directive 95/46/EC and the Framework Decision 2008/977/JHA on the protection of personal data processed in the framework of police and judicial cooperation in criminal matters.[39] Finally, the Regulation 2001/45/EC completes this legal framework, covering the processing of personal data by

---

[38]On this aspect we should note that the European Court of Human Rights has underlined that the concept of private life should be interpreted in a non-restrictive manner, for example as including a professional call. Cf. European Court of Human Rights, *Amann v. Switzerland*, Application No. 27798/95, 16 February 2000. In other case law, the Court has held that personal data as such irrespective of the format of conservation and/or storage, including in the concept communications, images, visual data and sounds collected through cctv. See, ECtHR, *von Hannover v. Germany*, N. 59320/00, of 24 June 2000; *Peck v. U.K.*, N. 44647/98, 28 January 2003; *Koepke v. Germany*, n. 420/07, 5 October 2010.

[39]Directive 95/46/EC of the European Parliament and of the Council of 24 October 1995 on the protection of individuals with regard to the processing of personal data and on the free movement of such data, in OJ L 281/31 (1995); Council Framework Decision 2008/977/JHA of 27 November 2008 on the protection of personal data processed in the framework of police and judicial cooperation in criminal matters, in OJ L 350 of 30.12.2008. On the relevance of the Data Protection Directive in the EU, see the recent judgment of the Court of Justice of the EU in Joined cases C-293/12, *Digital Rights Ireland Ltd v Minister for Communications, Marine and Natural Resources* and C-594/12, *Kärntner Landesregierung*, Judgment of 8 April 2014, in which the Court stated that data protection and privacy rights impede the massive retention of meta-data on tlc communications, invalidating the data retention directive.

EU institutions and bodies.[40] The core principles of the data protection regulation are the principle of consent and right of access to data of the data subject, the principle of specific, explicit and legitimate purpose for the collection of data (lawfulness), necessity and proportionality.[41]

The Data Protection Directive defines personal data as 'any information relating to an identified or identifiable natural person; an identifiable person is one who can be identified, directly or indirectly, in particular by reference to an identification number or to one or more factors specific to his physical, physiological, mental economic, cultural or social identity'.[42] Its scope covers 'the processing of personal data wholly or partly by automatic means' and by contrast does not cover 'processing operations concerning public security, defence, State security (including the economic well-being of the State when the processing operation relates to State security matters) and the activities of the State in areas of criminal law.'[43] The scope of the Framework Decision instead covers Member State activities in the context of the judicial cooperation in criminal matters, the former Third Pillar of EU law, and hence extends the data protection to that area, otherwise explicitly excluded from the scope of the data protection directive.

As anticipated, the deployment of drone technology in civil applications challenges these legal frameworks and the rights they protect: drones can collect and transmit specific information to the drone operator about what happens at sea, including visual data.[44] It has been argued that the application of the data protection directive to border surveillance is not clear

---

[40]Regulation (EC) No 45/2001 of the European Parliament and of the Council of 18 December 2000 on the protection of individuals with regard to the processing of personal data by the Community institutions and bodies and on the free movement of such data, in OJ L 8/1 (2001).
[41]Articles 6 and 7, Data Protection Directive.
[42]Article 2, Data Protection Directive.
[43]Article 3 and Article 13, Data Protection Directive.
[44]In the case of *Peck v. UK*, the ECtHR stated that: 'The monitoring of the actions of an individual in a public place by the use of photographic equipment which does not record the visual data does not, as such, give rise to an interference with the individual's private life'. European Court of Human Rights, 28 January 2003, Judgment, *Peck v. The United Kingdom*, Application No. 44647/98.

(Finn and Wright, 2012). Indeed, according to the Directive, the Member States can restrict the scope of the obligations and rights protected by the same, if necessary to protect, among others, national security, public security, and 'the prevention, investigation, detection and prosecution of criminal offences (...)'.[45] However, Frontex operations are covered by the Regulation 45/2001/EC, being an EU agency, as stated by Article 11 of the Frontex reformed Regulation. The Directive is currently in the process of being reformed by a General Data Protection Regulation. For this reason, the *Commissie Meijers* had suggested to postpone the enactment of the EUROSUR Regulation till the adoption of the General Data Protection Regulation, which has not been done [34].[46]

At the same time, activities of border surveillance are covered by data protection rights in specific rules, provided by the Frontex Regulation, as reformed by the Frontex Recast, and by the EUROSUR Regulation. More precisely, the EUROSUR Regulation defines rules for data protection, a sensitive issue since extensive surveillance carried out in the context of border surveillance potentially infringes upon data protection rights: the most sensitive issue concern the cooperation of Frontex and of with third country authorities in the framework of border surveillance. For this reason, '[a]ny exchange of personal data in the European situational picture and the common pre-frontier intelligence picture should constitute an exception. It should be conducted on the basis of existing national and Union law and should respect their specific data protection requirements.'[47] The European rules on data protection are applicable, in case more specific provisions do not apply.

Among the general provisions, Article 2(4) states that 'Member States and [Frontex] shall comply with fundamental rights, in particular the principles of non-refoulement and respect for human dignity and data protection requirements, when applying this Regulation'.

---

[45] Article 13, para. 1, letters a), c), d), DPD.

[46] While this chapter went to press, the General Data Protection Regulation was approved as Regulation (EU) 2016/679 of the European Parliament and of the Council of 27 April 2016 on the protection of natural persons with regard to the processing of personal data and on the free movement of such data, and repealing Directive 95/46/EC.

[47] Recital 13, EUROSUR Regulation.

Article 13 constitutes the general provision on data protection in EUROSUR. It provides that 'the European situational picture and the common pre-frontier intelligence picture may be used only for the processing of personal data concerning ship identification numbers.' And when the national situational picture is used for processing of persona data, those data shall be processed in accordance with the European data protection directive and the Framework Decision, and relevant national provisions on data protection.

Data protection requirements shall be applied also to the cooperation between Frontex and third parties, such as Union bodies, offices, agencies and international organisations.[48]

Another important provision is contained in Article 20 on cooperation of Member States with neighbouring third countries; it provides that '[a]ny exchange of personal data with third countries in the framework of EUROSUR shall be strictly limited to what is absolutely necessary for the purposes of this Regulation. It shall be carried out in accordance with Directive 95/46/EC, Framework Decision 2008/977/JHA and the relevant national provisions on data protection'.[49] Though strictly limited, exchange of personal data is still possible, and this inevitably triggers the question of the fate of those data, in the hands of third country institutions. The threat of the function creep is there.

12.6 Conclusion

The chapter has focused on the deployment of drone technology into border surveillance. It has shown the rationales for the metamorphosis of the drones, namely protecting human pilots from fatigue and environmental hazards, high flexibility in tasking, ability to cover remote areas, high potential in collecting information at a distance. However, their alleged economic benefits are counterbalanced by the high costs of this technology

---

[48] Article 18, EUROSUR Regulation, on cooperation of Frontex with third parties.

[49] Article 20, para. 4. See also para. 5, stating that: 'Any exchange of information under paragraph 1, which provides a third country with information that could be used to identify persons or groups of persons whose request for access to international protection is under examination or who are under a serious risk of being subjected to torture, inhuman and degrading treatment or punishment or any other violation of fundamental rights, shall be prohibited.'

and by the failures it currently displays. By highlighting policy and practices within the US, a state deploying drones in border surveillance since 2004, the chapter has shown that while drones per se are not more expensive than helicopters, their operations and maintenance are. Second, deploying drones requires a strong regulatory framework, in order to protect safety and prevent accidents, and to protect privacy: drones are powerful surveillance tools and their potential should not be underestimated.

Drones in border surveillance should be interpreted not simply as a new tool but potentially as a 'game changer', i.e., a new technology bearing the potential of radically changing the way a function—border policing—is performed. This radically affects the nature of such a function. For this reason, it is important that the regulatory approach to the deployment of drone technology into civil contexts is framed into a context of evidence- and risk-based regulation, mitigated by precautionary approaches, which could and should quickly take into account new issues emerging from the reality. From a policy perspective, the deployment of drone technology seems functional to a shift toward a quasi-militarised and intelligence-based border surveillance, shift which needs to be monitored in its practical implications, which can range from dehumanising border surveillance to strengthening the cooperation with third country authorities, in the perspective of 'moving the borders of Europe' and externalising border surveillance to the same third country. On the other side, while regulators and society should be aware of the surveillance powers of drones, which can cover remote and hazardous areas, the creation and the use of such a diffuse 'panopticon' can eventually also create or, at least, reinforce the obligations binding upon the beneficiaries of such an extended surveillance.

References

1. Alston, P. (2010) *Report of the Special Rapporteur on extrajudicial, summary or arbitrary executions* (Human Rights Council, Fourteenth Session, New York).
2. Thorp, A. (2011) Drone attacks and the killing of Anwar al-Awlaqi: Legal issues, *House of Commons Library Standard Note* SN06165 (House of Commons, London).

3. Rosén, F. (2014) Extremely stealthy and incredibly close: Drones, control and legal responsibility, *Journal of Conflict and Security Law*, **19**(1), 113–131.

4. European RPAS Steering Group (2013) *Roadmap for the Integration of Civil Remotely-Piloted Aircrafts Systems into the European Aviation System* (European Commission, Brussels).

5. Coeckelbergh, M. (2013) Drones, information technology, and distance: Mapping the moral epistemology of remote fighting, *Ethics and Information Technology*, **15**(2), 87–98.

6. European Commission (2012) *Towards a European Strategy for the Development of Civil Applications of Remotely Piloted Aircraft Systems (RPAS)*, SWD(2012) 259 final (European Commission, Brussels).

7. Gertler, J. (2012) *U.S. Unmanned Aerial Systems* (Congressional Research Services, Washington, D.C.).

8. Haddal, C. C., and J. Gertler (2010) *Homeland Security: Unmanned Aerial Vehicles and Border Surveillance* (Congressional Research Services, Washington, D.C.).

9. Ackerman, S. (2013) UN drone investigator: If facts lead to US war crimes, so be it, *Wired*, 29 January. Available at: http://www.wired.com/2013/01/un-drone-inquiry/.

10. Waever, O. (1995) Securitization and desecuritization. In: Lipschutz, R. D. (ed.) *On Security* (Columbia University Press, New York), pp. 46–86.

11. Bigo, D. (2000) When two become one: Internal and external securitizations in europe. In: Kelstrup, M., and Williams, M. C. (eds.) *International Relations Theory and the Politics of European Integration: Power, Security and Community* (London: Routledge).

12. Balzacq, T. (2008) The policy tools of securitization: Information exchange, EU foreign and interior politics, *Journal of Common Market Studies*, **46**, 75–100.

13. Besters, M. and F. W. A. Brom (2010) Greedy information technology: The digitalization of the European migration policy, *European Journal of Migration and Law*, **12**(4), 455–470.

14. Marin, L. (2011) Is Europe turning into a 'technological fortress'? Innovation and technology for the management of EU's external borders: Reflections on Frontex and Eurosur. In: Heldeweg, M. A., and Kica, E. (eds.) *Regulating Technological Innovation* (Palgrave Macmillan, London), pp. 131–151.

15. European Commission (2011) *Proposal for a Regulation of the European Parliament and the Council Establishing a European Border*

Surveillance System (EUROSUR), COM(2011) 873 final (European Commission, Brussels).

16. Booth, W. (2011) More Predator drones fly U.S.–Mexico borders, *Washington Post*, 21 December.
17. Harding, P. (2003) Eyes in the skies, *Richmond Times-Dispatch*, 30 October, p. F1.
18. Barry, B. (2012) *The Numbers Game: Government Agencies Falsely Report Meaningless Deportations and Drug Seizures as Victories*. Available at: http://www.ciponline.org/research/entry/numbers-game-government-agencies-falsely-report-meaningless-deportation.
19. Sternstein, A (2012) *Freeze Border Security Drones Purchases, IG Says*. Available at: http://www.nextgov.com/defense/2012/06/freeze-border-security-drone-purchases-ig-says/56212.
20. Lynch (2012) *EFF Demands Answers about Predator Drone Flights in the U.S.* Available at: https://www.eff.org/press/releases/eff-demands-answers-about-predator-drone-flights-us.
21. Thompson, R. M. (2013) *Drones in Domestic Surveillance Operations: Fourth Amendment Implications and Legislative Responses* (Congressional Research Service, Washington, D.C.).
22. Warren, S. D. and L. D. Brandeis (1890) The right to privacy, *Harvard Law Review*, **4**, 193–220.
23. Slemmons Stratford, J. and J. Stratford (1998) *Data Protection and Privacy in the United States and Europe*. Available at: http://www.iassistdata.org/downloads/iqvol223stratford.pdf.
24. Kenk, V. S., J. Križaj, V. Štruc and S. Dobrišek (2013) Smart surveillance technologies in border control, *European Journal of Law and Technology*, **4**(2).
25. Finn, R. L. and D. Wright (2012) Unmanned aircraft systems: Surveillance, ethics and privacy in civil applications, *Computer Law & Security Review*, **28**(2), 184–194.
26. Carrera, S. (2007) The EU border management strategy: FRONTEX and the challenges of irregular immigration in the Canary Islands, *CEPS Working Documents* (261).
27. Hayes, B., C. Jones and E. Toepfer (2014) *Eurodrones Inc* (Statewatch-TNI report, Washington, D.C.).
28. Statewatch (n.d.) *Drones over the Mediterranean*. Available at: http://www.statewatch.org/news/2013/may/02eu-drones-mediterranean.html.

29. European Commission (2014) *Communication from the European Commission to the European Parliament and the Council, A New Era for Aviation. Opening the Aviation Market to the Civil Use of Remotely Piloted Aircraft Systems in a Safe and Sustainable Manner,* COM(2014) 207 final (European Commission, Brussels).

30. Martin, M. (2013) *Trust in Frontex: The 2013 Work Programme.* (Statewatch, UK).

31. Gaiani, G. (2013) *Mare Nostrum: pro e contro della missione militare dell'Italia.* Available at: http://www.ispionline.it/sites/default/files/pubblicazioni/commentary_gaiani_22.10.2013_0.pdf.

32. Wall, T. and T. Monahan (2011) Surveillance and violence from afar: The politics of drones and liminal security-scapes, *Theoretical Criminology*, **15**, 239–254.

33. Moreno Lax, V. (2011) Seeking asylum in the Mediterranean: Against a fragmentary reading of EU member states' obligations accruing at sea, *International Journal of Refugee Law*, **23**(2), 174–220.

34. Commissie Meijers (2012) *Note on the Proposal for a Regulation Establishing the European Border Surveillance System* (COM (2011) 0873) (European Commission, Brussels).

Chapter 13

On the Disruptive Potential of 3D Printing

Pierre Delvenne and Lara Vigneron

Department of Political Science,
University of Liège, SPIRAL Research Centre,
Boulevard du Rectorat 7/29, Liège 4000, Belgique

pierre.delvenne@ulg.ac.be, Lara.Vigneron@ulg.ac.be

13.1 Introduction

A wide range of increasingly advanced manufacturing technologies is now emerging. They are changing basic assumptions about how products are designed and manufactured, and are to some extent re-dictating the terms governing who can successfully engage in manufacturing and where such production can be based [1]. These new manufacturing technologies, predominantly based on three-dimensional (hereafter '3D') printing technologies, today have a global spread. They have become usually associated to big claims like re-localising production processes in the West, creating a full range of high-tech jobs, or developing mass customisation market available to hardly anyone [2]. In other words, 3D printing technologies are expected to transform the world in the 21st century.

Embedding New Technologies into Society: A Regulatory, Ethical and Societal Perspective
Edited by Diana M. Bowman, Elen Stokes, and Arie Rip
Copyright © 2017 Pan Stanford Publishing Pte. Ltd.
ISBN 978-981-4745-74-1 (Hardcover), 978-1-315-37959-3 (eBook)
www.panstanford.com

However, these technologies do not form a monolithic block: Some have become the core of well-established or increasingly flourishing industries (i.e. in aerospace and biomedical sectors) while others are directly available to the end users (i.e. in personal fabrication). Either as established, or as emerging/future-oriented technologies, the challenges posed by 3D printing are likely to necessitate various innovative governance[1] interventions. In turn, governance agents will base their decisions on various narratives that structure present cognitive patterns. As the German phenomenologist Wilhelm Schapp puts it: 'We humans are always entangled in stories' [6:1]. For the last 20 years, cognitive research and neuroscience have kept confirming this statement, which is also true for the way we situate ourselves with regard to emerging technologies. It is now beyond dispute that narrative functions as a primary, universal and probably vital means of structuring information.

For 3D printing technologies, various narratives are at play. In the case of 3D printing technologies available to end users, 3D printing connects to a narrative of radical empowerment of individuals, either as manufacturers of customised products ('prosumers') in small-scale workshops offering (personal) digital fabrication or through dedicated enterprise services, or as enhanced by tissue regeneration and organ transplantations. This narrative needs to be connected to the increased valorisation of creative attitudes (also in terms of private appropriation of individuals' intellectual creativity) in enterprise culture, which is visible in the current interest in 'creative industries' [7]. In the case of industrial 3D printing technologies, 3D printing connects with another narrative of economic growth through technological innovation, which is largely state-driven. Although, for nearly 30 years, the sector of 3D printing has mostly been industry-driven, some states are now increasingly declaring a strategic interest in 3D printing. For example, President of the United States of America, Barack Obama, in his 2013 State of the Union Speech, declared:

---

[1] 'Governance' here means the broadening of government and the inclusion of more actors in collective choices involving science and technology. Governance is actually distributed between a number of actors as some definitions acknowledge: Governance can be discussed as the coordination and control of autonomous but interdependent actors either by an external authority or by internal mechanisms of self-regulation or self-control [3–4], including *de facto* governance arrangements that emerge and become forceful when institutionalised [5].

A once-shuttered warehouse is now a state-of-the art lab where new workers are mastering the 3D printing that has the potential to *revolutionize the way we make almost everything*. There's no reason this can't happen in other towns. So tonight, I'm announcing the launch of three more of these manufacturing hubs, where businesses will partner with the Departments of Defense and Energy to *turn regions left behind by globalization into global centers of high-tech jobs* [8] (our emphasis).

The 2013 UK Foresight report on the future of manufacturing stressed that technology will play a central role in driving change. It argued, for example, that

Some of the value being created in 2050 will derive from wholly unanticipated breakthroughs but many of the technologies that will transform manufacturing, such as additive manufacturing,[2] are already established or clearly emerging [9:20].

Therefore, it implies that manufacturing is entering a dynamic new phase, one in which technological advances such as 3D printing shall play a key transformative role.

Against the background of these narratives of big promises, this chapter seeks to describe a variety of contexts in which 3D printing technologies are expected to emerge over the next 15 years[3] and exert their so-called disruptive potential. In the Section 13.2, we first provide a brief introduction to 3D printing and we explain how it actually works. Next, in Section 13.3, we describe the paradigmatic change allowed by 3D printing in the industrial sector with a shift toward mass customisation. In particular, we focus on the biomedical sector (Section 13.1), which is an interesting case in point because of the important number of innovations and the growth of 3D printed biomedical parts, a trend that is expected to continue in the future.

To account for the dramatic, transversal, and transformative potential that 3D printing has in that whole sector, we first concentrate on 3D printing of biomedical instruments and implants for patients (Section 13.1) and, second, on additive bio-manufacturing of human tissues and organs (Section 13.1.2).

---

[2]Throughout this chapter, we will indistinctly make use of 'additive manufacturing technologies' and '3D printing'. The reason will be explained in Section 13.2.

[3]As we have just mentioned, 3D printing technologies do not form a monolithic block. In the technologies we describe in this chapter, some like biomedical instruments and implants are already part of routinised activities while some others, like bioprinting, are usually associated to a 15-year timeframe before it becomes a multi-billion US dollar industry.

Then, in the subsequent Section 13.4, we address the expectations raised by 3D printing to empowering users in non-industrial domains (such as in fabrication laboratories or with desktop 3D printers at home). In Section 13.5, we turn to discussing the impact of 3D printing on the governance actors and we raise important issues for further research in the political economy of 3D printing technologies.

The chapter posits that 3D printing, and its governance, are closely associated with more participatory means of manufacturing (and of decision-making, through various governance structures)—but that, as things currently stand, such openness and participation does not play out in practice. There is a distinction between the rhetoric and reality of 3D printing, as one might expect in the case of newly emerging technologies.

Before proceeding, we should like to make clear that we purposely chose to 'follow' the disruptive potential of 3D printing where it took us: from the work floor of a biomedical lab to the home desktop of users, from the instrument of a surgeon to the regulatory debate over bioprinted tissues, from the closed doors of the industry to the openness of a machine shop. As a result, we offer a rather unstructured account of the disruptive potential of 3D printing. But such an approach has to be taken as the main characteristic of our research objective. In turn, this observation leads to interesting speculations about the variety of governance interventions and strategies necessary at various levels to accompany the development of 3D printing technologies and, hopefully, maximise its positive upsides while minimising the possible side effects.[4]

13.2 A Brief Introduction to 3D Printing

In traditional manufacturing, parts are manufactured using processes such as injection moulding and casting, or subtractive manufacturing methods where material is trimmed. 3D printing

---

[4]In addition, like we do in this chapter, one has a tendency to distinguish 'large-scale' industrial additive manufacturing technologies on the one hand, and 'local', bottom-up, user-centred activities based on 3D printing on the other. How does this separation affect the analysis of the political economy of new additive manufacturing is an open question that we do not address here, but which deserves careful scrutiny in further research.

is an additive manufacturing process allowing the building of parts layer by layer, i.e. by *adding* one layer on top of the last-built layer. Although historical accounts of additive manufacturing technologies go back to late 19th century topography and photo-sculpture, 'the first significant work associated with modern photolithographic additive manufacturing systems only emerged during the 1970s' [10]. The first patent application was filed in 1980 by Hideo Kodama of Nagoya Municipal Industrial Research Institute, while the first working 3D printer was created in 1984 by Charles W. Hull of 3D Systems Corp. The first methods for rapid prototyping became available in the late 1980s and were used to produce models and prototype parts. By the mid-1990s, a community of academics, specialised suppliers and professional users developed, commercialised and adopted additive manufacturing technologies beyond prototype applications by manufacturing end products. Nowadays, 3D printing and additive manufacturing are interchangeable terms for all additive processes, independent of application.

Different technologies allow the layer-by-layer building process and they can roughly be divided into three categories depending on the initial state of the material used to build the part, i.e. liquid, solid or powder. The choice of a 3D printing technology and the material to be used is important because it relates to various factors such as the accuracy of the building process, the mechanical properties of the part, as well as other parameters like the number of materials, of colours and so on. Common 3D printed materials are plastic, metal, etc., but numerous 'unexpected' materials that are not belonging to industrial manufacturing field are currently in use such as food.

3D food printing is being used in several projects that are at various stages of development, like the Cornucopia project, with its Digital Fabricator, a 3D food printer that converts chosen ingredients into a delicious end product, CandyFab, printing very large objects out of pure sugar, or Modern Meadow, applying the latest advances in tissue engineering to culture leather and meat without requiring the raising, slaughtering and transporting animals. Very large objects have been also made possible in 3D building printing projects like concrete printing such as benches for the design of modern architecture buildings. While the price of printers for industrial use can reach the million US dollars, the price of printers for consumer use ranges from a few hundreds to a few thousand US dollars, an example of those being the RepRap, a self-replicating open source desktop printer.

The layer-by-layer building process has two important advantages over traditional manufacturing, allowing for a more cost-effective production that is better able quickly to respond to market changes [1]. The first advantage is that there is almost no constraint on the shapes that can be built, meaning that it enables the 3D printing of highly complex shapes. The second advantage relates to the better flexibility of 3D printing technology. Traditional manufacturing allows *mass production* by limiting cost when building the same part over and over again, but when different parts have to be manufactured it can become very costly, for instance by renewing the mould. 3D printing, by contrast, allows building each time a new part with no additional cost and with limited wastage. Thanks to these two advantages, the first wave of 3D printing applications showed a dominance of rapid prototyping. Compared to traditional methods, the flexibility, the relatively short time delay to get the part printed, and the potential lower cost made 3D printing technologies well suited for prototyping. Since then, the use the 3D printing of end product has grown fast thanks to the significant improvements in quality, reliability and repeatability of 3D printing building processes, making 3D printing a full manufacturing process like traditional ones. Combining these improvements with the two advantages of 3D printing, i.e. the absence of shape constraints and flexibility, it opens the opportunity to create different parts in mass by personalising or customising each part from the others, at reasonable cost compared to traditional manufacturing technologies. The second wave of 3D printing is thus being described as capturing value through *mass customisation*, both in the industrial and the non-industrial sectors [11].

13.3 3D Printing in the Industrial Sector

The shift from mass production to mass customisation has had major impacts in the industrial domain, as evidenced by the numerous applications developed in the automotive, aerospace [12] and biomedical [13, 14] fields. Big companies that usually worked on a business-to-business scheme currently offer online

3D printing services to their customers, using industrial 3D printers to print custom parts. The tech-guru Chris Anderson [2] highlights that 3D printing is suited to businesses in which even the production of specialised or customised products in low volumes is rendered economically viable. These critical evolutions in technologies of manufacturing are likely to increase the requirements in terms of competencies and expertise of the workers who will be expected to have higher levels of digital competence.

In the next section, we address the development and the potential of 3D printing in the biomedical sector, which appear to be one of the most promising and diverse in terms of subsequent applications. We illustrate such promises and actual product developments by first focusing on the personalised instruments and implants, before we concentrate on the more future-oriented case of additive bio-manufacturing.

13.4 3D Printing in the Biomedical Industry

13.4.1 3D Printing of Biomedical Instruments and Implants for Patients

The biomedical industry is currently demonstrating the potential of 3D printing owing to a number of developments in the area [15]. It is anticipated that the 3D printing of biomedical parts will be an important growth area and source of revenue for the biomedical industries. In the biomedical field, customised product often means that the product is personalised to the patient, therefore it is also called patient-specific product. Because of the similarity between the layer-wise representation of an object to be 3D printed, and the layer-wise visualisation of 3D medical images such as Computer Tomography (CT) and Magnetic Resonance Imaging (MRI), researchers and industrialists were quick to invest in the biomedical field, beginning of 1990s. At that time, 3D printing was used for printing anatomical parts visible on medical imaging, allowing surgical team members to comprehend the patient case and communicate more easily, and thus better to plan surgeries and improve surgical outcomes [16]. Over a decade ago, 3D printing models of skulls and interconnected vessels of conjoined twins helped for the highly complex planning of the surgery. 3D printed models were also recently used for the first Belgium's first full face transplant.

Nowadays, the biomedical use of 3D printing has drastically increased: Approximately ten of thousands of 3D printed metal implants are currently produced every year, while approximately 50,000 patients are treated every year with 3D printed surgical instruments [17]. Both 3D printed personalised surgical implants and instruments are designed based on the patient's anatomy, which is rendered visible by medical images, and on surgical planning [15]. Personalised implants (otherwise referred to as patient-specific implants) are required when standard implants cannot be used, often because of the complexity of the case. In bone cancer cases, the amount of bone that needs to be resected is sometimes so important because of the high proliferation of cancer cells, that no standard implants can be used and fixed to the remaining bone. Sometimes, these implants literally save one patient's life, as in the case of a baby treated with a 3D printed splint to hold his trachea open and to allow him to breathe. Patient-specific implants are also routinely used in cranio-maxillo-facial surgeries where personalised cranioplates, mandila implants, and those used in full-face transplants, greatly improve surgical outcomes but are also often essential for an aesthetic point of view in order to respect the remaining shape of patient's skull and face [16].

Personalised surgical instruments (otherwise referred to as patient-specific surgical guides or templates) are parts that indicate precisely surgeons where to cut, drill, how to reposition bones and implants during surgery. They are 3D printed parts designed pre-operatively such that they perfectly fit onto patient bones, hence explaining the reason this instruments should be personalised. Compared to standard instruments, personalised instruments are supposed to better reproduce the surgical planning that was considered as optimal for the patient by the surgeon.

While a few years ago, the use of 3D printed instruments and implants seemed 'futuristic', nowadays it has become commonplace in biomedical practices. It may even go further with manufacturing implants that fulfil more biological functions than the human body, thereby raising ethical and societal issues linked to human enhancement.

13.4.2 Additive Bio-Manufacturing

Due to its enormous potential in terms of health and the biomedical products' market, although still at its infancy, the field of additive bio-manufacturing, or bioprinting, is currently being paid a lot of attention [18, 19]. Bioprinting refers to a process by which living cells and a biocompatible scaffolds are printed to provide skin tissue, heart tissue, or even printed organs that could be suitable in the future for medical therapy and transplantation. It uses 3D printing to produce biologically relevant materials such as cells, tissues, and organs. This is based on the convergence of additive manufacturing technologies with tissue engineering, and it promises value capture through the manufacture of 3D tissues and organs. Building on the collective strengths in stem cells and regenerative medicine, materials science and engineering, 3D bioprinting could help to stimulate economic growth as well as to generate major health benefits.

One big expectation of the pharmaceutical industry relates to the possibility of testing drugs in clinical trials on 3D printed functional human tissues in order better to understand and predict a drug's effect on the body. Bioprinting functional preclinical human tissues would indeed help avoiding late-stage clinical trial failures for both toxicity and efficacy. These developments converge with well-known claims of animal rights movements that the use of human tissues is more accurate than animal tissues for toxicity testing. One NGO has argued that 'Drugs would be safer than they are now if the animal testing phase was eliminated' [20], and that animal testing has failed to predict numerous side effects of prescription medicine. A gradual phasing out of animal testing for pharmaceutical products, as currently envisioned by the European Commission's Joint Research Centre for cosmetics, may thus also benefit from progresses in additive bio-manufacturing of tissues.

Anyhow, the day when 3D bioprinted human organs are readily available is probably drawing closer. However, major challenges will need to be overcome in the first place. The barriers to full-organ printing are not just technical (this is expected to be very difficult to print blood vessels or vascular tissues); they are also economic and regulatory. The first organ-printing machine will cost hundreds of millions of dollars to develop, test, produce and market. Crucially, bioprinting tissues or organs for human consumption will probably involve lots of regulatory testing before any products can be brought to market. In the very near future, according to Gartner Group, bioprinting is expected to cause a global debate about regulating the technology or banning it for both human and nonhuman use [21]. If bioprinting is the way forward to safer drugs, pharmaceutical industry will have to provide the regulatory agencies, such as, for example, the American Food and Drug Administration and the European Medicines Agency, with sufficient evidence on the safety of such drugs when tested on 3D printed tissues.

The value of 3D printing in the global medical market is quite small at the moment, but with the rise of a richer middle class in the developing world and ageing populations in the west, the market is predicted to grow significantly in the coming years. Emerging markets in 'bioprinting' may thus transform political economies in both developed and developing countries, including the aspects related to economies of exchange in human tissues [22] human subjects for clinical trials [23] or illegal organ traffics.

13.5 3D Printing in the Non-Industrial Domains

With the democratisation of desktop machines and the rapid expansion of institutionalised spaces providing access to personal digital fabrication facilities, many more people outside the industrial world may have the opportunity to manufacture parts in collaborative environments. The empowerment of users equipped with 3D printers may arise through the act of creation itself, with a material engagement in our technological cultures [24] but also, more crucially, through hands-on experience altering our knowledge of technologies [25]. In this way, the expansion of 3D printing beyond the industrial sector encourages individuals to scrutinise the inner dynamics of the technologically complex world we live in [26], and to do so in 'real-life laboratories'—'wild' spaces in which the arrangements necessary for experimentation can be installed [27:3].

Today a variety of making, fabricating, fabbing, tinkering, assembling, prototyping, coding and manufacturing spaces or shops show promise of opening up opportunities for decentralised and collaborative engagements with technology. Such opportunities are not only related to material and technical experimentations, but also with economic, cultural, social and political consequences, and ultimately with conceptual and epistemological changes. With due attention to their differences, there is a common and shared rationale that supports an openness when approaching and thinking about technology [28]. From this standpoint, derives a multiplicity of potential pathways for empowerment through technology and democratisation of technology for broader social groups [28]. Spaces like FabLabs and maker- or hackerspaces have progressively included 3D printers as part of their services, which has greatly contributed to their popularity [29]. These spaces promote an open environment (working on the philosophy that everyone is welcome) and/or collaborative work. Within such spaces, the boundary between producers and consumers becomes increasingly blurred, thereby leading to the emergence of new business models.

Clearly, the mass customisation requires a new type of interaction or even collaborations between producers and consumers. The consumers interact with producers in the development process of the product sooner than after the market entry, when they do not become producers themselves. Such interaction may, even, contribute to better meet the needs of consumers. Design and production can much more easily be dissociated based on the principle 'ship the design, not the product', thereby allowing new actors, who may not have had production capacities in the past, to enter the market. It has been argued that do-it-yourself technologies are indeed revolutionary for so-called 'prosumption', for innovation, and for entrepreneurship [25].[5]

However, while users' material engagement in the design and manufacturing of products leads to some empowerment, centralised forms of industrial production are not likely to be overthrown by machine shops. Indeed, this might demand too much of these still fragile niches which have to handle the ambiguities between experimental freedom and socio-economic pressures [27]. Furthermore, it is to be expected that not everyone will be interested in customising products, not to mention the necessary skills, ranging from basic computing literacy to CAD

---

[5]Prosumption refers to individuals producing what they consume and the boundary between production and consumption of goods becoming blurred. Innovation refers to new types of physical goods that are manufactured/customised; and entrepreneurship refers to bringing new types of physical goods to market.

expertise, that are obvious assets for finding one's way in the 'makers community'. Even when the market of products manufactured in machine shops would become bigger, its expansion would most probably need public subsidies to be sustainable and could not easily extend to rural areas or developing countries lacking the necessary infrastructures. Related to that latter point, Fox [25] notes that because that kind of manufacturing is performed in countries with easy access to electricity supplies, plentiful water resources, and comprehensive transportation systems, the potential for expansion of personal fabrication in developing countries is limited.[6]

These observations tend to challenge big claims predicting that the 'disruptive innovation' coming from maker trends and platforms will greatly affect technologies, organisations, government, education and ultimately society [28]. This disruptive potential is fed by bold proclamations of 'we are all born makers' [30:13], or '(almost) anybody can make (almost) anything' [31], or in 'maker manifestos' [32] focused on buzzwords of 'make, share, give, learn, tool up, play, participate, support, change'. This phenomenon of the economics of techno-scientific promises is already well documented [33], as it happened with earlier technologies, for which similar claims were made. For example, in the US, the National Nanotechnology Initiative stressed the potential of 'nanotechnology research and development leading to a revolution in technology and industry' [34], which remained largely unrealised [35].

Politicians are embracing the hype too, like President Obama (2009) welcoming 'he promise of being the makers of things, and not just the consumers of things'. Other actors, like Deloitte, state that 'never before has it been so easy [for makers] to create or modify something with minimal technical training or investment in tools' [36:3]. Whether or not such imagined futures are achieved is a legitimate question, but prior to this one should consider the 'cost of hype' too. Indeed, the only potential of new technologies plays a tremendous role in shaping political-economic policies, public and private expenditures, institutional practices, and wider societal changes [37].

---

[6]Rather, he advocates the recourse to mobile factories that are 'engineered-to-order to meet the particular requirements of specific production needs in specific regions, including those without paved roads, mains electricity and piped water' [25:19].

As Nascimento stresses [28:2], the hype around the 'maker movement' covers a broad scope of people and communities with miscellaneous goals, from creative self-expression, technical curiosity, and commercial objectives up to ethical commitments and grassroots activism. For the most part, present maker discourses and practices are oriented towards the values of self-expression, knowledge sharing, community building, re-skilling, creativity, and innovation. Expectations for massive empowerment of users through broader inclusion of values, needs and expectations in AM technology may contain the seeds of further revolutionary changes, but most recent empirical works seem to attest the opposite: Shared machine shops are not new [38],[7] sharing is not happening [39],[8] hackerspaces are not open [40][9] and FabLabs are not the seeds of a revolution [27]. These findings go against the grain of the claims of greater collaborative governance, and therefore deserve further investigation.

13.6 Discussion

We have just outlined a series of developments with regards to the various ways (prosumers, grassroots activists, tinkerers) and different contexts (FabLabs, hackers- or makerspaces, machines shops) in which users are confronted with, and engaged by the innovative potential of 3D printing technologies. In recent years, as stressed by Söderberg,

---

[7]Referring to something 'new' immediately raises the questions: with respect to what and to whom is it new? For example, taking the case of the largely forgotten technology networks supported by the Greater London Council from 1983 to 1986, Smith [38] recalls us that movements for socially useful production are not new. He argues that, to some extent, state support for FabLabs and other workshops is a legacy of past activism.

[8]The case studies analysed in the paper revealed that knowledge sharing is not impeded by the barriers discussed elsewhere in literature such as motivational or technological impediments. Nevertheless, the cases showed that global open knowledge sharing was far from the norm, and sharing remains mainly local and personal.

[9]Sophie Toupin interviewed hacker/maker/geek women who decided to make their own spaces out of frustration with the politics of openness – or in other words, the practices of exclusion—in mainstream hackerspaces (see the editorial note of that special issue of the Journal of Peer Production).

the idea about engaging users in innovation processes has been adopted in policy documents about the 'knowledge society'. What invariably goes unmentioned in the official endorsements of user-initiated [or participatory] innovation is the acknowledgement that much of that creativity has been outlawed by intellectual property legislations [41:152–153].

Such as, for example, the Trade-Related Aspects of Intellectual Property Rights (TRIPS) agreement. Here lies the paradox of discursively encouraging users to take part in collaborative forms of engagement in innovation, while, at the same time, furthering the legal framework enclosing knowledge in ever more extended property rights. In other words, the coexistence of open or closed approaches to the governance of innovation creates tension in the very conception of the knowledge society.

States are actually entangled in contradictory measures with regard to the knowledge economy/society dyad. On the one hand, they aim at promoting the commodification of knowledge and its transformation from intellectual commons to intellectual property, for example, in the form of patent, copyright, and licenses [7]. On the other hand, they seek to protect the intellectual commons as a basis for competitive advantages and to develop a learning society and an informed public sphere [7]. Hence, there are various coexisting, competing governance strategies to manage the tension between intellectual commons and intellectual property, which, we argue, is especially salient in the case of 3D printing.

Governance strategies impregnated with the narrative of growth through additive manufacturing technologies also uncover another aspect that relates to a 'Catching up' repertoire. In the quote mentioned in the introduction, Obama places the USA resolutely ahead of their competitors in the global race towards knowledge, excellence, and growth because the USA could take the necessary measures to 'turn regions *left behind* by globalisation into global centers of high-tech jobs' [8]. This suggests that, if policy actions were not been taken in due time to take the best advantage of the potential of 3D printing technologies, other countries may, in due course, *overtake* the USA as a top competitive region.[10] Given the expectations and/or fears that

---

[10]By a large margin (38%), the US currently leads the world in the adoption of industrial additive-manufacturing technology, followed by Japan (9.7%), Germany (9.4%) and China (8.7%) [23].

'Catching up' evokes among policymakers worldwide, the narrative serves these actors as a flexible discursive resource to make sense of, and shape, their collective futures, and thus their identities.[11] In this case, the narrative relies on the observation of a threatening present (an increased international competition) that would lead to an apocalyptic future (being left behind by globalisation) 'if nothing were done' to exploit the potential of 3D printing. Yet, however inescapable this future may be described, the very existence of such a narrative presupposes that the political community it tries to reach is actually able to do something to thwart it [42], *if embracing the disruptive potential of new technologies is the favoured route* out of economic crisis toward renewed economic growth.

Reaping the benefits of 3D printing is not for anyone, though. Contrary to certain portrayals of the technology, not everyone can access a 3D printer, not everyone is interested in mass customisation, and not everyone can 'start in his garage and become the new Mark Zuckerberg'.[12] Nor is it a possible future for any country to turn additive manufacturing technologies into a central axis of its economy [25]. The emergence of additive manufacturing technologies is not a general 'dynamic' made out of thin air. Rather, it is always rooted in material conditions and performed and managed by identifiable actors who attempt to steer the potential of new technologies to their own advantage. Therefore, as for AM technologies as for other emerging technologies, in spite of a rhetoric of openness and wide democratisation, there are structural biases favouring certain directions and distributions of innovation over others. Provided 3D printing technologies effectively help to re-localise production processes in the 'West', it may mean a dramatic increase of the competition in the high-costs countries, whereas the creation of a new range of high-tech jobs, if any, is much likely to widen gap with non-highly skilled workers. Such considerations deserve attention, as policymakers worldwide are entangled in cornucopian imaginaries of abundant creative resources best exploited with the help of technologies, with the promise of a better world at the end of day. (What does 'better' mean, anyway?).

---

[11]On these ideas and the catching up repertoire, see [43].
[12]This assertion was Chris Anderson's, at the occasion of a lecture on the makers movement, given at the University of Liège, Belgium, on the 5th of November 2013.

In this chapter on the disruptive potential of 3D printing, we contend that openness and participation does not (yet?) play out in practice. The key questions then become: If 3D printing does not entail user-driven and collaborative approaches to governance, why is that and what does it involve instead? As we have shown, 3D printing technologies do not form a monolithic block. Crucially, the potential for distributed, participatory and collaborative governance of 3D printing only concerns one part of additive manufacturing technologies: personal fabrication relating to and directly engaging with end users. For industrial 3D printing technologies, our findings point to a market-oriented, business-as-usual governance strikingly similar to what is to be found in the literature on other technologies such as biotechnology, particularly the role of marketisation and an enlarged international regime of intellectual property rights [44]. The strategic interest in commercial outcomes of biotechnologies has led to the emergence of a 'global bioeconomy' [45, 46], in which the latent value of biological materials and products offers the opportunity for economic growth. The same seems to be happening with the global political economy of 3D printing. For emerging technologies too (we may even say 'for emerging technologies in particular'), the system of production of norms in advanced capitalism exacerbates and generalises global competition—among states and among individuals. In this context, even collective and participatory experiments enabled by new technologies take place against the backdrop of a neoliberal context striving users for becoming the entrepreneurs of their selves or for (economically more than socially) valuing their own creative potential. Thus, rather than allowing the reinforcement of social ties and a greater inclusiveness, material engagements with 3D printing technology in a FabLab are scattered into radically individualised actions in which sharing is the exception rather than the rule [39].

While makers have initially been driven by open-source technology and a do-it-yourself ethos, there are increasingly institutionalised forms of organisation of their community, as the number of FabLabs established with the support of local or national governments currently attests. This shift towards the mainstream leads to struggles within the maker community, as users-oriented 3D printing technologies seem to work as technologies of power '"governing at a distance", seeking to create locales, entities and persons able to operate a regulated autonomy' [47:173]. Against the proliferation of norms and policies further enforcing the enclosure of knowledge and creativity, it is urgent to find new ways enabling the sharing of knowledge and the re-articulation of *truly collective* actions. Recent works in 'commons studies', particularly Dardot and Laval [48], provide powerful theoretical tools to take the Common as a political principle that defines a new regime for struggling against the domination of neoliberalism. Instead of a being resurgence of communist ideas, the concept of 'Common' points at new ways of contesting and going beyond capitalism by governing the resources (natural resources, creativity, knowledge commons) away from both the State and the market [48:16–17]. Concrete governance propositions to secure the Common as a general political principle are also advanced, such as, for example, opposing the right of use to the property right, instituting the common company, or founding the social democracy on the basis of Common [48].

For 3D printing technologies like for other emerging technologies discussed in this book, there are predominant political-economic factors intertwined with strategic practices of designing or using technology to constitute, embody or enact economic and political goals that deserve careful scrutiny in further empirical research.

Acknowledgments

This work was supported by the Walloon Fonds de la Recherche Scientifique.

References

1. Norwegian Board of Technology (2014) *Made in Norway? How Robot, 3D-printers and Digitalisation Bring New Opportunities for Norwegian Industry* (Norwegian Board of Technology, Oslo).
2. Anderson, C. (2006) *The Long Tail: Why the Future of Business is Selling Less of More* (Hyperion, New York).
3. Mayntz, R., and F. Scharpf (1995) Steuerung und Selbstorganisation in staatsnahen Sektoren. In: Mayntz, R., and F. Scharpf (eds.) *Gesellschaftliche Selbstregelung und politische Steuerung* (Frankfurt a/Main and New York, Campus), pp. 9–38.

4. Benz, A. (2007) Governance in connected arenas–Political science analysis of coordination and control in complex rule systems. In: Jansen, D. (ed.) *New Forms of Governance in Research Organizations. From Disciplinary Theories towards Interfaces and Integration* (Heidelberg, Springer), pp. 3–22.
5. Kooiman, J. (2003) *Governing as Governance* (London, Sage).
6. Schapp, W. (2004) *Im Geschichten verstrickt: Zum Sein von Mensch und Ding* (Frankfurt am Main, Klostermann).
7. Jessop, B. (2005) Cultural political economy, the knowledge-based economy and the state. In: Slater, D., and A. Barry (eds.) *The Technological Economy* (Routledge, London), pp. 144–166.
8. Obama, B. (2013), *State of the Union* (White House, Washington, D.C.).
9. Foresight (2013) *The Future of Manufacturing: A New Era of Opportunity and Challenge for the UK* (The Government Office for Science, London).
10. Bártolo, P. J., and I. Gibson, (2011) History of stereolithographic processes. In: Bártolo, P. J. (ed.) *Stereolithography: Materials, Processes and Applications* (Heidelberg, Springer), pp. 37–56.
11. Robinson, D., and A. Lagnau (2015) Additive manufacturing in transition: Characterising distinct value-capturing pathways in the move from rapid prototyping to mass customization (unpublished manuscript).
12. Costa Santos, E., M. Shiomi, K. Osakada, and T. Laoui (2006) Rapid manufacturing of metal components by laser forming, *International Journal of Machine Tools & Manufacture*, **46**, 1459–1468.
13. Mironov, V., V. Kasyanov, and R. Markwald (2011) Organ printing: From bioprinter to organ biofabrication line, *Current Opinion in Biotechnology*, **22**, 667–673.
14. Campbell, P., and L. Weiss (2007) Tissue engineering with the aid of inkjet printers, *Expert Opinion on Biological Therapy*, **7**(8), 1123–1127.
15. Giannatsis, J., and V. Dedoussis (2009) Additive fabrication technologies applied to medicine and health care: A review, *The International Journal of Advanced Manufacturing Technology*, **40**(1–2), 116–127.
16. Rengier, F., A. Mehndiratta, H. von Tengg-Kobligk, C. M. Zechmann, R. Unterhinninghofen, H. U. Kauczor, and F. L. Giesel (2010) 3D printing based on imaging data: Review of medical applications, *International Journal of Computer Assisted Radiology and Surgery*, **5**, 335–341.

17. Wohlers (2013) *Additive Manufacturing and 3D Printing State of the Industry. Annual Worldwide Progress Report* (Wohlers Associates, Fort Collins).
18. Bártolo, P. J., C. K. Chua, H. A. Almeida, S. M. Chou, and A. S. Lim (2009) Biomanufacturing for tissue engineering: Present and future trends, *Virtual Physics Prototypes*, **4**(4), 203–216.
19. Melchels, F. P. W., M. A. N. Domingos, T. J. Klein, J. Malda, P. J. Bártolo, and D. W. Hutmacher (2012) Additive manufacturing of tissues and organs, *Progress in Polymer Science,* **37**(8), 1079–1104.
20. Safe Medicines (undated) *Facts.* Available at: http://www.safermedicines.org/faqs/faq01.shtml.
21. Gartner (undated), Latest News. Available at: http://www.gartner.com/newsroom/id/2658315.
22. Waldby, C., and R. Mitchell (2006) *Tissues Economies. Blood, Organs and Cell Lines in Late Capitalism* (Duke University Press, Durham).
23. Petryna A. (2009) *When Experiments Travel. Clinical Trials and the Global Search for Human Subjects* (Princeton University Press, Princeton).
24. Hommels, A., J. Mesman, and W. Bijker (2014) *Vulnerability in Technological Cultures* (MIT Press, Cambridge).
25. Fox, S. (2014) Third wave do-it-yourself (DIY): Potential for prosumption, innovation, and entrepreneurship by local populations in regions without industrial manufacturing infrastructure, *Technology in Society*, **39**, 18–30.
26. Pinch, T., and W. Bijker (1987) *The Social Construction of Technological Systems* (MIT Press, Cambridge).
27. Dickel, S., J. P. Ferdinand, and U. Petschow (2014) Shared machine shops as real-life laboratories, *Journal of Peer Production*, **5**, 1–9.
28. Nascimento S. (2014) Critical notions of technology and the promises of empowerment in shared machine shops, *Journal of Peer Production*, **5**, 1–4.
29. Walter-Herrmann, J., and C. Büching (2013) *FabLab: Of Machines, Makers and Inventors* (Transcript Verlag, Bielefeld).
30. Anderson, C. (2012) *Makers. The New Industrial Revolution* (Crown Business, New York).
31. Gershenfeld, N. (2005) *FAB: The Coming Revolution on Your Desktop: From Personal Computers to Personal Fabrication* (Basic Books, Cambridge).

32. Hatch, M. (2014) *The Maker Movement Manifesto: Rules for Innovation in the New World of Crafters, Hackers, and Tinkerers* (McGraw-Hill, New York).

33. Joly, P.-B. (2010) On the economics of techno-scientific promises. In: Akrich, M., Y. Barthe, F. Muniesa, and P. Mustar (eds) *Débordements-Mélanges offerts à Michel Callon* (Paris, Presse des Mines).

34. National Science and Technology Council (2012) *The National Nanotechnology Initiative. Research and Development Leading to a Revolution in Technology and Industry. Supplement to the President's 2013 Budget* (NSTC, Washington D.C).

35. Thoreau, F. (2013) *Embarquement immédiat pour les nanotechnologies responsables. Comment poser et re-poser la question de la réflexivité?* (unpublished PhD dissertation, University of Liège, Belgium).

36. Deloitte Center for the Edge (2014) *A Movement in the Making*. Available at: http://dupress.com/articles/a-movement-in-the-making/.

37. Birch, K., L. Levidow, and T. Papaioannou (2010) Sustainable capital? the neoliberalization of nature and knowledge in the European "knowledge-based Bio-economy", *Sustainability*, **2**, 2898–2918.

38. Smith, A. (2014) Technology networks for socially useful production, *Journal of Peer Production*, **5**, 1–9.

39. Wolf, P., P. Troxler, K. Y. Kocher, J. Harboe, and U. Gaudenz (2014) Sharing is sparing: Open knowledge sharing in fab labs, *Journal of Peer Production*, **5**, 1–11.

40. Toupin, S. (2014) Feminist hackerspaces: The synthesis of feminist and hacker cultures, *Journal of Peer Production*, **5**, 1–9.

41. Söderberg, J. (2010) Misuser inventions and the invention of the misuser: Hackers, crackers and filesharers, *Science as Culture*, **19**(2), 151–179.

42. Claisse, F., and Delvenne, P. (2015). Building on anticipation: Dystopia as empowerment, *Current Sociology*, **63**(2), 155–169.

43. Van Oudheusden, M., N. Charlier, B. Rosskamp, and P. Delvenne (in preparation), *Flanders Ahead, Wallonia Behind (But Catching Up): Reconstructing Communities through Science, Technology, and Innovation Policymaking*.

44. Tyfield, D. (2012) *The Economics of Science: A Critical Realist Overview* (Routledge, London).

45. European Commission (2010) *The Knowledge Based Bio-Economy (KBBE) in Europe: Achievements and Challenges* (Clever Consult BVBA).
46. Organization for Economic Cooperation and Development (2009) *The Bioeconomy to 2030: Designing a Policy Agenda* (International Futures Programmes, Paris).
47. Rose, N., and P. Miller (1992) Political power beyond the state: Problematics of government, *The British Journal of Sociology*, **43**(2), 173–205.
48. Dardot, P., and C. Laval (2014) *Commun. Essai sur la révolution au XXIème siècle* (La Découverte, Paris).

Chapter 14

Advanced Materials and Modified Mosquitoes: The Regulation of Nanotechnologies and Synthetic Biology

Diana M. Bowman,[a] Elen Stokes,[b] and Ben Trump[c]

[a]*Sandra Day O'Connor College of Law and
School for the Future of Innovation in Society, Arizona State University,
111 E Taylor St, Phoenix, Arizona 85004, USA*
[b]*Birmingham Law School, Edgbaston, Birmingham, B15 2TT, UK*
[c]*Department of Health Management & Policy, And the Risk Science Center,
School of Public Health, University of Michigan, 1415 Washington Heights,
Ann Arbor, Michigan, USA*
Diana.Bowman@asu.edu, e.stokes@bham.ac.uk, bdtrump@umich.edu

14.1 Introduction

As this edited volume highlights, the policy and regulatory radars have long been attuned to the challenges and opportunities of technological developments. New technologies and their high-tech derivative products present regulators with a host of difficult, though not always unfamiliar, questions. The first of these usually centres on whether the technology and its products fall within the remit of existing regulatory frameworks. If they do, then

---

Embedding New Technologies into Society: A Regulatory, Ethical and Societal Perspective
Edited by Diana M. Bowman, Elen Stokes, and Arie Rip
Copyright © 2017 Pan Stanford Publishing Pte. Ltd.
ISBN 978-981-4745-74-1 (Hardcover), 978-1-315-37959-3 (eBook)
www.panstanford.com

subsequent questions are likely to include whether the regulatory frameworks are adequate, proceeding to questions of how regulation ought to develop if the current level of intervention is deemed inappropriate. Intuitively, one might expect that new technologies warrant new regulatory responses, especially where they give rise to new commercial applications, new exposure scenarios, new types or magnitudes of risk, or new problems of uncertainty. Experience has shown, however—across time, jurisdictions and technologies—that the most favoured regulatory approach is often the one that is already in place [1–4].

This chapter explores the widespread tendency among policymakers and safety regulators to manage a new technology by defaulting to existing regulatory measures. There may be good reasons for this. For example, in the early stages of a technology's development, it can be difficult to predict the variety of uses to which the technology will be put, the commercial trajectories and possibilities, or the potential human and/or environmental risks. There would be little point, therefore, in undertaking a radical overhaul of the regulations before the issues and challenges involved are better understood. Moreover, existing regulations provide an element of certainty in an otherwise rapidly changing and contested technological environment [5]. The pros and cons of a particular technology may be hotly disputed, but the application of existing regulations implies a degree of continuity and stability.

Although such an approach is readily adopted for some new technologies, it can have its limitations. For instance, it may result in a failure to address broader social issues arising outside the narrow confines of the legislation—legislation that, in areas of health and environmental protection, tends to be directed at the reduction of known risks. And what of other issues, not necessarily governed by the numerous regulatory regimes already in place? These issues, such as the public acceptance of a new technology, or the distributional effects of a particular technological development, also raise important questions in need of examination, even if they do not fall squarely within the province of existing regulation. Yet, they are in danger of being overlooked given that they are peripheral to the core business of regulation.

The purpose of this chapter is to highlight some of the implications of regulating technological advances using well-established rules and principles. Drawing on examples of nanotechnology and synthetic biology, it illustrates the process of having recourse to existing regulations. The chapter further argues that the process entails more than the simple transference of rules and principles from one technology context to another. It involves the transference of other, ingrained elements of the regulatory frameworks in question. In other words, these are situations in which a new technology—such as nanotechnology or synthetic biology—is governed by the content as well as the operational practices, assumptions and priorities of existing regulations [6].

The chapter proceeds in four parts. Section 14.2 begins to unpack the incidence and effects of resorting to established regulations. It focuses on recent regulatory responses to nanotechnologies, drawing on specific examples from the European Union (EU) and Australia to demonstrate the propensity—certainly in the initial stages of technological development—to rely on existing, generically worded regulatory measures. It goes on to discuss steps taken more recently to amend existing regulations as they apply to nanotechnology-based products. These recent amendments are in contrast to the initial period of regulatory stasis, characterised by the automatic application of regulations pre-dating the emergence of nanotechnologies. It is important to bear in mind, however, that none of the amendments mark a significant departure from existing regulations—they simply involve a little tweaking.

Section 14.3 turns its attention to developments in the field of synthetic biology. It identifies some of the major issues currently facing safety regulators, and again highlights the tendency in the first instance to defer to existing regulatory structures. One of the notable consequences of this, we argue, is that opportunities to subject evolutionary technological developments to scrutiny may be more limited because the assumption is that they are covered by the existing regulations *per se*. Such an assumption can lead to a failure to assess the merits and effectiveness of said regulations, either on their own terms or as applied to the products of synthetic biology.

Section 14.4 focuses on a specific application of synthetic biology: the engineered *Aedes aegypti* mosquitoes. Pioneered by Oxford Insect Technologies (or 'Oxitec'), the mosquitoes are sex-selected and engineered to make eventual reproduction prohibitively difficult. Such techniques have tremendous potential to improve living conditions and save lives, by reducing the incidence of mosquito-transmitted disease. The technology's initial field trials in Grand Cayman have demonstrated highly successful results; however, a number of questions have been raised in relation to the mosquito. They include, but are not limited to

(1) the ability of existing governance structures to manage the potential risks associated with the technology as suggested by the current scientific state of the art;
(2) the uncertainty surrounding the impact that genetically modified (GM) mosquitoes may have on the environment in the short and longer term; and
(3) the ethical concerns regarding the deployment of organisms in developing regions with little, or no, stakeholder involvement and/or public consent.

Section 14.5 pulls together these threads and concludes that, while it is inevitable that existing regulations will play an important role in determining how best to manage new technologies, it may not be enough simply to allow the automatic extrapolation of established regulatory rules and approaches. Instead of presuming that former provisions offer an adequate response, we argue that the regulatory future would be better served by a case-by-case approach that pays greater attention to the strengths, but also the weaknesses of previously adopted regulations.

14.2 Regulations: Past and Present

The incidence and effects of relying on existing regulatory instruments and arrangements become more pronounced in the case of evolutionary technologies, which combine new scientific practices with old to create new or improved versions of familiar products and processes. Since evolutionary technologies tend, in the first instance, to bring about incremental improvements in existing materials, devices and systems, the question facing policymakers is not whether regulations are in need of large-scale reform. Rather, it is whether current regulations are broad enough adequately to deal with these more sophisticated materials and variants. Applications of nanotechnology and synthetic biology offer excellent cases in point. Both are examples of technologies with huge promise across a range of commercial sectors. At the same time, both raise a number of concerns about potential risks from human and/or environmental exposure. Moreover, both involve layers of uncertainty and complexity, not just in terms of their potentially harmful consequences but also in terms of the interests and interactions of relevant stakeholders—all are to some degree open-ended and contingent rather than fixed and final. This is especially so as a new technology begins to emerge, when the facts are uncertain and disagreements about the appropriate path of development are more or less unavoidable.

These factors alone make applications of nanotechnology and synthetic biology *bona fide* objects of regulation. Yet, rather than introduce entirely new, technology-specific regulatory regimes, the emphasis in many jurisdictions—at least to date—has been on the capacity of existing regulations to manage the challenges associated with such technological progress. Hence, these are examples in which the emergence of new, evolutionary technologies has occurred largely within the bounds of what might be described as an 'inherited regulatory environment' [6:95].

14.2.1 Common Narratives of Early-Stage Emerging Technology Risk Governance

Consequently, despite claims by some commentators that nanotechnologies are essentially 'unregulated' [7], the reality is somewhat more sobering. As government and independent reviews have repeatedly shown us [8–12], the commercialisation of nanotechnology-based products and processes has occurred under the auspices of pre-existing and sector-specific statutory frameworks, which are further underpinned by the common law. These frameworks cover an extensive range of uses and types of product, such as industrial chemicals, foods and therapeutic goods, and each is typically guided by a number of high-level principles and policies. They include, for example, pre-market safety assessment, burdens of proving safety, safety evaluation and determination, and post-market surveillance.

Central to any single national regulatory framework are formal legislative measures established by government that set out the rights and responsibilities of individuals and firms, and articulate the powers of regulatory agencies, along with their objectives and operational remit. Formal statutory instruments include, for example, acts, directives and regulations which operate at the national as well as supranational level. In areas of occupational health, public health and the environment, these frameworks will often set overarching goals such as 'safety' or a 'high level of protection', which would apply to all applications of technology, new or not. Underpinning these broad goals is secondly, a wide range of policies and practices that go into implementation. The aim here is to ensure that all agreed higher level principles, such as pre-market safety testing and review for example, can be effectively carried out. Implementation activities can include risk assessment, testing and monitoring, and enforcement. These practices are fundamental when considering the operation and adequacy of current regulatory arrangements for nanotechnologies. Accordingly, the question is not whether evolutionary technologies, such as nanotechnologies, are regulated, but instead how effective existing regimes—including their implementation practices—are for ensuring human and environmental health and safety.

The sheer breadth of the provisions, in terms of both coverage and implementation, has led to the now familiar conclusion that current regulations are adequate to manage nanotechnologies [10]. Although many commentators have questioned the extent to which current regulations are 'fit for purpose' [9, 13–16], the weight of opinion is that those measures, which were designed with more conventional technologies in mind, are already broad enough to cover new nano-enhanced products and processes.

14.2.2 Nanotechnology Regulation in the EU and Australia

This has been the position of the European Commission in its various reviews of EU regulation as applied to nanotechnology.

For example, following its first regulatory review, the Commission concluded that '[c]urrent legislation covers in principle the potential health, safety and environmental risks in relation to nanomaterials' [10:6]. The Australian Government [17] echoed similar sentiments after its own independent review of the key federal regulatory regimes [12]. So too did various regulatory agencies and government departments in the UK, after they were asked to review current regulation as it applied to nanotechnologies [18].

There are exceptions to this approach, and in spite of the initial policy stance of the European Commission and the Australian Government, other policymaking bodies have started to turn their attention to questions of amendment. Going against the prevailing view that existing regulations are sufficient, the European Parliament and the Australian Government (more recently), for example, have introduced modifications to the regulations so that they are better able to address the various issues raised by nanotechnologies. The European Parliament has introduced nano-specific requirements into legislation in a range of sectors, including: food additives (Regulation (EC) No 1333/2008[1]); food information for consumers (Regulation (EU) No 1169/201[2]); cosmetic products (Regulation (EC) No 1223/2009[3]); waste electrical and electronic equipment (Directive 2012/19/E[4]); restrictions on hazardous substances in electrical and electronic equipment (Directive 2011/65/EU[5]); biocidal products (Regulation (EU) 528/2012[6]); and food for infants (Regulation (EU) No 609/2013[7]).

It is important to note that these are not in the form of whole legislative measures designed especially for nanotechnology; rather, they are smaller, nano-specific provisions inserted into existing legislation as and when it came up for review. It is further worth bearing in mind that these nano-specific provisions are of a type that is consistent with previous ways of doing things:

---

[1]Article 12.
[2]Article 2(t), 18(3) and 18(5).
[3]See Preamble 29- 31, and 65; Articles 2, 13(f), 16 and 19(g)(ii).
[4]Article 8(4).
[5]See Preamble 16.
[6]Article 3.
[7]Articles 2 and 9(2).

they are informational in nature, meaning that they introduce nano-specific labelling requirements and/or nano-specific data disclosure obligations (see, for example, Regulation (EC) No 1223/2009). Both labelling and data reporting have a long history in EU safety regulation [10], so they are tried and tested strategies—and, importantly, they come with their own set of well-established rationales and assumptions [19].

In Australia, the Government has introduced additional administrative provisions for 'new' nanoscale industrial chemicals, the *status quo* has largely remained intact as regards the key federal regulatory frameworks [20]. For example, despite the Therapeutic Goods Administration's diverse activities in relation to its oversight of therapeutic goods containing nanomaterials, the Administration continues to assert that it has the necessary tools to ensure the safety of any such products [21, 22]. And while new administrative measures may have been introduced for 'new' nanomaterials under the National Industrial Chemicals Notification and Assessment Scheme, 'existing' materials that are reformulated at the nanoscale continue to be deemed to be 'existing' for the purposes of regulatory oversight [20]. This approach is similar to that in the United States (US), where existing regulations have been deemed by the regulatory authorities to be sufficient overall. The US Environmental Protection Agency has, however, used its existing powers under Toxic Substances Control Act of 1976 to issue 'significant new use rules' for several nanomaterials [23–25]. As such, we can conclude that at this time, there have been some attempts to modify existing frameworks as they apply to applications of nanotechnology, while the dominant model comprises existing rules and procedures.

That is not to say, of course, that regulatory activity has not taken place outside the realm of legislation. Early responses from industry suggested a preference for self-regulation—not instead of, but on top of, formal, state-led regulation. This is reflected by the numerous codes of conduct and risk management frameworks that have emerged from industry over recent years [26]. In a further response to calls for the clarification of industry obligations and for inter-jurisdictional consistency, voluntary standard-setting bodies at international, regional and national levels have played an important role in crafting the practical standards by which to regulate the technology [27]. One area in which standard-setting bodies have been particularly active is in relation to the definitional aspects of nanotechnologies, reflecting concerns that ambiguity surrounding the meaning of terms such as 'nanomaterial' and 'nanoscale' could create regulatory uncertainty and incoherence not only among stakeholders but between nation states and trading regions. With high-profile trade disputes between the US and EU fresh in the memories of policymakers [28, 29], there was initial optimism that the definitions agenda would offer a vehicle for transatlantic convergence in the management of nanotechnologies.

This was short-lived, however, with clear differences beginning to materialise as the EU and US have forged distinct paths on nanotechnology, both in policy and practice. A case in point here is the adoption by the European Commission on a definition of 'nanomaterial' [30]. The quest for uniformity has proved difficult even within the EU's jurisdictional boundaries. For instance, two pieces of EU legislation to include provisions on nanotechnologies—the Cosmetic Products Regulation (Regulation (EC) No 1223/2009) and the Food Labelling Regulation (Regulation (EU) No 1169/2011)—contained definitions of 'nanomaterial' that are different from that recommended by the Commission [30]. It is important to note that the Commission's Recommendation is not legally binding, but that its 'nanomaterial' definition is intended to be incorporated into any legislation that makes reference to the technology. Again, this provides a good example of regulatory activity outside the confines of legislation—except that, on this occasion, the non-binding provision may well end up having legal bite as a result of subsequent legislative uptake. This approach can be contrasted to the position of the US; the federal government has resisted calls for the adoption of a uniform definition across agencies.

So far, the picture that emerges is one of well-established legislative frameworks interspersed with some examples of amendment and, more prominently, a host of non-legislative initiatives designed to support the effective functioning of existing regulatory regimes. In other words, activity to date has either focused on updating existing legislative measures so that they make specific reference to 'nanotechnology' or 'nanomaterials', or on assisting regulating and regulated actors in carrying out existing legislative obligations—for example, through the provision of guidance, and codes of practice. These activities undoubtedly improve the clarity and coherence of the regulation of nanotechnologies.

Yet, by focusing (understandably enough) on existing regulatory arrangements, less attention has been paid to questions and concerns falling outside the terms of the legislation. For example, although there is agreement among policymakers that the regulation of nanotechnologies should be addressed in an open and inclusive manner, attempts to initiate stakeholder and public engagement have been relatively few and far between. And even though public participation has been identified as a core objective of the policy process [31], there are differing views as to how this ought to play out. In some arenas, it is deemed to entail direct public involvement through upstream deliberation [32], while in others it is interpreted more passively as the implementation of transparency measures to encourage openness on the part of governments and industry [33].

The crucial point here is that, since engagement activities are not usually within the scope of legislation, their implementation are often either non-existent, or disparate and contradictory. As yet, orchestrated attempts to make decision processes more inclusive are largely absent from the policy agenda. Miller and Scrinis [34] also make the point that even when stakeholders such as non-governmental organisations (NGOs) have been engaged, their impact on the policy process has been negligible at best. What this suggests is that when newly emerging technologies become subject to existing legislative regimes, they inherit the strengths of those regimes but also their weaknesses. One such weakness is a lack of joined-up thinking in addressing wider public involvement and participation in the regulatory policy in this area.

14.3 Synthetic Biology: The Next Evolutionary Technology

Synthetic biology—which has been defined as 'apply[ing] standardised engineering techniques to biology and thereby create organisms or biological systems with novel or specialised functions to address' [35:17])—has been in the peripheral vision of scientists, policymakers and regulators for a number of years. But while many have recognised its potential to raise key questions of regulation, there has so far been limited political and public debate on the matter [36–38].

14.3.1 Synthetic Biology: Early Steps to Regulation and Governance

Following news in May 2010 that a research team at the J Craig Venter Institute in the US had successfully created the first synthetic self-replicating life form [39, 40], synthetic biology no longer belonged to the realms of science fiction. It had, instead, become a scientific and regulatory reality. On the same day, President Barrack Obama wrote to the *Presidential Commission for the Study of Bioethical Issues* requesting that it undertake 'as its first order of business' a study of the implications of synthetic biology and issue recommendations regarding 'any actions the Federal government should take to ensure that America reaps the benefits of this developing field of science while identifying appropriate ethical boundaries and minimising identified risks' [35:17; 41, 42]. There could be no clearer sign that synthetic biology was to be placed high on the policy and regulatory agenda.

Currently (February 2017), and to our knowledge, there is no legislation in any jurisdiction that deals specifically with synthetic biology. The general consensus among policymakers, however, is that uses of synthetic biology will fall within the remit of existing regulations and that those regulations are, once again, 'fit for purpose'. Given that synthetic biology, at least in terms of its regulation, is conceived of as an extension to established genetic modification technologies (Organisation for Economic Co-operation and Development (OECD) and Royal Society 2010), regulators are already assumed to have a wide range of regulatory instruments at their disposal. The UK's Royal Academy of Engineering (RAEng), for instance, reported that

> At present the official view in the UK is that the majority of synthetic biology research will be covered by current GMO

regulations and that there is no need for any new regulations relating specifically to synthetic biology at present. However, this may not be the case in the future [43:48].

In other words, the emerging technology is once again assumed to come within the regulatory framework for the technology from which it evolved. This is despite the fact that there are unanswered questions about the suitability of existing regulatory arrangements for dealing with potential risks associated with synthetic biology. For example, is has been pointed out that

> The long term impact of these fabricated systems on the environment is unknown and effective regulation is not evolving as rapidly as the number of potential applications [44:4].

14.3.2 Synthetic Biology: The Challenges for Regulators

The challenges facing regulators are many. For example, concerns have been expressed about the 'dual-use' potential of synthetic biology—meaning its potential to be used for good but also bad purposes—and the capacity of existing regulations to cope with the range of possible scenarios. While issues surrounding biosafety and biosecurity have received some attention, questions about whether current regulations are sufficiently agile to enable regulators to identify, monitor and respond to unimagined futures [45], or futures at the margins of those anticipated by initial regulations, remain unanswered. Biosafety concerns relate to the accidental release of synthetic organisms, which could have unintended detrimental effects on human health or the environment [46]. Virus leaks from laboratories are not unheard of, and as synthetic biology looks set to increase the ease with which viruses are produced, the prospect of unintended releases is also likely to rise [47]. Well-known examples include the escape of smallpox in the UK in 1978 and of severe acute respiratory syndrome (SARS) in Singapore in 2003 and in China in 2004 [48].

Therefore, the primary issue regarding synthetic biology appears relate to unexpected outcomes of any accidental release. The deliberate release of such novel organisms is also cause for concern, since the growth of the Internet means that the tools for 'doing' synthetic biology are becoming increasingly accessible. This makes it difficult to gain a handle on the nature and extent of possible biosecurity threats, especially those emerging from lessor- or un-regulated markets. It only exacerbates and intensifies the anxiety that synthetic biology could lead to the production of organisms that are radically different from those that currently exist, with new and unpredictable properties [43]. This apprehension is summarised by the Royal Society and OECD, as follows:

> Taxonomy can act as a guide to pathogenicity, but it is less useful for synthetic biology. Novel organisms would present a particular challenge because of the lack of prior experience. Identifying sequences with pathogenic properties is also difficult, and conventional tools may no longer be appropriate [48:32].

In addition to biosafety and biosecurity concerns, synthetic biology raises a range of other ethical and social issues. They include, for example, concerns relating to the scope of protection that shall be afforded to such organisms under current intellectual property rights frameworks. Specific issues here relate to the scope of protection of foundational technologies, such as methods for producing synthetic DNA [43], the potential monopolisation of such patents by a small number of actors, and the potential patent thickets that may emerge. Ethical concerns relate to the distribution of harms and benefits associated with synthetic biology, and to the very idea of creating 'life' ab initio in the laboratory.

Such questions and concerns are not new, nor are they necessarily unique to synthetic biology (see, in the context of nanotechnology, [49]). But their re-appearance would suggest that there are unresolved tensions in existing regulatory frameworks. For example, they bear some resemblance to previous debates about the role of genetic modification in allowing scientists to 'play God' by interfering with nature. They bring to the fore foundational questions about the meaning of life, the distinction between natural and artificial, the existence and importance of such categorisations, and rights and obligations both to pursue progress and to refrain from doing so. These arguments commonly arise independently of issues of regulatory coverage, but they frequently reflect concerns about the intrinsically permissive nature of the current regulatory set-up. Typically, the starting point of regulation is the potential of new technology to open up markets and create wealth through the formation of competitive and dynamic knowledge societies. The prioritisation of these objectives over other political, ethical, cultural and moral goals can itself be scrutinised, and the emergence of new technologies such as synthetic biology provide the very opportunity to unpack established regulatory positions [19].

As these clusters of examples should illustrate, the practice of defaulting to former regulations may prove to be problematic because they can only be as receptive and responsive to new technological challenges as the regime permits. Novel types of threat, lacking information and inadequate inquiry into the assumptions that underpin new policy agendas, may limit the extent to which those regulations confront the problems directly.

14.4 Synthetic Biology in the Environment

One current area of synthetic biology research is the focused population control of *Aedes aegypti* mosquitoes. This species of mosquito is the carrier of infectious tropical diseases such as dengue fever, malaria, yellow fever, and chikungunya. To date, methods of mosquito population control have focused primarily on the use of physical barriers to prevent insect-human contact (i.e. mosquito nets) or chemical treatments to kill adult and larval mosquitoes alike. This has included, for example, aerial spraying, sticky ovitraps which trap mosquitoes when they land, and other chemically treated traps. In recognition of the limitations of these current approaches, Oxitec, a UK-based research company, has focused its efforts on the development of a biological alternative by genetically altering the mosquito itself. Specifically, Oxitec has engineered a strain of male *Aedes aegypti* (the 'OX513A' mosquito) so that its offspring cannot survive and reproduce without access to the antibiotic tetracycline, which is not commonly found in nature. This differs from previous mosquito genetic modifications due to a systems biology approach to the genetic alteration. In the OX513A, the tetracycline dependency is built-in to the larva's ability to the grow and survive.

14.4.1 Genetically Modified Mosquitoes as Population Control: Initial Trials

The logic behind the creation of the male *Aedes aegypti* OX513A mosquito is as follows: while the release of such mosquitoes into the environment would temporarily contribute to a spike in the *Aedes aegypti* population, the modified mosquito would, over time, reduce the local mosquito population as the modified mosquitoes breed with native, unmodified mosquitoes. Additionally, the release of male *Aedes aegypti* would, at least in theory, ensure that those mosquitoes would decrease disease transmission. If successful, this technology would help control a constant vector of disease transmission without the use of environmentally destructive chemicals or the need for constant vigilance and human involvement in barrier technologies.

Oxitec's OX513A mosquito was first deployed in field trials in the Grand Cayman in 2009. Repeated trials there (held in 2009 and 2010) were considered to be successful: approximately 3.3 million mosquitoes were released over a five-month period and were able competitively to breed with the local mosquito population. Specifically, Oxitec noted that by the end of the Grand Cayman trial, 88% of all mosquito eggs found on the island were the offspring of the genetically modified Oxitec mosquitoes, and that the overall *Aedes aegypti* population had declined by 80% over a three month period [50]. This success spurred additional environmental trials in regions of Brazil and Panama in 2014.

Yet, despite these positive results, the company noted some unintended negative consequences. These included approximately 3.5% of the modified mosquitoes living to adulthood, allowing for the potential for continued mosquito population growth and eventual dengue transmission [51]. This has the potential to negate any benefits of the OX513A in the medium to longer term. An additional concern was raised by the NGO GeneWatch UK, which argued that little to no warning was given to the public prior to the widespread mosquito releases [52]. Of particular concern to GeneWatch UK were the moral and ethical issues associated with selling such a technology to developing countries with endemic dengue and malaria. It noted, for example, that local policy- and decision-makers were under considerable pressure to agree to trials of the mosquito in order to combat the endemic threat of dengue without a full consideration of the long-term risks involved [53].

14.4.2 Critical Response to Modified Mosquito Field Trials

To illustrate, GeneWatch UK focused on the ongoing field trials of the mosquito in Panama in 2014. Its principal concern was the apparent failure of Oxitec to conduct an adequate risk assessment of the release of its genetically modified mosquitoes. Oxitec's OX513A mosquito falls within the definition of a 'living modified organism' under the Cartagena Protocol on Biosafety [54], to which the UK and Panama are parties. In the UK, the requirements for export (e.g. from the UK to Panama) are contained in the Regulation on the transboundary movement of genetically modified organisms (Regulation (EC) No 1946/2003, hereafter referred to as 'the Regulation'), which is an EU-wide measure of legislation that applies to genetic modification generally. It does not contain any provisions specifically on synthetic biology. Instead, as explained above, the assumption is that existing regulatory frameworks will apply, even though they were designed before synthetic biology became practically or commercially viable.

Article 4 of the Regulation stipulates that the exporter (in this case, Oxitec) shall issue written notification to the competent authority in the host country (Panama) before the first intentional transboundary movement of a genetically modified organism (the OX513A mosquito) intended for deliberate release into the environment. The notification shall contain, *inter alia*, the name and contact details of the exporter, the name and identity of the genetically modified organism, information about its intended use, and—importantly, for our purposes—a risk assessment report (Annex I to the Regulation). The Regulation also makes it clear that there can be no intentional transboundary movement of a genetically modified organism unless and until the host country has given its consent (see Article 5).

GeneWatch UK has been highly critical of Oxitec for seeking to conduct the experimental release of hundreds of thousands of modified mosquitoes without (what GeneWatch UK considered to be) a thorough risk assessment as required under the Regulation [55]. Oxitec's Research Director responded by noting that the company is required to 'provide the necessary information to allow the scientists and technical advisers to governments in affected countries to decide whether this is right for their country' [56]. The question for GeneWatch UK was whether that necessary information has been provided. GeneWatch UK submitted a request to the UK's Department for Environment, Food & Rural Affairs (Defra) [57] for a copy of Oxitec's risk assessment and of evidence of consent from Panama. Defra is reported to have told GeneWatch UK that those documents did not exist, which GeneWatch interpreted as a 'clear breach' of the Regulation [58]. Moreover, GeneWatch UK argued that Oxitec's alleged failure to provide risk assessment in compliance with EU requirements meant that 'the Panamanian authorities and public had no means to hold Oxitec to account for the information it provided in its role as exporter', and that 'the public in Panama may be exposed to unnecessary and unacceptable risks to human health and the environment' [58:3].

14.4.3 Synthetic Biology and Mosquitoes: Health Concerns and Regulatory Challenges

As this suggests, the development and release of the OX513A mosquito brings into question the adequacy of existing regulatory instruments as they apply to new applications of technology. Even if the legislative text is broad enough to encompass synthetic biology, there appear to be unresolved issues at the level of implementation. In recognition of the tensions at play in this context, GeneWatch UK called upon Oxitec to conduct a risk assessment in line with the Regulation. Previously, in 2012, GeneWatch UK obtained copies of Oxitec's risk assessment, which, GeneWatch UK said, failed to comply with the Regulation's notification procedure [59]. It highlighted the importance of a new risk assessment, in order to review the efficacy and implications of the mosquitoes in a manner befitting the Regulation. This, said GeneWatch UK, would provide invaluable data on potential risks and benefits of the novel technology to other prospective client nations. GeneWatch UK called for adequate testing prior to the release of genetically modified material, so that full consideration could be given to how this could impact the host nation. Without such information, GeneWatch UK argued, it would be difficult for a nation's governing authority to make a fully informed decision about the welfare of their constituents, or truly to understand the risks that may result from the technology's use. This raises obvious challenges for the Article 5 (see above) requirement that, before there is transboundary movement of such an organism, a host country must have given its consent.

One of the concerns related to the previously untested release of living GM organisms is the potential for undesirable consequences to ecosystems and population health more generally [55]. Specifically, Oxitec described a small rate of failure in their controlled releases, including (i) roughly 3.5% of modified mosquitoes survived to adulthood and (ii) a small number of modified biting female mosquitoes were released into the wild (approximately 200 per million males as of February 2014) [51, 55]. The nature and full extent of negative effects remain largely unknown and under-researched [60], which only serves to emphasise the importance of data collection under the Regulation.

A further complicating factor to the regulation of the OX513A mosquito is the novelty of this form of insect population control. Existing regulation and governance structures are geared to review the risks and benefits of chemical and barrier systems that have been in use for several decades. As such, governance structures in developing and developed countries alike would be strained to offer proper guidance on the deployment of the mosquito. Under the Cartagena Protocol, an Ad Hoc Technical Expert Group has issued additional guidance on the risk assessment and management of environmental releases of 'living modified' mosquitoes [61]. This is a welcome step. It may still be prohibitively difficult, however, for regulators adequately to manage, let alone mitigate, potential risks from the release of the mosquitoes, particularly where there is significant political and social pressure quickly to resolve the spread of diseases like dengue or malaria. It also fails to address situations in which exporting companies are confident that they have already complied with the letter of the law, as set out in the various regulations. The additional guidance is, after all, just that: guidance. It does not introduce new and targeted requirements into the legislation. That is not to say that its lack of legal 'bite' is necessarily problematic. What it does highlight, however, is that, for voluntary initiatives to work, there needs to be at least some degree of consensus among stakeholders that those initiatives are appropriate.

A related concern of GeneWatch UK was the general lack of stakeholder engagement in the deployment and trial of the OX513A mosquitoes. Recent guidance issued by the World Health Organisation (and others) notes, in relation to genetically modified mosquitoes, that 'adequate plans for communication and engagement are important even at the early field-testing stage' [62]. It further recommends that investigators (which would include potential exporters)

> work cooperatively...with the host communities to avoid miscommunications and misunderstandings that could undermine trust and transparency [62:xxiv].

The guidance also notes that, although there may be some overlap between engagement activities and regulatory requirements, 'ethical issues and responsibilities are generally broader than just those activities specifically mandated by administrative law or organisational policies' [62:xxiii]. In other words, the mere fact that relevant legislation exists in this area provides no guarantee that broader social and ethical issues will be properly taken into account. GeneWatch UK has specifically pushed for increased transparency and attention to issues of consent in the selected host nations prior to any field trials or release of the mosquitoes. It also highlighted the need for citizen participation in decision-making processes, following the notable absence of such engagement activities leading up to the release of the modified mosquitoes in Panama in mid-2014 [58].

14.4.4 Synthetic Biology and Mosquitoes: Looking Forward

In sum, the release Oxitec's OX513A mosquito in trials in Grand Cayman, Brazil, and Panama since 2009 raises questions regarding the ethical dimensions of the environmental release of such an organism from both the standpoint of transparency and consent. Questions over the adequacy of existing governance structures in terms of the regulatory instruments themselves, risk assessment, risk management and enforcement, have similarly been raised in parallel with these broader societal issues. The purpose of reviewing this case is not to pass judgement on the efficacy or benefits of an application of synthetic biology. Rather, it is to illustrate the need for careful consideration of the distributional effects of a new technology being put to use outside the remit of relevant legislation—legislation that is said to 'cover' applications of synthetic biology. It is to highlight that, while synthetic biology may in principle be governed by existing legislative measures in the EU, this does not eliminate the need for close scrutiny of the wider, perhaps unforeseen, implications of the technology—especially for the countries and communities acting as hosts.

One thing is clear: synthetic biology has already arrived and is no longer belongs to the realms of science fiction. Its emergence in the laboratory and its uses in the environment are likely to strain existing regulatory arrangements, while also raising ethical and moral concerns associated with its deployment. This is particularly true, we would argue, in cases where the emerging technology has been designed to help address a challenge faced primarily by developing countries, such as dengue fever. Many of these countries, which are likely to become the 'test beds' for early trials, may have more limited capacity effectively to oversee and manage such tests, as the GeneWatch UK studies have indicated.

Therefore, it is incumbent on those working in the field, and those charged with safeguarding human and/or environment health and safety, to start assessing the adequacy of existing governance structures for near-term synthetic biology applications. Their adequacy needs to be addressed not just in terms of 'coverage' but also in terms of their potential effects within and, importantly, outside their jurisdictional scope. Existing regulation and governance conventions are a useful starting point from which to review and assess the risks of synthetic biology applications, but new governance approaches may be warranted to ensure that the most appropriate and effective risk and ethics practices are being upheld.

14.5 Conclusions

This chapter has identified a series of broad themes that have come to pervade regulatory approaches to nanotechnologies and synthetic biology. One is that, the initial response to these new technological developments is often to have recourse to existing regulatory frameworks. There is nothing inherently problematic with this practice, unless—as this chapter has indicated—it involves an unquestioning acceptance of existing regulations and their adequacy. As the example of nanotechnology has shown us, new technologies and their products will, at least in the first instance, be shepherded into the regulatory regime used to govern the technology from which it evolved, or the conventional counterparts of its new products. This pattern looks set to be replicated for the 'next' evolutionary technology, synthetic biology, despite the lessons that policymakers should be able to draw from nanotechnologies even today.

There may be very good reasons for continuing to rely on existing regulations, not least because the fields discussed can be affected by scientific uncertainty and contested ideas about appropriate paths of development. For example, where a regulatory regime is well established, it can make sense, from the point of view of market stability, to continue with regulatory 'business as usual'. But, at the very least, we would argue that new technologies provide useful opportunities for policymakers and regulators to reflect on existing regulations—not only because a new technology might throw up new kinds of issue, but also because there may be scope to improve the regulations as they apply to technologies generally, new and old. A more proactive approach may involve efforts to ensure that the social benefits of such technologies are maximised without overlooking the need continually to reflect on the adequacy of current measures, in terms both of their rules and of their underlying assumptions. This might involve the more widespread use of guidance materials by industry, or the introduction of various forms of information-sharing by and between relevant stakeholders. Examples of these approaches have been used in relation to nanotechnologies and synthetic biology, albeit in a variety of different forms. The question is whether and to what extent they will be implemented, given that many of them operate at the fringes of the regulatory regimes.

The challenges facing regulators ought not to be underestimated, and in a climate of scientific, not to mention political and financial, uncertainty, the practice of resorting to default strategies will hold considerable appeal. But, as has previously been argued, a lack of scientific evidence should not in itself stop governments and regulators from considering alternative regulatory paths for these evolutionary technologies [63]. Moreover, reliance on existing regulations can have its drawbacks, such that it will be necessary to reflect hard not only on the constituent rules but also on more entrenched aspects of the regulatory landscape. We would argue for a more proactive stance; one which attempts to overcome the limitations of existing regulations as they apply generally as well as to technological evolutions. The emergence of nanotechnologies and synthetic biology offers a perfect opportunity to do just that.

References

1. Bowman, D. M., and G. A. Hodge (2007) A small matter of regulation: An international review of nanotechnology regulation, *Columbia Science and Technology Law Review*, **8**, 1–32.

2. Friedman, D. (2011) Does technology require new law? *Harvard Journal of Law & Public Policy*, **25**(1), 71.

3. Hodge, G. A., D. M. Bowman, and A. D. Maynard (eds.) (2010) *International Handbook on Regulating Nanotechnologies* (Edward Elgar, Cheltenham).

4. Ludlow, K., D. M. Bowman, J. Gatof, and M. G. Bennett (2015) Regulating emerging and future technologies in the present, *NanoEthics*, **9**(2), 151–163.

5. Stokes, E. (2013) Demand for command: Responding to technological risks and scientific uncertainties, *Medical Law Review*, **21**, 11–38.

6. Stokes, E. (2012) Nanotechnology and the products of inherited regulation, *Journal of Law & Society*, **39**(1), 93–112.

7. Breyer, H., (2006) *European Parliament, Debate No. 4*. (European Parliament, Brussels).

8. Davies, J. C. (2006) *Managing the Effects of Nanotechnology* (Project on Emerging Nanotechnologies, Washington, D.C.).

9. Frater, L., E. Stokes, R. Lee, and T. Oriola (2006) *An Overview of the Framework of Current Regulation Affecting the Development and Marketing of Nanomaterials* (Cardiff University, Cardiff).

10. European Commission (2008) *Regulatory Aspects of Nanomaterials* (European Commission, Brussels).

11. Taylor, M. (2006) *Regulating the Products of Nanotechnology: Does FDA Have the Tools It Needs?* (Project on Emerging Nanotechnologies, Washington, D.C.).

12. Ludlow, K., D. M. Bowman, and G. A. Hodge (2007) *A Review of the Possible Impacts of Nanotechnology on Australia's Regulatory Framework* (Monash University, Melbourne).

13. Council for Science and Technology (2007) *Nanosciences and Nanotechnologies: A Review of Government's Progress on its Policy Commitments* (CST, London).

14. Royal Commission on Environmental Pollution (2008) *Novel Materials in the Environment: The Case of Nanotechnology* (RCEP, London).

15. Stokes, E. (2009) Regulating nanotechnology: Sizing up the options, *Legal Studies*, **29**(2), 281–304.

16. Lee, R., and S. Vaughan (2010) REACHing down: Nanomaterials and chemical safety in the European Union, *Law, Innovation and Technology*, **2**(2), 193–217.

17. Australian Office of Nanotechnology (2008) *National Nanotechnology Strategy Annual Report 2007–08* (Australian Government, Canberra).

18. Chaudhry, Q., J. Blackburn, P. Floyd, C. George, T. Nwaogu, A. Boxall, and R. Aitken (2006) *Final Report: A Scoping Study to Identify Gaps in Environmental Regulation for the Products and Applications of Nanotechnologies* (Defra, London).

19. Stokes, E. (2011) You are what you eat: Market citizens and the right to know about nano foods, *Journal of Human Rights and the Environment*, **2**, 178–200.

20. Bowman, D. M., and K. Ludlow (2013) Assessing the impact of a 'for government' review on the nanotechnology regulatory landscape, *Monash Law Journal*, **38**(3), 168–212.

21. Therapeutic Goods Administration (2009) *A Review of the Scientific Literature on the Safety of Nanoparticulate Titanium Dioxide or Zinc Oxide in Sunscreens* (Australian Government, Canberra).

22. Therapeutic Goods Administration (2008) *Nanotechnology and Therapeutic Products. Government of Australia* (Australian Government, Canberra).

23. Preston, C. J., M. Y. Sheinin, D. J. Sproat, and V. P. Swarup (2010) The novelty of nano and the regulatory challenge of newness, *NanoEthics,* **4**(1), 13–26.

24. Monica Jr, J. C., and J. C. Monica (2009) Examples of recent EPA regulation of nanoscale materials under the toxic substances control act, *Nanotechnology Law & Business,* **6**, 388.

25. Hull, M., and D. M. Bowman (eds.) (2014) *Nanotechnology Risk Management: Perspectives and Progress* (2nd edition) (Elsevier, London).

26. Meili, C., and W. Widmer (2010), Voluntary measures in nanotechnology risk governance: The difficulty of holding the wolf by the ears. In: Hodge, G. A., D. M. Bowman, and A. D. Maynard (eds.) *International Handbook on Regulating Nanotechnologies* (Cheltenham: Edward Elgar), pp. 446–461.

27. Miles, J. (2010) Nanotechnology captured. In: Hodge, G. A., D. M. Bowman, and A. D. Maynard (eds.) *International Handbook on Regulating Nanotechnologies* (Cheltenham: Edward Elgar), pp. 83–106.

28. Newell, P. (2003) Globalization and the governance of biotechnology, *Global Environmental Politics,* **3**(2), 56–71.

29. Sheldon, I. M. (2002) Regulation of biotechnology: Will we ever 'freely' trade GMOs?, *European Review of Agricultural Economics,* **29**(1), 155–176.

30. European Commission (2011) *Recommendation on the Definition of Nanomaterial, 2011/696/EU, 2011 OJ L275/38* (European Commission, Brussels).

31. Renn, O., and M. Roco (2006) *Nanotechnology Risk Governance* (International Risk Governance Council, Geneva).

32. Rogers-Hayden, T., and N. Pidgeon (2007) Moving engagement "upstream"? Nanotechnologies and the Royal Society and Royal Academy of Engineering's inquiry, *Public Understanding of Science,* **16**(3), 345–364.

33. Stebbing, M. (2009) Avoiding the trust deficit: Public engagement, values, the precautionary principle and the future of nanotechnology, *Journal of Bioethical Inquiry,* **6**(1), 37–48.

34. Miller, G., and G. Scrinis (2010) The role of NGO in governing nanotechnologies: Challenging the 'benefits versus risks' framing of nanotech innovation.' In: Hodge, G. A., D. M. Bowman, and A. D. Maynard (eds.) *International Handbook on Regulating Nanotechnologies* (Cheltenham: Edward Elgar), pp. 409–445.

35. Presidential Commission for the Study of Bioethical Issues (2010) *New Directions: The Ethics of Synthetic Biology and Emerging Technologies* (PCSBI, Washington, D.C.).

36. Calvert, J., and P. Martin (2009) The role of social scientists in synthetic biology, Science & Society Series on Convergence Research, *EMBO Reports*, **10**, 201–204.

37. Caruso, D. (2008) *Synthetic Biology: An Overview and Recommendations for Anticipating and Addressing Emerging Risks* (Center for American Progress, Washington, D.C.).

38. Serrano, L. (2008) Synthetic biology: Promises and challenges, *Molecular Systems Biology*, **3**, 158.

39. Gibson, D. G., J. I. Glass, C. Lartigue, V. N. Noskov, R. Y. Chuang, et al. (2010) Creation of a bacterial cell controlled by a chemically synthesized genome, *Science*, **329**(5987), 52–56.

40. Pennisi, E. (2010) Synthetic genome brings new life to bacterium, *Science*, **328**, 958–959.

41. Presidential Commission for the Study of Bioethical Issues (2010) *Letter from President Barack Obama to Dr. Amy Gutmann* (PCSBI, Washington, D.C.).

42. Bhattacharjee, Y. (2010) U.S. panel weighs guidelines for synthetic biology, *Science*, **329**, 264–265.

43. Royal Academy of Engineering (2009) *Synthetic Biology: Scope, Applications and Implications* (RAEng, London).

44. Lloyds of London (2009) *Synthetic Biology: Influencing Development* (Lloyds, London).

45. Samuel, G., M. Selgelid, and I. Kerridge (2009) Managing the unimaginable: Regulatory responses to the challenges posed by synthetic biology and synthetic genomics, *European Molecular Biology Organization Reports*, **19**(1), 7–11.

46. Bhutkar, J. (2005) Synthetic biology: Navigating the challenges ahead, *Journal of Biolaw & Business*, **8**(2), 19–29.

47. Andrianantoandro, E., S. Basu, D. K. Karig, and R. Weiss (2006) Synthetic biology: New engineering rules for an emerging discipline, *Molecular Systems Biology*, **2**(1), DOI 10.1038/msb4100073.

48. Organisation for Economic Co-Operation and Development and Royal Society (2010) *Symposium on Opportunities and Challenges in the Emerging Field of Synthetic Biology: A Synthesis Report* (OECD and Royal Society, Paris).

49. Lee, M. (2010) Risk and beyond: EU regulation of nanotechnology, *European Law Review*, **35**(6), 799–821.
50. Harris, A. F., A. R. McKemey, D. Nimmo, Z. Curtis, I. Black, et al. (2012) Successful suppression of a field mosquito population by sustained release of engineered male mosquitoes, *Nature Biotechnology*, **30**(9), 828–830.
51. Naik, G. (2011) Scientists tweak bugs to zap disease, *The Wall Street Journal*, 11 April. Available at: http://www.wsj.com/articles/SB10001424052970204505304577003792892772710.
52. Pollack, A. (2011) Concerns are raised about genetically engineered mosquitoes, *The New York Times*, 30 October. Available at: http://www.nytimes.com/2011/10/31/science/concerns-raised-about-genetically-engineered-mosquitoes.html?_r=0.
53. GeneWatch UK (2010) *Oxitec's Genetically-Modified Mosquitoes* (GeneWatch UK, London).
54. Secretariat of the Convention on Biological Diversity (2000) *Cartagena Protocol on Biosafety to the Convention on Biological Diversity* (United Nations, Montreal).
55. Wallace, H. (2013) *Genetically Modified Mosquitoes: Ongoing Concerns* (TWN Biotechnology and Safety Series, New York).
56. World Health Organization (2009) Mosquito wars, *Bulletin of the World Health Organization*, **87**(3) 161–244.
57. Department for Environment, Food & Rural Affairs (2004) *Environmental Protection, England: The Genetically Modified Organisms (Transboundary Movements)* (HM Government, Statutory Instrument No. 2692, London).
58. GeneWatch UK (2014) *Failures of the Transboundary Notification Process for Living Genetically Modified Insects* (GeneWatch UK, London).
59. GeneWatch UK (2012) *Oxitec's Genetically Modified Mosquitoes: Ongoing Concerns* (GeneWatch UK, London).
60. Sustainable Pulse (2014) *Oxitec 'Negligent' Over GM Mosquito Release in Panama*. Available at: http://sustainablepulse.com/2014/02/12/oxitec-negligent-gm-mosquito-release-panama/.
61. United Nations Environment Programme (2010) *Final Report of the Ad Hoc Technical Expert Group on Risk Assessment and Risk Management Under the Cartagena Protocol on Biosafety. UNEP/CBD/BS/COP-MOP/5/INF/15* (UNEP, New York).

62. World Health Organization and Foundation for the National Institutes of Health (2014) *The Guidance Framework for Testing Genetically Modified Mosquitoes* (WHO, Geneva).
63. Bowman, D. (2008) Governing nanotechnologies: Weaving new regulatory webs or patching up the old?, *NanoEthics*, **2**(2), 179–182.

Index

actor constellations 191–192, 195
actor strategies 191–192
actors
 constellation of 191, 208
 non-state 11, 223, 226, 229–230, 277
 types of 230
additive bio-manufacturing 337, 341, 343
additive manufacturing 8, 337, 339
additive manufacturing technologies 337, 339, 343, 348–350
Aedes aegypti 360, 370–371
AM/3D printing 12
anticipation 18, 23, 25, 29, 63–64, 78–80, 123, 142, 146–147, 167, 187–189
anticipation and tentative governance 5
anticipatory practices 188–189, 194–195, 201, 203, 207–208
artifacts 93, 95, 105
 technological 102–103, 106
autonomy 107–108

beverages sectors 161, 168–169, 175
bioethics 67, 72, 90
biomedical field 341

biomedical industry 341, 343
bioprinting 337, 343–344
biotechnology 21, 24–25, 98, 101, 124, 269, 350
border surveillance 299, 303–307, 309, 311–319, 321–325, 327–330
 increased militarisation of 319
business association activities 261, 264–265, 267–269, 271, 273, 275, 280
business associations 11, 45, 229, 260, 264–267, 269, 277–281
businesses, nanotechnological 41

cancer 80
care 9, 43, 90–95, 97, 99–100, 102, 104–108, 121, 261–262, 280
 duty of 91
care and techno-science 89–108
care-full re-embedding 102–107
Cartagena Protocol on Biosafety 372
chemical industry 265, 269–271, 280, 282–283

chemical safety 224, 242, 248
chemical substances 261–263, 283–284, 287
chemicals 171, 221, 226, 242–243, 261–263, 273, 285
civil society 223, 226
civil society actors 28, 122, 222
co-evolution 18–20, 76, 149
  reflexive 19–21, 28
co-evolution of science 9, 17, 19
co-evolution of technology 20–21, 29
co-regulation 36, 40, 229, 260, 277–278
co-regulation of nanomaterials 259–260, 262, 264, 266, 268, 270, 272, 274, 276, 278, 280, 282, 284, 286
code of conduct 25, 132–135, 137, 150, 164, 260, 266, 364
competence 117, 138, 231
compliance 38, 47, 105, 125, 172, 219, 267, 300, 373
  rule 43, 276
consent 67, 326–327, 372–376
constructive technology assessment (CTA) 3, 25, 63, 78, 116, 144, 160, 167–168, 179
consumers 47–48, 50–51, 82, 122, 162, 165, 174, 176, 219–220, 228, 235, 345–346, 363
contestation 20, 22–24, 46, 123–124
corporate social responsibility 122
cosmetic products 39, 42, 47–50, 235, 280, 363

CTA, *see* constructive technology assessment

data protection 310, 312, 325–329
data protection requirements 328–329
de facto governance 18, 21, 193–194, 207
de facto governance patterns 23, 28, 162
derived no-effect levels (DNELs) 44
disembedded futures 102, 107–108
disembedding 97–99, 101, 106
DIY, *see* do-it-yourself
DNELs, *see* derived no-effect levels
do-it-yourself (DIY) 345, 350
drinking water 168–171, 173, 176, 178–179
drinking water companies 172, 179
drinking water sector 170, 174–175, 177
drone surveillance 312
drone technology 11, 299–307, 309–310, 312–315, 317–318, 321–322, 324–325, 327, 329–330
drones 11–12, 299–304, 306–312, 314–330
  armed 303–304

economics of techno-scientific promises 346

electric vehicles 143
electronics 195–196, 198, 206, 226
ELSI, *see* ethical, legal and social issues
engineered nanomaterials 52, 222
engineering ethics 67
environmental regulation 125
epistemological uncertainty 74–75
ethical, legal and social issues (ELSI) 11, 21, 132, 136
ethical reflexivity 10, 131–138, 140, 142, 144, 146–148, 150
ethicists 132, 134, 138, 148–149
ethics 108, 120, 132, 134, 138–139, 146
expectation building 189, 192–195, 203, 207
expectation dynamics 195, 207–208
expectations
 construction of 96, 98
 governance of 188, 190, 192, 198, 200–201, 203, 206–208
 production of 95–96, 189
 societal 11, 18
 sociology of 100, 103, 193
 technological 95
experimental sociology 68
experiments
 traditional 64
 uncontrolled 65, 71

fictive scripts 143–144, 150
futures 91, 93–94, 97, 103, 106, 368
 singular 93–94, 97, 100, 102–103, 105–108

genetic modification (GM) 1, 360, 367, 369–370, 372
GM, *see* genetic modification
GoodNanoGuide 241–242, 245, 249
governance
 anticipatory 47, 132, 188, 207, 209
 dynamic 126
 modes of 2, 9, 188, 190–191, 194, 204, 208, 225, 244
 new forms of 229
governance actors 224, 338
governance approaches, responsive 37, 45
governance arrangements 2, 8, 10, 17, 38–39, 163–164, 166, 189, 191, 225–226, 229, 231–232, 244, 246, 248
governance of emerging technologies 115–126, 219–220, 222, 226, 228, 230, 232, 234, 236, 238, 240, 242, 244, 246, 248
governance of governance 21
governance tools, anticipatory 192, 201
graphene 189, 195–200, 202–208
 possible applications of 197, 199
graphene actors 190, 204
graphene expectations 195–197, 201–202
graphene flagship 203

graphene market   204–205
graphene oxide   206
graphene researchers   200, 202
graphene scientists   190, 196, 201

hazardous substances   43–44, 363
human enhancement   19, 63, 342
human health   8, 44, 62, 236, 238, 245, 261, 273, 276, 285–286, 368, 373
human tissues   337, 343–344

implants   337, 341–342
indeterminacy   3, 23, 74–75
informed consent   67, 72, 82
innovation, disruptive   195, 199, 346
innovation actors   187, 192–193, 208
innovation chains   144
innovation governance   91, 159–161, 164, 167, 169, 179–180, 348
interactions, user-producer   160

judicial cooperation   326–327

learning
  mutual   277, 279
  normative   77

learning-by-anticipation   73, 77–79, 83
learning-by-doing   64, 78–79, 83
learning-by-experimentation   63, 77, 79, 83

manned aircraft   304, 308, 315, 318
manufactured nanomaterials (MNs)   225, 236–238, 243–245, 248
meta-regulation   40–41
MNs, *see* manufactured nanomaterials
modified organisms   123, 226, 372
moral labour   8, 26, 115–126, 133
  division of   10, 116–117, 119, 121–123, 125–126, 149–150
mosquitoes   360, 370–375
  modified   357–358, 360, 362, 364, 366, 368, 370–372, 374–376, 378

nano-based products   37
nano market   220
Nano Reference Values (NRVs)   44–47, 53
nanoethics   79
nanomaterials   38, 44–52, 103–104, 220–221, 228, 234–236, 238, 241, 243, 247, 259–264, 266, 268–280, 282–287, 363–365

definition of 41, 49, 272–273, 285, 365
handling of 242, 283
new generations of 223, 227
potential risks of 262, 270
nanomaterials occupational health and safety (OHS) 260–261, 268–272, 274–275, 277, 280–282, 286
nanomaterials OHS
co-regulation of 260–261, 281
context of 265, 267
regulation of 260–261, 275
nanomaterials risk assessment 274, 279, 283, 285
nanoparticles 4, 22, 39, 43–44, 46, 62, 66, 80–83, 245, 271, 280, 283, 286
engineered 44, 46, 220
titanium dioxide 80–81
nanosciences 2, 124, 132, 135, 220–221, 226, 247
nanosensors 103–105
nanotechnologies
emergence of 359, 378
governance of 224–227, 233, 235, 237, 239, 241, 243–245, 248–249
novel 170
regulation of 226
regulatory challenges of 222, 224
regulatory models for 40
nanotechnologists 117
nanotechnology-based products 10, 234, 361
nanotechnology developments 124, 135, 221–222, 227
nanotechnology enactors 98
nanotechnology regulation 244, 362

nanotechnology research 132, 219, 227, 240, 346
nanotechnology risk 239, 248
nanotechnology risk governance 239–240, 248
nanotechnology risk regulation 249
nanotechnology stakeholders 135
new technologies
development of 18, 37
governance of 10, 89
introduction of 66, 73
managing 26
NGOs, *see* non-governmental organisations
non-governmental organisations (NGOs) 21, 26, 125, 231–232, 239–240, 242–244, 246–248, 343, 366
normative issues 69–70, 75–77
novel sensor technologies 172, 175–176, 179
novel sensors 172–173, 176–177
NRVs, *see* Nano Reference Values

occupational health and safety (OHS) 259, 261, 275, 278, 280
occupational safety 228, 234
OHS, *see* occupational health and safety
over-regulation 9, 40, 221

Predator drone 307–308, 322
printers 338–339, 341, 344–345, 349
printing 8, 12, 335–346, 348–350
printing technologies 335–340, 347–351

real time technology assessment 25
reflexivity 6, 8, 20, 131–133, 135, 137–142, 145, 148–150, 167–168
 articulation of 138, 144–145, 149
reflexivity of scientists 132, 138, 150
regulation
 data protection 325–327
 nanotechnological 46
regulatory effectiveness 43, 276, 281
regulatory oversight strategies 220–221
remotely piloted aerial systems (RPAS) 299, 302
remotely piloted aircraft (RPA) 316, 318–319
responsible nanosciences 41, 135, 222
Responsible Research and Innovation (RRI) 2, 10, 25–26, 37–38, 40, 53, 90, 92, 94–95, 108, 116, 121, 124, 131–132
RPA, see remotely piloted aircraft
RPAS, see remotely piloted aerial systems

RRI, see Responsible Research and Innovation

Safety of Life at Sea (SoLaS) 322
SAR, see search and rescue
scientific knowledge 90, 132, 269, 279
search and rescue (SAR) 321–322, 324
sensor applications 161, 171, 176, 179
sensor technologies 169, 174, 176
 nano-based 10, 159
sensors 103, 168–177, 315, 324
small and medium-sized enterprises (SMEs) 271, 273, 284, 286
SMEs, see small and medium-sized enterprises
social experimentation 66–68
social experiments 64–72, 80–84
 characteristics of 69, 71
social license 36–37, 40
social order 116, 119–120
social sciences 5, 20, 35, 68, 89, 131, 136, 299
social scientists 68, 132, 134, 138, 144, 148–149
societal benefit 134, 319
societal challenges 26, 198, 203
societal embedding 19, 25, 64, 159, 163–164, 167, 176, 178
societal embedding
 articulation 163–164, 169, 171–172, 174–175

societal impacts 131, 136
socio-technical dynamics 144–145, 167
socio-technical scenarios 145–146
soft regulation instruments 260, 271–272
SoLaS, *see* Safety of Life at Sea
strategic niche management 68–69
stretch 163, 171, 173–174
stretch strategies 163–164, 166, 169–170, 174–175, 179
sunscreens 63, 66, 80–82
surveillance 301, 305, 307, 310, 312–314, 317, 320, 322–325, 328
surveillance tools, common application of 314
synthetic biology 8, 22, 100, 357, 359, 361, 366–373, 375–378
synthetic biology applications 136, 376

TA, *see* technology assessment
techno-science 89–92, 94, 96, 98, 100, 102, 104, 106, 108
technological developments evolutionary 359
fast moving 37
technological evolution 378
technological innovation 4, 6, 35–36, 40, 89, 92, 103, 105, 139, 163, 312, 336
technology, dual-use 300, 325
technology assessment (TA) 63, 78
technology regulation, models of 39
tentative governance 5
TGAs, *see* transnational governance arrangements
TiO_2 nanoparticles 80–82
tissues 339, 343
transnational governance arrangements (TGAs) 223, 225–226, 228–233, 244
transparency 51–52, 54, 174, 284, 318, 375–376

UAVs, *see* unmanned aerial vehicles
unmanned aerial vehicles (UAVs) 8, 299–300, 302, 304, 309, 317

workers, health and safety of 262–263